Environmental Pollution Control Engineering

C.S. Rao
Ph.D. (Houston)
Director,
Sardar Patel Renewable Energy Research Institute
Vallabh Vidyanagar, Gujarat, India
(Formerly Professor of Chemical Engineering,
Birla Institute of Technology & Science, Pilani)

JOHN WILEY & SONS
New York Chichester Brisbane Toronto Singapore

First published in 1991 by
WILEY EASTERN LIMITED
4835/24, Ansari Road, Daryaganj
New Delhi 110 002, India

Distributors:

Australia and New Zealand:
JACARANDA WILEY LTD.
PO Box 1226, Brisbane, Milton Old 4064,
Australia

Canada:
JOHN WILEY & SONS CANADA LIMITED
22 Worcester Road, Rexdale, Ontario, Canada

Europe and Africa:
JOHN WILEY & SONS LIMITED
Baffins Lane, Chichester, West Sussex, England

South East Asia:
JOHN WILEY & SONS (PTE) LIMITED
05-04, Block B, Union Industrial Building
37 Jalan Pemimpin, Singapore 2057

Africa and South Asia:
WILEY EASTERN LIMITED
4835/24, Ansari Road, Daryaganj
New Delhi 110 002, India

North and South America and rest of the World:
JOHN WILEY & SONS, INC.
605 Third Avenue, New York, NY 10158, USA

Library of Congress Cataloging in Publication Data

Rao, C.S.
Environmental pollution engineering/by C.S. Rao.
p. cm.
Includes bibliographical references and index.
1. Pollution. 2. Pollution—India. I. Title.
TD191.5.R33 1991 91-10923
628.5—dc20 CIP

ISBN 0-470-21763-4 John Wiley & Sons, Inc.
ISBN 81-224-0301-8 Wiley Eastern Limited

Printed in India at Prabhat Press, Meerut.

Preface

The systematic pollution of our environment is one of the biggest hazards that humanity faces today. People are becoming increasingly aware of the threat posed by pollution and governments are enacting legislations aimed at protecting the environment. India is no exception. Recognising the need for trained technical manpower in this area, more and more universities and institutions are offering courses related to environmental pollution and its control. Thus, a text book dealing with the air, water and land pollution at a level suitable for an advanced undergraduate or a first year post-graduate student in engineering, which emphasises the fundamentals of the origin and control of pollution, is very much required. Specialized books dealing with a particular aspect of environmental pollution have been published in the United States and in Europe, but these serve mostly as reference books and moreover they are not within the financial reach of an average Indian student. The material in the present book is designed to meet the above mentioned need.

The main emphasis in the book is on topics related to air and water pollution—the origin of various pollutants, their effects on man and on the environment, and the methods available to control them. The fundamental aspects are stressed and due importance is given to their application to the design of pollution control equipment. In Chapter 1, an overview of the impact of man and his activities on the environment is given with a view to making the reader aware of the magnitude of the problem of pollution.

The topical Chapters 2 through 6 deal with air pollution aspects—classification and origin of air pollutants and their harmful effects, air pollution modelling, air quality monitoring, general methods and equipment available for controlling air pollutants, and finally the control of specific gaseous pollutants like sulphur and nitrogen oxides.

Chapters 7 through 9 are devoted to water pollution aspects—sources and classification of water pollutants, sampling and analysis of various pollutants and the methods available for the treatment of wastewaters. Solid waste management is of growing importance, particularly in urban areas, and Chapter 10 is devoted to its discussion. Data dealing with conditions in India are given wherever possible to help the reader gain an insight into the country's specific pollution problems.

A discussion on noise as the fourth pollutant is left out as it is considered beyond the scope of the task for which the book is designed. The material presented in the book is suitable for a one semester course covering about 45 lecture hours. It was tested extensively by the author for a number of years in the class room when he taught a course in environmental pollution control for final year students of engineering at the Birla Institute of Technology and Science (BITS), Pilani. The book can also be useful to practising engineers, research scientists and administrators who have an interest in seeking solutions to problems of environmental pollution. With the omission of a few topics, the material in the book can also serve as an introductory course to first year engineering or science students. Environmental pollution control is an active and creative field; new methods of measurement and control are constantly being introduced. Only regular perusal of the current literature will enable the reader to keep abreast of the developments in this rapidly changing field. It is hoped that the book will ultimately serve to develop in the readers a proper concern for our deteriorating environment and encourage them to become actively involved in matters related to the environment.

The author completed a large part of the work on the book when he was Professor of Chemical Engineering at BITS, Pilani. He wishes to thank the authorities of BITS for providing the necessary library facilities and many of his former colleagues and students for their valuable suggestions. Finally, the author records his deep sense of gratitude to his father and teacher Prof. C. Venkata Rao without whose guidance this work would have been impossible.

Vallabh Vidyanagar
Gujarat
March, 1991

C.S. Rao

Contents

Preface iii

Chapter 1 IMPACT OF MAN ON THE ENVIRONMENT: AN OVERVIEW **1**

The Biosphere 1
The Hydrologic Cycle 3
The Nutrient Cycles 4
Consequences of Population Growth 10
Energy Problem 14
Pollution of Air, Water and Soil 17
Towards a Solution 25

Chapter 2 AIR POLLUTION: SOURCES AND EFFECTS **30**

Definition and Concentrations 30
Classification and Properties of Air Pollutants 34
Emission Sources 38
Major Emissions from Global Sources 41
Importance of Anthropogenic Sources 44
Behaviour and Fate of Air Pollutants 46
Photochemical Smog 50
Effects of Air Pollution 55
Health, Vegetation and Materials Damage in India 70
Air Pollution Laws and Standards 71

Chapter 3 METEOROLOGICAL ASPECTS OF AIR POLLUTANT DISPERSION **77**

Temperature Lapse Rates and Stability 77
Wind Velocity and Turbulence 85
Plume Behaviour 90
Dispersion of Air Pollutants 94

Solutions to the Atmospheric Dispersion Equation 97
The Gaussian Plume Model 101

Chapter 4 AIR POLLUTION SAMPLING AND MEASUREMENT 122

Types of Pollutant Sampling and Measurement 122
Ambient Air Sampling 122
Collection of Gaseous Air Pollutants 124
Collection of Particulate Pollutants 126
Stack Sampling 132
Analysis of Air Pollutants 142
Sulphur Dioxide 142
Nitrogen Oxides 146
Carbon Monoxide 148
Oxidants and Ozone 150
Hydrocarbons 151
Particulate Matter 152

Chapter 5 AIR POLLUTION CONTROL METHODS AND EQUIPMENT 157

Control Methods 157
Source Correction Methods 157
Cleaning of Gaseous Effluents 159
Particulate Emission Control 160
Gravitational Settling Chambers 163
Cyclone Separators 169
Fabric Filters 175
Electrostatic Precipitators 182
Wet Scrubbers 188
Selection of a Particulate Collector 198
Control of Gaseous Emissions 200
Absorption by Liquids 200
Adsorption by Solids 219
Combustion 229

Chapter 6 CONTROL OF SPECIFIC GASEOUS POLLUTANTS 238

Control of Sulphur Dioxide Emission 238
Desulphurization of Flue Gases 243
Dry Methods 243

Wet Scrubbing Methods 248
Control of Nitrogen Oxides 257
Modification of Operating Conditions 260
Modification of Design Conditions 263
Effluent Gas Treatment Methods 263
Carbon Monoxide Control 268
Control of Hydrocarbons 269
Mobile Sources 270

Chapter 7 SOURCES AND CLASSIFICATION OF WATER POLLUTANTS 277

Water Resources 277
Origin of Wastewater 282
Types of Water Pollutants and Their Effects 286
Oxygen Demanding Wastes 286
Disease Causing Agents 298
Synthetic Organic Compounds 299
Plant Nutrients 302
Inorganic Chemicals and Mineral Substances 303
Sediments 305
Radioactive Substances 305
Thermal Discharges 306
Oil 307
Water Pollution Laws and Standards 308

Chapter 8 WASTEWATER SAMPLING AND ANALYSIS 313

Sampling 313
Methods of Analysis 315
Determination of Organic Matter 315
Determination of Inorganic Substances 319
Physical Characteristics 323
Bacteriolcgical Measurements 325
Water Quality Standards 328

Chapter 9 WASTEWATER TREATMENT 333

Basic Process of Water Treatment 333
Primary Treatment 334
Pretreatment 334

Sedimentation 335
Flotation 347
Secondary (Biological) Treatment 348
Activated Sludge Process 349
Trickling Filters 361
Sludge Treatment and Disposal 363
Advanced Wastewater Treatment 368
Removal of Suspended Solids 368
Removal of Dissolved Solids 370
Nitrogen Removal 380
Phosphorus Removal 383
Advanced Biological Systems 385
Chemical Oxidation 387
Recovery of Materials from Process Effluents 389

Chapter 10 SOLID WASTE MANAGEMENT **396**

Sources and Classification 396
Public Health Aspects 399
Methods of Collection 399
Disposal Methods 402
Potential Methods of Disposal 408

APPENDICES

A. Conversion Factors and Constants **415**

B. Physical Properties of Fluids **418**

Notation Index **420**
Index **427**

1. Impact of Man on the Environment: An Overview

1.1 THE BIOSPHERE

The environment, which supports life and sustains various human activities, is widely known as the biosphere. The biosphere is a shallow layer compared to the total size of the Earth and extends to about 20 km from the bottom of the ocean to the highest point in the atmosphere at which life can survive without man-made protective devices. It is estimated that the biosphere contains at least 3.3×10^5 species of green plant, 9.3×10^5 species of animal and 8×10^4 species of bacteria and fungi known to science. The total number of existing species is certainly much more and may be close to three million. The essential requisites of life for all these species, namely, light, heat, water, food and habitats are supplied by the biosphere.

Since the biosphere is very complex and large, it is usually divided into smaller units or ecosystems. All ecosystems can be divided into two parts known as the biotic (living) and the abiotic (nonliving) components. Abiotic substances are the basic elements such as phosphorus and nitrogen, and the compounds found in the environment. The biotic category can be subdivided into three functional groups:

(i) producers—the autotrophic (self-nourishing) organisms, largely the green plants and algae;

(ii) consumers—the heterotrophic (other-nourishing) organisms, chiefly all animal life, including mammals, fish, insects and birds; these ultimately depend upon consumption of producers to sustain life;

(iii) decomposers—heterotrophic organisms, chiefly the bacteria and fungi that break down complex compounds from waste materials—including dead producers and consumers—to again make the chemical components available to producers.

A typical example of an operating ecosystem is the pond. The abiotic substances are the water, the nutrients, oxygen, carbon dioxide, etc. The producers in the ecosystem (Fig. 1.1) are the large rooted plants and the free floating minute plants, usually algae, called phytoplankton. These store energy and liberate oxygen. The primary consumers are benthos, or bottom forms, and zooplankton with little or no swimming ability. The phytoplankton are consumed by zooplankton, which are in turn eaten by

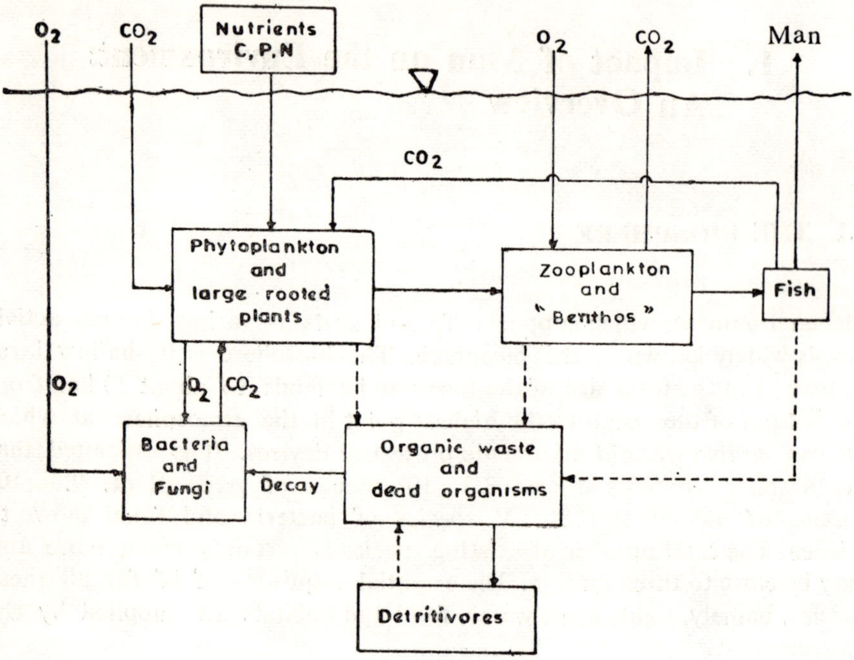

Fig. 1.1 An aquatic ecosystem.

large aquatic life such as fish. Other consumers are the insects, frogs, man, etc., and a category called detritivores which live on the organic wastes. All of these forms produce organic waste and dead organisms. The decomposers, bacteria and fungi, utilize the organic carbon and generate CO_2 which in turn is used by the algae. Additional CO_2 is provided from the atmosphere and through the respiration of fish. In a healthy system, the availability of the nutrients carbon, phosphorus and nitrogen is sufficiently small so as to limit the production of algae and a dynamic equilibrium is maintained throughout the ecosystem. Chemical elements circulate between the organisms and the environment through pathways comprising the natural cycles, the most important of which are probably (1) the hydrologic cycle and (2) the biogeochemical cycles of (i) carbon, (ii) nitrogen, (iii) phosphorus and (iv) sulphur. In the unpolluted natural environment, these cycles operate in a balanced state with little variation thereby contributing to the stability of the whole biosphere.

Ever since he appeared on the face of the Earth, man has exploited and modified the environment to his advantage in many ways. Until recently, however, these changes have represented only a small perturbation in a large system and have not seriously threatened the homeostatic mechanisms, called the feedback loops, which tend to maintain the system in a stable condition. Now unfortunately man is capable of inducing large enough perturbations in the ecosystems, which can permanently upset the balanced state of the natural cycles.

1.2 THE HYDROLOGIC CYCLE

The hydrologic or water cycle is probably the most important of all the natural cycles in the biosphere. More than 97 per cent of water in the biosphere is found in the oceans; the remaining 3 per cent is found on the continents and in the atmosphere. But more than 70 per cent of this latter portion is locked in glaciers and icecaps. The water on which humans

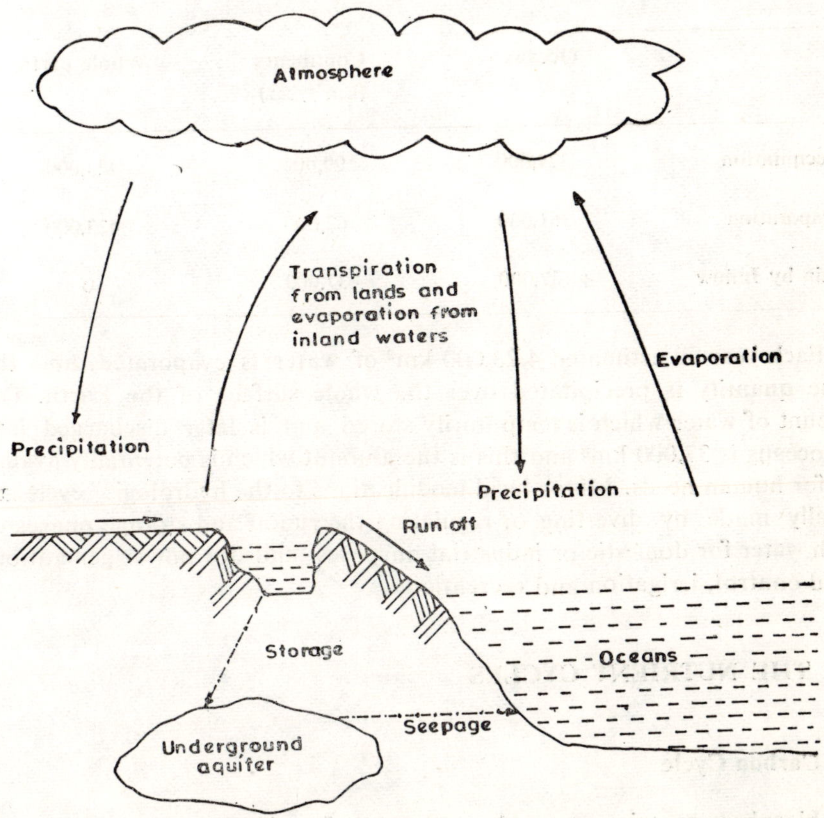

Fig. 1.2 The hydrologic cycle.

depend so heavily—lakes, streams and ground water—accounts for less than 1 per cent of the total supply, and it is this water which is currently being used and reused in many parts of the world.

The hydrologic cycle of the biosphere depends on the reciprocity of evaporation and precipitation. Liquid water on the Earth goes into the atmosphere as vapour by evaporation and transpiration of the plants. The vapour is returned to Earth as rain or snow (precipitation). Fig. 1.2 illustrates the complete hydrologic cycle. Most evaporation occurs over the oceans but some oceans lose more water by evaporation than they gain by precipitation. The difference is made up by runoff and seepage from the continents, over which there is more precipitation than evaporation. The continents lose more than 50 per cent of the precipitation through evaporation and the remainder is temporarily stored in lakes and rivers or as ground water which is later discharged into the oceans. The global cycle can be summarized as shown in Table 1.1.

Table 1.1 Water balance of the earth

	Oceans	Continents (km^3/year)	Whole earth
Precipitation	324,000	99,000	423,000
Evaporation	361,000	62,000	423,000
Gain by Inflow	+ 37,000	−37,000	0

Each year an estimated 4,23,000 km^3 of water is evaporated, and the same quantity is precipitated over the whole surface of the Earth. The amount of water which is temporarily stored and is later discharged into the oceans is 37,000 km^3 and this is the amount which is potentially available for human needs. Minor local modifications to the hydrologic cycle are usually made by diverting or regulating the runoff and storage phases of fresh water for domestic or industrial purposes and for power generation, flood control, irrigation and recreation.

1.3 THE NUTRIENT CYCLES

The Carbon Cycle

The biosphere contains a complex mixture of carbon compounds in a dynamic equilibrium of formation, transformation and decomposition. The dynamics of the carbon cycle is presented in Fig. 1.3.

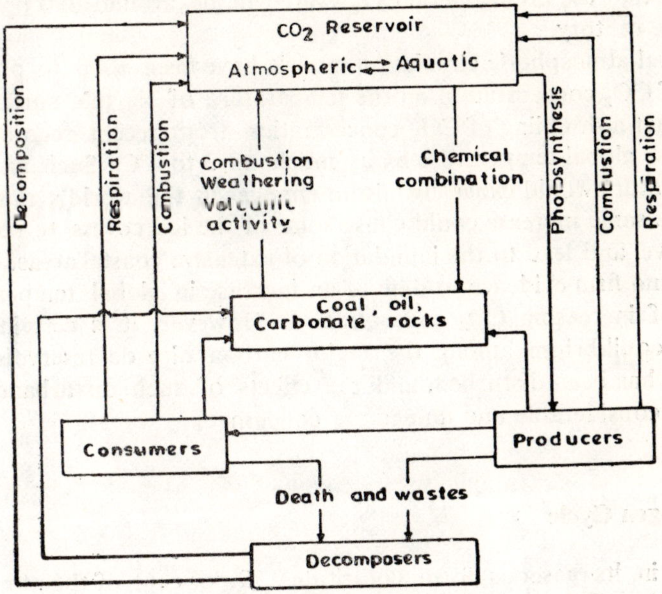

Fig. 1.3 The carbon cycle.

The producers, through photosynthesis, reduce carbon dioxide from the atmosphere to organic carbon. This then passes through consumers and decomposers, then usually re-enters the atmosphere through respiration and decomposition. Additional return from producers and consumers occurs through the nonbiological process of combustion. In addition to the atmosphere, another major reservoir of carbon is the ocean, which stores more than 50 times as much as the atmosphere. Interchange of carbon dioxide between the atmosphere and the ocean takes place through diffusion and the oceanic reservoir tends to regulate the atmospheric CO_2 concentration.

Apart from the daily production and consumption of carbon (in the form of CO_2), the Earth has significant reserves of bound carbon in the form of inorganic deposits such as limestone and organic fossil fuel deposits consisting of mainly coal and petroleum. Due to combustion of fossil fuels, weathering and dissolution of carbonate rocks, and volcanic activity, some of the bound carbon returns to the atmospheric aquatic reservoir as carbon dioxide or carbonic acid.

Combustion of large amounts of fossil fuels can increase the amount of CO_2 in the atmosphere. Studies have shown that plants tend to grow faster in a CO_2-enriched atmosphere[1], but this benefit is offset by denudation of forests by man thereby decreasing nature's ability to remove atmospheric

CO_2. As a result, a detectable increase in the concentration of atmospheric CO_2, from 290 ppm in the nineteenth century to the current level of about 325 ppm, has been observed. If the present trend continues, it has been estimated that the atmospheric CO_2 level will be around 380 ppm by the end of the century.

General atmospheric circulation models have been used to predict the impact of CO_2 concentration on the temperature of Earth's surface. They suggest that a doubling of CO_2 concentration from recent recorded levels might raise global temperature by as much as 3 to 4°C. Such an increase in temperature would cause significant changes in the world's rainfall patterns. The same increase could cause some of the ice covers to melt, raise the sea level and lead to the inundation of extensive coastal areas. There is, however, no firm evidence to date of an increase in global temperature as a result of increasing CO_2 concentration. However, it is certain that the dynamic equilibrium among the major carbon dioxide reservoirs of the biosphere has been disturbed, and the effects of such disturbances are a matter of considerable and immediate concern.

The Nitrogen Cycle

Nitrogen in its gaseous form constitutes 79 per cent of the atmosphere. However, it cannot be used directly by most forms of life. It must first be "fixed" before it can be utilized by plants and animals. By fixation, nitrogen is converted into its chemical compounds, largely nitrates (NO_3) and ammonia (NH_3). The fixation of nitrogen takes place through both physicochemical and biological means although the latter is by far the much bigger contributor. The biological fixation is limited to a few, but abundant organisms like the free living bacteria *Azetobacter* and *Clostridium*, nodule bacteria on leguminous plants like *Rhizobium*, and some blue-green Algae. These are the keys to the movement of nitrogen from the atmospheric reservoir into the cycle shown in Fig. 1.4.

The nitrates are assimilated to form amino acids, urea, and other organic residues in the producer, consumer and decomposer cycles. The amino acids and urea are then converted to ammonia through a process called "ammonification". To complete the cycle, denitrifying bacteria convert the ammonia into nitrites, then into nitrates, and then back into gaseous nitrogen. In this way, under normal circumstances, the total amount of nitrogen fixed equals the total amount returned to the atmosphere as gas.

Man has interfered with this natural cycle by industrially fixing nitrogen. This includes production of nitrogen fertilizers and oxidation of nitrogen during fossil fuel combustion. Most of the excess nitrogen is carried off into rivers and lakes and ultimately reaches the ocean. This increased runoff has greatly increased the productivity in many aquatic environments and has contributed to the process of eutrophication.

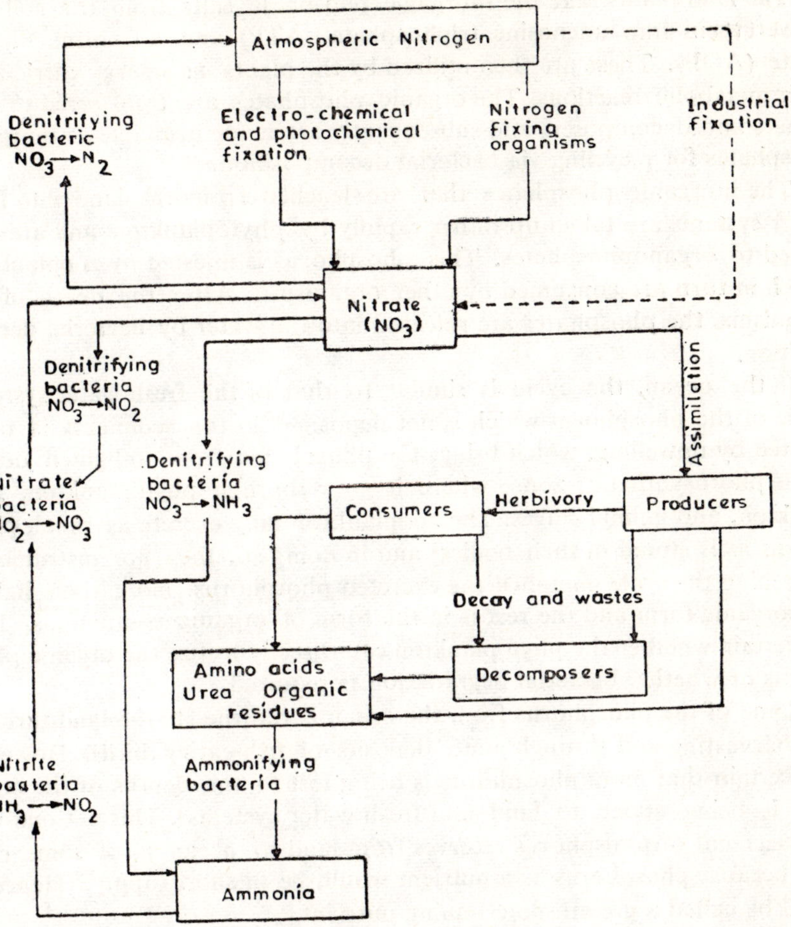

Fig. 1.4 The nitrogen cycle.

The Phosphorus Cycle

Phosphorus plays an important role in the growth of living tissue, for it is involved in the metabolic processes of energy transfer. It is quite different from other major elements of the biosphere in that it does not form gaseous compounds at normal temperatures and pressures, and hence it cannot return to the atmosphere. The main reservoirs of phosphorus on land are rock and natural phosphate deposits. Rain and other natural processes cause phosphorus to be released to the soil and much of it is fixed in the soil or absorbed on to soil particles. But some of it is lost to the water bodies such as lakes and streams and eventually ends up in the ocean, where the phosphorus is deposited partly in shallow sediments and partly in the deep zone.

The land plants take the inorganic phosphate salts from the soil and convert them into adenosine triphosphate (ATP) and adenosine diphosphate (ADP). These are then utilized by the plants as energy carriers for their metabolic reactions. The organic phosphates are transferred to consumers and decomposers and subsequently are made available as inorganic phosphates for recycling via bacterial decomposition.

The inorganic phosphates that are leached from the land into fresh water systems are taken up rather rapidly by phytoplankton and are converted to organophosphates. This phosphorus is injested by zooplankton, which in turn are consumed by other organisms. After the death of the organisms, the phosphates are released into the water by bacterial decomposition.

In the ocean, the cycle is similar to that of the fresh water systems. Some of the phosphorus which is not deposited in the sediments is recirculated by upwelling, which brings the phosphates from unlighted depths to the photosynthetic zone, where it goes through phytoplankton, zooplankton, and animal stages. The zooplankton may excrete as much phosphorus as is stored in their bodies[2] and in doing so, they are instrumental in keeping the cycle going. Of the excreted phosphorus, more than half is in inorganic form and the rest is in the form of organic compounds. It is not certain whether the phytoplankton can directly utilize the organic phosphorus or whether bacterial degradation is required.

Some of the phosphorus from the ocean is returned to the land through fish harvesting and through guno deposits of fish-eating birds. However, it is certain that more phosphorus is being lost to the depths of the ocean than is being added to land and fresh water systems[3]. This net one-way displacement of phosphorus reserves from land to ocean is of some concern because phosphorus as a nutrient would be in short supply. Hence, it could be called a growth-determining nutrient.

Lack of phosphorus in the soil causes soil infertility, and to replenish the lost phosphorus, synthetic fertilizers are now being frequently used. Because of the accelerated use of fertilizers and synthetic detergents, the phosphorus cycle is substantially affected. Phosphorus pollution has contributed to the eutrophication of many water bodies and it may also have an adverse effect on the natural food chains.

The Sulphur Cycle

Sulphur, like nitrogen, is a basic constituent of proteins in plants and animals. It is found in the biosphere in a wide variety of forms. Sulphur dioxide (SO_2) and hydrogen sulphide (H_2S) are the important gaseous forms, and the sulphate ion (SO_4^{2-}) is the common form found in water and soil. The distribution of sulphur in the environment is shown in Fig. 1.5.

Sulphate ion is reduced after being absorbed from the soil by plants and bacteria, and ultimately incorporated as the sulphydryl group (—SH) in proteins. Some sulphates are reduced under anaerobic conditions directly to sulphides, including H_2S, or to elemental sulphur by a class of bacteria known as *Desulfovibrio* bacteria, found largely at ocean bottom. The hydrogen sulphide thus produced escapes as a gas into the atmosphere and replenishes the sulphur lost by precipitation. In the presence of oxygen, H_2S is rapidly oxidized to sulphates by bacteria of genus *Thiobacillus*. Even in the absence of oxygen, several types of bacteria such as *Chlorobacteriaceae* and *Thiorhodaceae* oxidize H_2S to elemental sulphur.

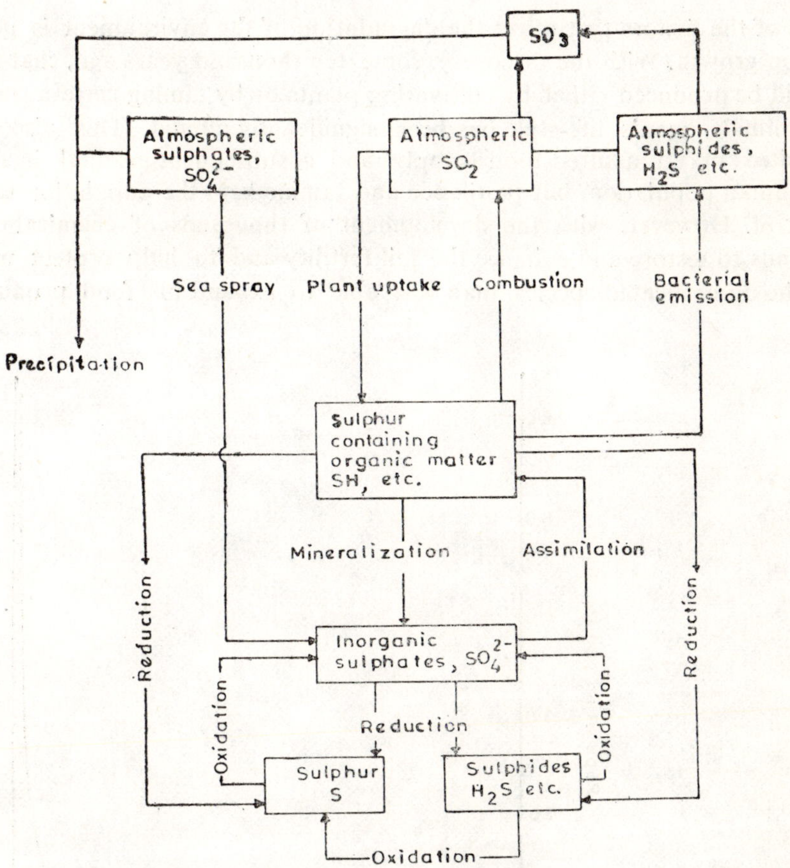

Fig. 1.5 The sulphur cycle

Atmosphere receives sulphur through bacterial emission (H_2S), fossil fuel burning (SO_2), wind-blown sea salts (SO_4^{2-}) and volcanic emissions (H_2S, SO_2, SO_4^{2-}). Most of the sulphur in the form of SO_2 or H_2S is converted to sulphur trioxide (SO_3) which dissolves in water droplets to form sulphuric acid. The sulphates and the acid then precipitate with rain.

The sulphur cycle is overloaded due to burning of fossil fuels at an ever increasing rate. As a result, the SO_2 emitted into the atmosphere this way constitutes a significant fraction of total global sulphur transport. This increased amount of sulphur is changed mostly to the form of sulphuric acid in rain water causing adverse ecological effects.

1.4 CONSEQUENCES OF POPULATION GROWTH

One of the factors that affect the degradation of the environment is population growth. With the discovery, some ten thousand years ago, that food could be produced either by cultivating plants or by taming certain species of animals, man's life-style has been significantly altered. This discovery resulted in an assured food supply and a subsequent gradual increase in human population, but pestilence and famine kept the population under control. However, with the development of thousands of chemical compounds to restore and enhance the soil fertility and to help protect many of the domesticated species, man was able to expand his food producing

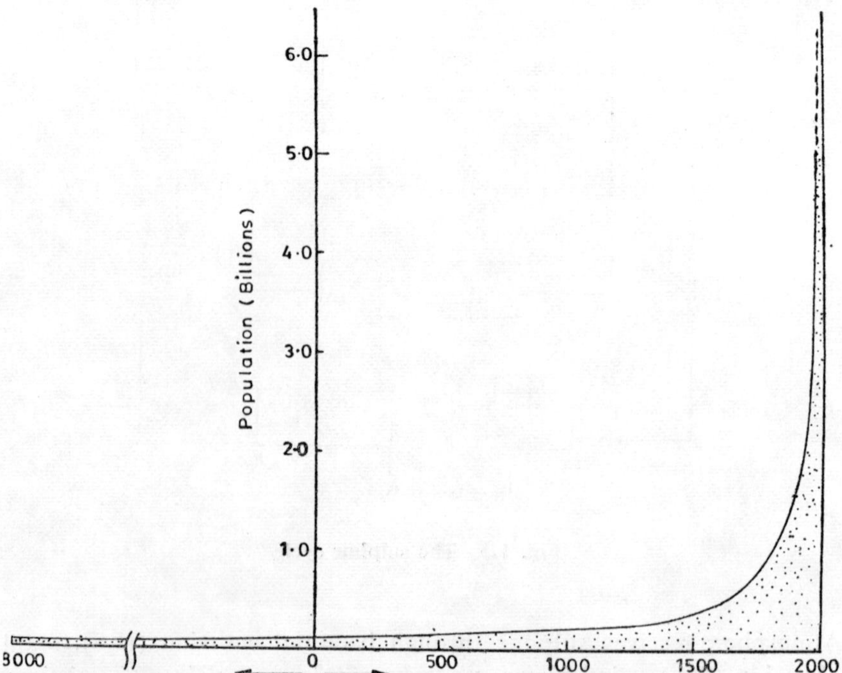

Fig. 1.6 World's population growth.

capacity significantly. These activities together with the birth of modern medicine have resulted in an explosive growth of population with inevitable consequences.

To assess the impact of population on the environment, it is first necessary to consider the growth of world population, the general trend of which is shown in Fig. 1.6. This arithmetic plot shows a slow rate of increase until about 500 years ago, when the numbers began to increase more rapidly. The growth is not only exponential in nature but the rate of growth has also increased. Table 1.2 shows the population change alongwith the doubling time at various stages since 8000 B.C.

Table 1.2 Population growth and doubling time (ref. 4)

Year	Population (Millions)	Doubling time (Years)
8000 B.C.	5.32	
		10×10^2
4000 B.C.	86.5	
		64×10^2
A.D. 0	1.33×10^2	
		8.3×10^2
1650	5.45×10^2	
		2.4×10^2
1750	7.28×10^2	
		1.6×10^2
1800	9.06×10^2	
		1.2×10^2
1900	16.10×10^2	
		87
1950	24.0×10^2	
		36
2000 (estimated)	62.70×10^2	

The doubling time ranges from 6400 years—a very stable population in equilibrium with its environment—to 36 years, the value predicted for the present time. The world population reached ~545 million by 1650. Advances in food production and industry lowered the death rate, and this further shortened the time needed for the population to double. By 1850 the population had crossed one billion, and by 1930, it had doubled again. In 1976 world's population crossed the four billion mark, with a doubling time of 46 years. This dramatic growth has coincided with the development of cheap sources of energy (coal, petroleum, natural gas, and hydroelectric power) and with the scientific and industrial revolution.

The simplest measure of population growth is the annual percentage increase which relates easily to doubling times :

$$T_d \simeq \frac{70}{r}$$

where r is the per cent rate of growth, and T_d is the doubling time in years. The rate of growth of population of a region can be predicted by the equation :

$$r = b - d + m$$

where b is the birth rate, d is the death rate, and m is the net migration rate into the region. For example, world's annual birth and death rates in 1973 were 3.3 per cent and 1.3 per cent, respectively. Since $m = 0$, the rate of growth of world population becomes 2.0 per cent, which gives a doubling time of about 35 years.

The annual 2.0 per cent increase in world population does not seem particularly large until the characteristics of exponential growth are examined carefully. With the doubling time of 35 years, and at 2.0 per cent increase, a population of 4 billion in 1976 becomes 16 billion in 2046 and 64 billion by the year 2116. The uniqueness and abruptness of population explosion is self-evident. Obviously this unlimited growth is impossible in a finite system such as the Earth and environmental constraints will eventually restrict any further increase. The cause for alarm, however, is that the population explosion poses a grave threat to the environment in relation to human settlements; this can be clearly elucidated in the Indian context.

India occupies 2.5 per cent of the land area of the world while about 15 per cent of the population of the world reside in this country. The population of India was 684 million in 1981 and is growing exponentially at a rate of slightly less than 2 per cent per year with a doubling time of about 35 years. The steep increase in population is attributed to the combined influence of a steep reduction in the death rate and an almost stable level of fertility at 40 per thousand of population. This is evident from Fig. 1.7 where the birth and death rates during successive decades from 1901 to 1971 are plotted.

As is seen from the plot, the growth of population during the decade 1961–1971 has been the highest. This rapid population growth has been interacting with a fixed arable land area and thus contributing to a massive rural→urban migration. The rate of urban growth has constantly risen from 1901 to 1971. For instance, about 10 per cent of the population lived in urban areas in 1911; this has almost doubled by 1971. Most of this increase has occurred since 1931, with the largest per cent increase occurring in the decade 1941–1951. It is estimated that by the end of the century, about one-third of the population of India will be urbanized. This continuous migration of people from countryside to city is causing serious social problems like over-crowding, increase in crime rate, and unemployment.

Similarly, the disproportionate need for social services due to rapid urbanization leads to many dangerous situations, particularly regarding sanitation and health.

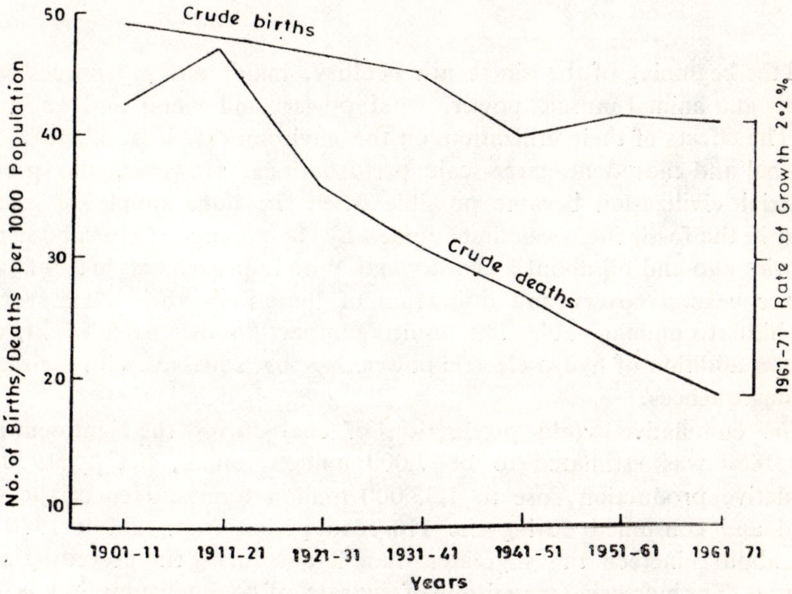

Fig. 1.7 Different patterns of change in birth and death rates in India (ref. 5, reproduced with permission of National Book Trust, India, New Delhi, 1977).

The role of population growth in creating a disharmonious impact on the environment was thoroughly analyzed by Ehrlich and Holdren[6]. As human population increases, the absolute minimum requirements of minerals and fossil fuels also increases, and even provision for these requirements at fixed per capita consumption entails an increase in per capita use of energy. This is due to the use of lower-grade ores after concentrated ores are exhausted, deeper drilling for petroleum and extended distribution networks. All these activities increase the per capita impact on the environment. Similarly, increase in food production results in an increase in per capita energy consumption which is disproportionate to the population fed because of the need to obtain and distribute water, fertilizer and pesticides.

A study conducted by the SCEP (Study of Critical Environmental Problems) group[7] points out that the demand for minerals, energy and space are exponentially increasing at a rate of 5 to 6 per cent per year, and if this trend continues, the next doubling of population will increase the environmental impact sixfold. Thus, it is certain that the systems of the biosphere will collapse under the weight of exponentially growing population unless control measures are implemented successfully.

1.5 ENERGY PROBLEM

Until the beginning of the nineteenth century, man's energy sources were human and animal muscle power, wind power, and wood and vegetable oils. The effects of their utilization on the environment, if any, were only marginal and caused no large-scale perturbations. However, the present industrial civilization became possible when the huge supply of energy stored in the fossil fuels was first tapped by the mining of coal about nine centuries ago and oil about a century ago. With improvements in technology for discovery, recovery and utilization of these fuels, their use grew on a scale hitherto unimaginable. The resulting impact on the environment, with the later addition of hydro-electric power, has been massive with far-reaching consequences.

The cumulative world production of coal during the eight centuries upto 1860 was estimated to be 7,000 million tonnes, but by 1970 the cumulative production rose to 1,33,000 million tonnes. Hence, the coal mined and consumed during the 110-years period from 1860 to 1970 has been about nineteen times greater than it was during the preceding eight centuries. The increasing magnitude of the rate of coal consumption is also evident from the fact that during the 30 years period from 1940 to 1970, the amount of coal mined was approximately equal to that produced during all the preceding history. However, in the 1960s, coal lost its place as the dominant source of energy to oil due to the cheap availability and convenience in use of the latter.

The world cumulative production of oil upto 1970 was about 233×10^9 barrels. Of this amount, the first half required 103 years (from 1857 to 1960) to produce in contrast to only 10 years (from 1960 to 1970) required to produce the remaining half. From 1890 to 1970, the oil production grew exponentially at a rate of 6.94 per cent per year with a doubling time of about ten years. The world reserves of crude oil were estimated to range from 1350×10^9 barrels to 2100×10^9 barrels. Starting in the mid-1960s, the rate of discovery of new oil reserves has slowed down appreciably. The present indications[8] are that the continuing increase in world oil demand combined with the decline in reserve additions may result in the occurrence of the peak in world oil production sometime in the period 1990 to 2010 after which it will decrease steadily.

Since oil will not be able to meet the world energy demand much longer, the only fossil fuel which seems to remain a major source of energy much beyond the year 2000 is coal. The world's resources of minable coal (50 per cent of the coal actually present) are estimated to be

7.6×10^{12} metric tonnes. The production of coal is increasing exponentially at a rate of 3.6 per cent per year, and if this growth is maintained, the resources will be exhausted in about 120 years. More realistically, Hubbert[8] has estimated that the peak production will occur sometime between the years 2100 and 2150 and that the middle 80 per cent of the resources will be consumed in a period of about 300 years, from 2000 to 2300. However, if the production of coal is accelerated further because of shortages in petroleum, it will last only 100 to 200 years as a major energy source.

India's Fossil Fuel Scene

India's energy requirements are met from traditional sources such as firewood, dung cakes, and agricultural wastes, and modern commercial sources such as coal, oil and hydro-electric power. Associated gas and nuclear energy have also a small share as energy sources in the primary energy consumption. Though the consumption of modern forms of energy has been growing rapidly during the past three decades, the non-commercial fuels still play a dominant role in India's energy scene and still account for about 50 per cent of the total energy consumption. The use of non-commercial energy sources on a large-scale should be discouraged as far as possible from the point of view of the environmental damage. Felling of trees for fuel leads to deforestation, soil erosion and finally desertification. Firewood consumption is about 65 per cent of the total non-commercial energy sources in India, and it has been estimated that only 23 per cent of the country is now under forest cover as against 33 per cent recommended in the National forest policy. As vast areas are deforested, people are forced to use cowdung for fuel, thereby depriving the soil of a natural fertilizer. This in turn is forcing the farmers to depend more and more on chemical fertilizers leading to adverse effects on the environment.

In India the requirements of commercial energy are closely related to the issues of population expansion and economic development. Growth in population increases the demand for commercial energy manyfold because of the urgency of economic development to improve the living standard of the vast majority of population. For example, the coal supply industry has been steadily expanded and modernized, and its output has increased nearly four-fold in the last three decades, from 37 million tonnes in 1953-54 to about 160 million tonnes in 1984-85. Similarly, the intensive and extensive explorations for oil and gas—both on- and off-shore—have led to major discoveries of oil and gas reserves and a spectacular increase in indigenous production. It can be seen from Fig. 1.8 that the oil production has increased tremendously, from 0.45 million tonnes in 1960-61 to 21 million tonnes in 1982-83. Now, petroleum accounts for 50 per cent of the total commercial energy consumption in India and its products are used in diverse sectors like transportation, industry and agriculture. At present the proven recoverable reserves of oil and gas are placed at 471 million tonnes and 420 million cubic meters, respectively. These figures are likely to increase with new discoveries particularly in the offshore basins of Godavari and Cavery on the East Coast. However, with increasing demands on petroleum, this may be the scarcest energy source in the long term.

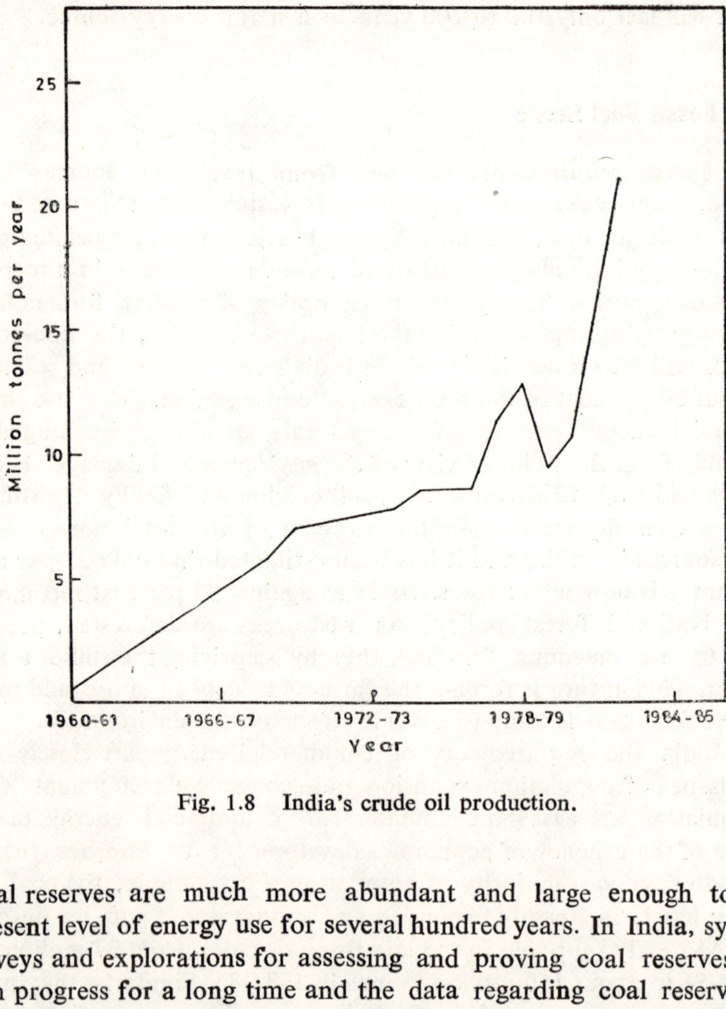

Fig. 1.8 India's crude oil production.

Coal reserves are much more abundant and large enough to meet the present level of energy use for several hundred years. In India, systematic surveys and explorations for assessing and proving coal reserves have been in progress for a long time and the data regarding coal reserves are fairly reliable. The assessment of 1978 places the coal reserves, occurring at the depth of up to 600 metres and in seams of 1.2 metres and above, at 86,428 million tonnes.

At present the production of coal is increasing exponentially at a rate of 6.6 per cent per year and this rate has to be maintained if the target of around 420 million tonnes for the year 2000 has to be met. However, at this rate of increase the coal reserves are expected to last for only 65 years,

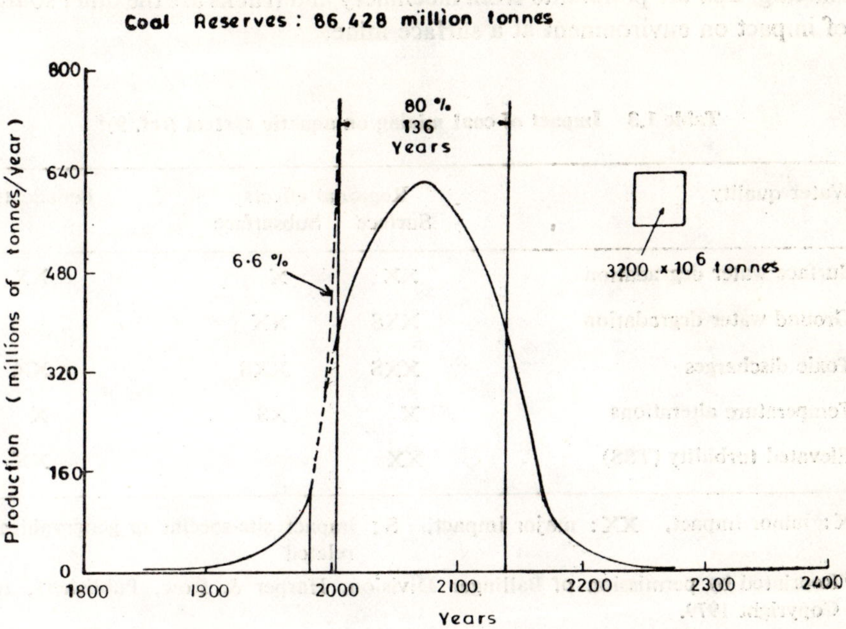

Fig. 1.9 A possible trend in coal consumption.

counting from 1974 when the production was 78 million tonnes. More realistically, the expected trend is likely to be as shown in Fig. 1.9, where the coal production will attain a peak sometime between the years 2060 and 2080 and the middle 80 per cent of the reserves will be exhausted in a period of 136 years.

1.6 POLLUTION OF AIR, WATER AND SOIL

Impact of Coal Mining

Coal mining requires a lot of land, whether the coal is mined on surface or underground. Surface or open-cost mining takes up large land areas, and nearly 75 per cent of the increased production in India is sought to be achieved through progressively increasing the share of open-cast mines. In addition to destroying large land areas, unreclaimed mine sites pollute aquatic streams with runoff of mine drainage containing toxic soluble metal salts and sediment. The potential impact of coal mining on aquatic ecosystems is shown in Table 1.3. Reclamation efforts have been limited and generally ineffective even in Europe and U.S.A. Also, noise, dust from blasting, and air pollutants from mechinery and trucks are the other sources of impact on environment at a surface mine.

Table 1.3 Impact of coal mining on aquatic system (ref. 9)*

Water quality	Regional effects Surface	Regional effects Subsurface	Benefication
Surface water degradation	XX	X	XX
Ground water degradation	XXS	XX	
Toxic discharges	XXS	XXS	XX
Temperature alterations	X	XS	X
Elevated turbidity (TSS)	XX		XX

X: minor impact, XX: major impact, S: impact site-specific or geographically related

In underground mining, the main problems are the acid mine drainage and the health and safety of the miners. Coal mines have a high record of death and injury from explosions, fires and cave-ins. About 80,000 workers have died so far in U.S.A. alone in coal mine accidents; the miners inhale coal dust which may result in pneumoconiosis—popularly known as black lung. It has been estimated that 10 to 15 per cent of the miners suffer from this disease. It leads to disability and often results in death because no cure is known for this disease. These are the results of analysis carried out in U.S.A. where coal mining is more mechanised than in India. Even though proper statistics about the fatalities in coal mines in India are not available, it can be concluded that the fatalities in India are not likely to be less than the U.S. figures.

Fossil Fuel Related Pollutants in the Environment

When coal and other fossil fuels are burned or processed, a variety of pollutants are emitted into the atmosphere. These include, particulates, trace elements, a variety of hydrocarbons, and sulphur and nitrogen compounds. Probably the most important pollutant emitted by coal combustion is sulphur dioxide. SO_2 is oxidized to H_2SO_4 and some of the H_2SO_4 in turn reacts with trace elements emitted alongwith SO_2 or other substances present in the atmosphere to form submicron-size sulphate compounds. Recent studies indicate that it is the aerosol or 'mist' containing H_2SO_4

and metallic sulphates, and not the SO_2 gas itself, that is the major ingredient in sulphur pollution. It impairs visibility, damages vegetation and materials, and harms human health. Small-particle air pollution is a major unsolved environmental health hazard relating to coal and other fossil fuels.

Another environmental problem of great concern, which is related to sulphur in fossil fuels, is the acid rain. There is some speculation that tall smoke stacks aid the long-range transport of sulphur (up to 1000 km), making the acid rain problem a regional one. This has been corroborated by a series of studies in Sweden where vast tracts of forest and lakes have been or are being destroyed. Acidity kills off fish, bacteria and algae[10], and the aquatic ecosystem collapses into sterility leaving a clear but ultimately dead lake. There is also evidence that rainfall in the industrialized regions of U.S.A. has a significantly lower pH than that elsewhere in the country.

Table 1.4 Coal-related environmental effects (ref. 9)*

Environmental problem	Geographic scale
Carbon dioxide from combustion	Hemispheric/global
SO_2 emissions from power plants	Regional/hemispheric
Particulate emissions from power plants	Local/regional
NO_x emissions from combustion	Local/regional
Surface mining and discharges	Local/regional
Waste heat from power plants	Local/regional
Cooling water discharges	Local
Sludge disposal	Local

*Reprinted by permission of Ballinger Division, Harper & Row, Publishers, Inc., Copyright 1979.

The environmental effects of acid rain probably include increased leaching of nutrients such as calcium from soil, changes in metabolic rates of organisms which depend on acid or base catalysts, and corrosion of basic materials such as limestone and marble[11].

The adverse effects of using coal may vary widely in the geographic scale. Table 1.4 lists some important sources of coal-related environmental effects and the geographic scope of these effects.

Impact of Urban Areas

The combustion of fossil fuels and their products is responsible for a sizable amount of anthropogenic air pollution, and this problem is particularly acute in urban areas. Motor vehicles in urban areas produce hydrocarbons, CO, NO_x, and lead particles. The problem is further aggravated by the formation of secondary pollutants in the atmosphere due to complex chemical reactions. Photochemical smog, the product of such reactions, gives rise to the familiar haze (particularly in summer) and to vegetation damage, eye irritation, and respiratory distress.

In urban areas, the most noticeable effect of air pollution is the reduction in visibility. This results from the scattering of light by airborne liquid and solid particles, particularly in the submicron range from 0.1 to 1 μm, leading to safety hazards. Air pollution also affects urban climate with respect to increased fog formation and reduced solar radiation. It is estimated that urban areas receive 15 to 20 per cent less total solar radiation than rural areas due to pollution. The weather is also affected because a city forms a heat island, i.e., the temperature within an urban area may be many degrees higher than that in the surrounding rural area. In addition, pollutants such as SO_2 in sulphurous smogs and ozone in photochemical smogs may intensify the greenhouse effect over urban regions, leading to even higher ambient temperatures.

Environmental Impacts of Hydro-electric and Nuclear Energy

Hydro-electric Energy

The world's potential production of hydro-electric power is roughly half the amount of power now being produced by fossil fuels. The areas with the largest potential capacities of hydro-electric power are the industrially developing regions of Africa, South America and South East Asia where combined capacities represent about 60 per cent of the world total. The total hydro-electric potential of India is estimated to be around 41,000 MW, but so far only 11 per cent of the total potential has been realised. This indicates that dam construction is going to be a major development activity. Power schemes presently under construction would enable utilization of further 7 per cent of the total potential.

Although regarded as clean and harmless, hydroelectric plants have come in for much criticism recently by environmentalists. Construction of large dams adversely affects the hydrology, the terrestrial system and the aquatic systems of the basin. The hydrology of the basin is altered downstream of a dam, especially if the dam is completely closed during the filling period. Thereafter, the volume of discharge and the current velocity downstream are obviously altered resulting in a reduction in river turbidity. The salinity of water flowing into the river below the reservoir increases as the basin evaporation rates rise. This in turn affects crop lands using river water for irrigation in downstream areas, and ultimately the soil is ruined for crop cultivation.

There is also the possibility that large reservoirs may trigger seismic movements, especially in high fault regions. Since silting reduces the life-span of a reservoir, a silt trap is built into the design, where possible. If the sediment piles against the dam, it may be periodically flushed out, with disastrous effects on the ecology of the river downstream. When the initial flooding takes place, there is a biogeochemical enrichment of the water, usually resulting in an explosive growth of weeds. These weeds, in addition to transpiring large amounts of water into the atmosphere, prevent algal photosynthesis and, hence, lead to serious depletion of fisheries.

Nuclear energy

Nuclear energy now supplies about 3 per cent of India's electricity and an ambitious target of 10,000 MW of nuclear power has been set for the year 2000, which is a ten-fold increase from the installed capacity of 1095 MW as of 1983. The latest estimates of the total uranium reserves in India are of the order of 67,000 tonnes of U_3O_8. In addition, about 13,000 tonnes of U_3O_8 are estimated to be present in the monazite sands. India is reputed to have one of the largest reserves of thorium in the world. The known reserves in the form of monazite sands are estimated at 363,000 tonnes of thorium oxide. The uranium reserves are not ample; they can barely support the thermal reactor programme of 10,000 MW. However, with the development of fast breeder reactor programme based on uranium or thorium, the total capacity can grow to many hundreds of thousands of MW.

Nuclear power has significant advantages with regard to air pollution, and in normal operations, radioactive emissions can be kept to very low levels. The land use advantage with nuclear power is also significant, since coal surface mining requires about 17 times as much land as uranium ore mining to produce an equivalent amount of electricity[12]. Yet the nuclear power has been a source of much controversy in recent years. Fears have been expressed about environmental contamination by radioactivity, particularly that arising from disposal of radioactive wastes. Concern has also been shown about the technological and sociological problems involved in the handling of plutonium when fast breeder reactor technology becomes established.

Mining, milling and refining of the uranium and thorium ore may result in considerable radiation hazard to the workers. The most significant hazard in uranium mining is through the inhalation by workers of radon gas and its decay products. This is known to increase the incidence of cancer among miners. The major threat to the environment from radio-active pollution, however, occurs during the phases of reactor operation, chemical reprocessing of spent fuel, and waste disposal. Under normal working conditions the public and the workers in the industry are exposed to low level background radiation that can cause both somatic and genetic damage. Fission byproducts of nuclear reactors can escape in liquid and gaseous effluents. Some of these fission products, their half-lives and environmental effects are summarized in Table 1.5. Strontium-90, for example,

Table 1.5 Environmental effects of some fission reactor emissions

Radioisotope	Half-life	Environmental effect
Sr—90	25 years	Collects in bones, may cause leukemia
I—131	8 days	Concentrated in thyroid
Kr—85	10.4 years	Not chemically reactive, so not concentrated
Cs—137	30 years	Deposited in the soft tissue, irradiates the whole body

with a half-life of about 25 years is one of the most dangerous radioactive pollutants. Because of its chemical similarity to calcium, it is deposited in the bones. Apart from the somatic effects such as leukemia and bone cancer, it has been reported to have significant genetic effects on mice.

The greatest risk of radioactive contamination exists when spent reactor fuel is taken to a chemical reprocessing plant. The fuel is dissolved in chemical solutions and the unburned uranium and plutonium are recovered from the radioactive waste. Thus, the bulk high level radioactive wastes are reduced to a form suitable for long term storage. The waste material is usually stored in steel containers in underground concrete vaults. But in view of the growing quantity of waste, a more efficient and less expensive method must be found for disposal. Table 1.6 gives an idea of the magnitude of the nuclear wastes generated in India. Several other problems are related to nuclear power. Hijacking, sabotage and diversion of fissionable material from power plants, reprocessing plants, or when these are in transit, are some of the potential problems. These are bound to become important in the near future as more and more power plants are constructed, especially if they are breeder reactors based on a plutonium fuel cycle.

In addition to the radioactive pollution problems, nuclear power plants create about 50 per cent more thermal pollution than fossil fuel plants. Thermal pollution is a serious hazard to aquatic systems because the availability of dissolved oxygen is greatly reduced and this is detrimental to aquatic life. Moreover, the life cycle and natural processes of many aquatic life forms are so closely geared to water temperature that any artificial changes in temperature can completely disrupt the normal life patterns.

Table 1.6 Estimated nuclear wastes generated in India (ref. 15)

	Year	
	1982	2000
Installed capacity (MW)	800	10,000
Primary solid waste (m^3)	1700	107,000
Low-level waste concentrates (m^3)	2500	77,100
Intermediate level wastes (m^3)	650	19,900
High level wastes (m^3)	350	8,000

Chemical Pollution

Modern technological developments have multiplied the hazards to which human beings are exposed. Nearly 5 million chemicals have been synthesized in the world in the last 40 years and some 50,000 to 70,000 chemicals are used extensively in millions of different commercial products without the availability of proper toxicological information on the majority of chemicals[13]. Various industries like steel, non-ferrous metals, fertilizers, and petroleum, in addition to emitting the common pollutants, are sources of toxic pollutants like lead, cadmium, zinc, arsenic, nickel, beryllium and mercury. Many environmentalists believe that these toxic metals pose a greater threat to the environment than SO_2, NO_x, CO and other gross contaminants. The present levels of these metal concentrations in the atmosphere are quite small compared to the total suspended matter; however, it is not quite apparent whether these levels are necessarily small relative to the amounts required to produce adverse health effects.

Arsenic occurs in phosphate rock used to make fertilizers and detergents and may thus end up in both rural and urban waterways. Episodes in Sweden and Japan have pointed to dangers arising from the discharge of large quantities of mercury into the environment. In Sweden, the use of methylmercury in a seed dressing had led to a drastic decrease in wild bird populations. Near Minimata in Japan, over 40 people died between 1953 and 1960 from mercury poisoning after eating contaminated fish and many others suffered serious neurological damage. Most of these toxic elements are slowly accumulated in the body, due to their very slow excretion. Therefore, a low exposure over a long period can result in building up of highly toxic levels. In addition to their toxicity, they also produce in the victim symptoms of a deficiency-related disease[14]. A typical case is the effect of excess zinc in animals. The symptom of zinc toxicity is iron deficiency anemia.

The worldwide production and use of chemical compounds have increased tremendously since the second world war. Much of this growth can be attributed partly to the needs of a growing population and partly to the development of new compounds for the sake of 'progress'. The environmental impact of these chemical interventions has only slowly become apparent and is a cause for much concern. Thousands of chemical compounds are released into the environment and many of these compounds resist decay and are biologically non-degradable. Indiscriminate use of pesticides like DDT, dieldrin and endrin is of great concern because these chemicals are passed on from one organism to another through all the links of the food chain and can ultimately be accumulated in the fat of human body. Pesticide residues now appear in all varieties of foods like milk, fish, fruits and vegetables. The pesticides also exhibit specific effects on species other than the pests for which they are solely intended. For example, the egg shells of many birds have become thinner thereby reducing hatching success. In several other species also these effects now seriously threaten reproductive capabilities.

The pollutional characteristics of industrial waste waters are of extremely varied nature since these contain suspended matter, dissolved solids, toxic metals, chemicals, strong acids, alkalies, oils, and dyes. These substances may render the water corrosive and toxic to fish. They also deplete the dissolved oxygen, impair the biological activity and destroy aquatic life. In addition to industrial wastes, large quantities of domestic sewage are discharged into rivers and lakes, leading to depletion of dissolved oxygen. Fertilizers, detergents and various biologically active materials find their way, in increasing quantities, into rivers, lakes and finally the ocean where the environment of many of the aquatic inhabitants is adversely affected.

Oil pollution is an almost inevitable consequence of the dependence of a rapidly growing population on oil-based technology. The production, distribution and use of large quantities of oil the world over results in oil contamination of our aquatic systems, primarily the ocean. Oil spills resulting from both accidents and routine discharges are the major problems arising from offshore drilling operations. These oil spills affect coastal waters and estuaries, which are valuable marine breeding grounds. The shipping of petroleum and its refining also have significant environmental consequences. In 1978 some 323 million tonnes of crude oil were shipped across the Arabian Sea to India and Far East from the Gulf countries. This intensive shipping of oil is one of the important reasons of oil pollution in the seas around India.

The effects of oil pollution include reduction of light transmission across the water surface and consequent slowing of photosynthesis by aquatic algae, depletion of dissolved oxygen, damage to water birds whose feathers become coated, smothering of shoreline plants, and the toxicity of many petroleum components. Some aromatic substances exhibit carcinogenic effects and these, such as 3,4-benzpyrene, are occasionally concentrated by shellfish.

Ozone Depletion

The Ozone layer serves as a shield, protecting the Earth's surface from most of the ultraviolet radiation found in the Sun's radiation. However, this layer may be depleted by reactions involving a variety of compounds, which reach the stratosphere. Of particular concern are the water vapour and nitrogen oxides released by high altitude aircraft, nitrous oxide produced by the action of bacteria in soils (amounts of which have become significant with the large scale use of nitrate fertilizers), and chlorofluoro-hydrocarbons which are widely used as aerosol spray propellants and refrigerants. It is postulated that the combined effect of aircraft engine effluents, nitrate fertilizers and halogens would cause considerable depletion of the ozone layer and could result in an increase in ultraviolet radiation reaching the earth leading to crop damage and marked rise in cases of skin cancer. The recent reports of the ozone hole in the atmosphere over the antarctica is a cause for great concern.

Solid Wastes

Solid wastes disposal in urban areas is a serious problem. Due to rapid increase in urban population, the amount of solid wastes that must be regularly collected, transported and ultimately disposed off has increased tremendously in recent years. For example, greater Bombay generates more than 3000 tonnes of garbage daily as against some 300 tonnes nearly 50 years ago. As a result, in large parts of our cities subhuman conditions prevail as the garbage removal machinery is inadequate to cope with the tasks of sanitation and cleanliness. In addition, uncontrolled dumping of garbage destroys the beauty of the surroundings, and burning of these wastes causes air pollution due to release of a variety of pollutants depending on the waste composition and combustion conditions.

1.7 TOWARDS A SOLUTION

Many of the environmental pollution problems can be solved with the available technology, but unfortunately the efforts being made are badly managed and inadequate in India. In order to ensure environmental protection, the Government of India has made some administrative and legislative arrangements. A department of environment has been created with a wide range of responsibilities. Similarly, many state level organisations are being set up. There are more than 200 central and state laws in existence today to protect the environment, but these laws can be interpreted one way or another. The major central laws include the Wild Life Protection Act of 1972, the Water (Prevention and Control of Pollution) Act of 1974, the Air (Prevention and Control of Pollution) Act of 1981 and the new environment (Protection) Act of 1986.

Presently, many large-scale projects related to energy and industry have to go through the environment appraisal with a view to evaluating environmental implications and incorporating necessary safeguards to protect and improve the environment. If the government is determined to use the technical know-how, legislative checks and management skills for controlling pollution, there is a reasonable hope for the goal of environmental protection.

There are many uncertainties surrounding environmental problems. These uncertainties can be reduced in part by better scientific information which can identify and deal with specific pollution problems. Most of the pollution problems in real situations are integrated and, hence, a systems approach to environmental control is needed. In addition, greater public awareness of the environment is necessary and should be encouraged. It should be clear that with the present rate of growth of pollution, industrialisation and urbanisation, there is an urgent need for a holistic thinking on the management of our resources and environment.

REFERENCES

1. B. Bolin, "The le", *Sci. Amer.*, Vol. 223, No. 3, 1970.
2. L.R. Pomeroy, H.M. Mathews and H. Shik Min, "Excretion of phosphate and soluble organic phosphorus compounds by zooplankton", *Limnol. Oceanog.*, Vol. 8 (1963), pp. 50–55.
3. G.E. Hutchinson, *A treatise on limnology*, Vol. I, *geography, physics, chemistry* (Wiley, New York), 1957.
4. E.S. Deevey, *Sci. Amer.*, Vol. 203, No. 5, 1960.
5. S.N. Agarwala, *Population*, 2nd ed. (National Book Trust, India, New Delhi), 1977, p. 18.
6. P. Ehrlich and J.P. Holdern, "Impact of population growth", *Science*, Vol. 171 (1971), pp. 1212–17.
7. SCEP, *Man's impact on the global environment*: assessment and recommendation for action (MIT Press, Cambridge, Mass.), 1970.
8. M. King Hubbert, "The energy resources of the earth", *Sci. Amer.*, Vol. 224 (1971), pp. 61–70.
9. Hans H. Landsburg, *Energy : the next twenty years*, Report sponsored

by the Ford Foundation (Ballinger Pub. Co., Cambridge, Mass.), 1979.

10. S. Oden, "The acidification of air and precipitation and its consequences on the natural environment" (in Swedish), *Ecol. Comm. Bull.*, Vol. 1, Trans. by Transla, No. TR-1172, 1968.
11. C.E. Likens, F.H. Bormann and N.M. Johnson, *Environment*, Vol. 14, No. 2 (1972), pp. 37-38.
12. R. Ramanna, *Inevitability of atomic energy in power programme, energy policy for India,* edited by R. Pachauri (MacMillian Co. of India Ltd., Delhi), 1980, p. 221.
13. *The State of India's environment* 1984-85, *the second citizen's report* (Centre For Science and Environment, New Delhi), 1985.
14. A. Furst, *Environmental geochemistry in health and disease* (Geol. Soc. Amer., Boulder), 1971, pp. 109-130.
15. H.N. Sethna, *Waste management policy and its implementation in India* (IAEA, Vienna), CN-43/133.

SUGGESTIONS FOR FURTHER READING

1. Edward J. Kormandy, *Concepts of ecology,* 2nd Ed. (Prentice-Hall, Englewood Cliffs, N.J.), 1976.
2. Paul R. Ehrilich, Anne Ehrilich and John P. Holdren, *Human ecology-problems and solutions* (W.H. Freeman & Co., San Fransisco), 1973.
3. H.L. Penman, "The water cycle", *Sci. Amer.*, Vol. 223, Sept. 1970.
4. C.C. Delwiche, "The nitrogen cycle", *Sci. Amer.*, Vol. 223, Sept. 1970.
5. Arthur Baurne, *Pollute and be damned* (J.M. Dent & Sons Ltd., New York), 1972.
6. Richard A. Falk, *This endangered planet* (Random House, New York), 1971.
7. J.Y. Wang, W.R. Hemmer and F.B. Tuttle, *Exploring man's environment* (Cummings Publishing Co., California), 1973.
8. L.K. Coldwell, *Environment : a challange to modern society* (Doubleday Anchor Books, New York), 1971.
9. Barry Commoner, *The closing circle, nature, man and technology* (Knopf, New York), 1971.
10. D.H. Meadows, D.L. Meadows, J. Randers and W.W. Behrens, *The limits to growth* (Universe Books, New York), 1972.
11. D.M. Gates, "The flow of energy in the biosphere", *Sci. Amer.*, Vol. 224, No. 3 (1971), pp. 88–100.
12. B.P. Cloud (Ed.), *Resources and man* (W.H. Freeman & Co., San Fransisco). 1969.

13. C.S. Rao, "Impact of exponential growth on India's fossil fuel reserves", *Energy Management*, Vol. 3 (1979), pp. 187–192.
14. R. Carson, *Silent spring* (Houghton Mifflin, Boston), 1963.
15. C. Walker, *Environmental pollution by chemicals* (Hutchinson, London), 1971.
16. P.D. Henderson, *India: the energy sector* (Oxford Univ. Press, New Delhi), 1975.

PROBLEMS

1. The population of India according to 1981 census was 684 million and it is growing exponentially at a rate of 2.0 per cent per year. If it continues to grow at that rate, when will the population be doubled? What will be the population in the years 2000 and 2050? (2016)
2. Several types of population curves are observable in nature. One is the exponential growth curve. What are the other types? Sketch them and briefly discuss.
3. How does the age structure help in making detailed population projections? Discuss.
4. What is replacement level? If the population of India reaches replacement level, will it mean reaching zero growth in population?
5. What environmental factors will influence future growths of population in India? Explain.
6. Consider India of the future which is over-crowded (double the present population), polluted and threatened with starvation of fossil fuels energy. Write your own scenario for that India.
7. The number of years (T) required to deplete a quantity of reserves (R), when the present rate of consumption is P_0 and the consumption grows exponentially at the rate r is given by the equation,

$$T = \frac{1}{r} \ln \left(\frac{rR}{P_0} + 1 \right)$$

   Derive the above equation.
8. The following table gives the data on India's crude oil production.

Year	1961	1967	1969	1972	1975	1980	1981	1982	1983
Production (million tonnes per year)	0.45	4.6	5.8	7.4	8.5	9.4	10.5	16.2	21.0

(a) Is the production growing exponentially? If so what is the rate of growth per year? (6.2%)
(b) The proved recoverable reserves of oil are estimated to be 471 million tonnes. At the above growth rate, how long will the reserves last? (till 2003)
(c) Assume that the amount of reserves is doubled because of new discoveries and new technology. At the same growth rate, how long will the reserves last? (till 2012)

9. Barry Commoner, in his book, *The closing circle,* presents his four laws of ecology:

(a) Everything is connected to everything else.
(b) Everything must go somewhere.
(c) Nature knows best.
(d) There is no such thing as a free lunch.

Discuss briefly each law.

10. Which do you think is a greater danger to man: pesticides used by farmers and householders, or toxic substances discharged into the air and water by industrial plants? Explain.

2. Air Pollution: Sources and Effects

2.1 DEFINITION AND SCALES OF CONCENTRATION

Air pollution may be defined as any atmospheric condition in which certain substances are present in such concentrations that they can produce undesirable effects on man and his environment. These substances include gases (sulphur oxides, nitrogen oxides, carbon monoxide, hydrocarbons, etc.), particulate matter (smoke, dust, fumes, aerosols), radioactive materials and many others. Most of these substances are naturally present in the atmosphere in low (background) concentrations and are usually considered to be harmless. The background concentrations of various components of dry air near sea level and their estimated residence times are given in Table 2.1. Thus, a particular substance can be considered an air pollutant only when its concentration is relatively high compared with the background value and causes adverse effects. For example, sulphur dioxide, if present in the atmosphere in concentrations greater than the background value of 2×10^{-4} ppm and causes measurable effects on humans, animals, plants, or property, then only it is classified as an air pollutant.

The concentration of a pollutant in the atmosphere can be expressed in a number of ways involving units of weights or volume per unit weight or volume of air. Four concentration scales are generally used to describe the concentrations of either gaseous or particulate pollutants.

The first is the mass concentration, ω_p, defined as the ratio of the mass of pollutant to the mass of air plus mass of pollutant:

$$\omega_p = \frac{m_p}{m_a + m_p} \tag{2.1}$$

where m_p is the mass of the pollutant and m_a is the mass of pure air in a given volume of air-pollutant mixture. The second concentration scale is the volume concentration, y_p, defined as the ratio of the volume of pollutant to the volume of air plus volume of pollutant:

$$y_p = \frac{v_p}{v_a + v_p} \tag{2.2}$$

The third concentration scale is the volume concentration in parts per million (ppm), y_{ppm}:

$$y_{ppm} = y_p \times 10^6 \tag{2.3}$$

Table 2.1 Composition of clean, dry atmospheric air

Component	Concentration* (ppm)	Estimated residence time (ref. 1)
Nitrogen	78.09×10^4	Continuous
Oxygen	20.94×10^4	Continuous
Argon	93×10^2	Continuous
Carbon dioxide	3.2×10^2	2–4 years
Neon	18	Continuous
Helium	5.2	~ 2 million years
Krypton	1.0	Continuous
Xenon	8×10^{-2}	Continuous
Carbon monoxide	1×10^{-1}	0.5 years
Methane	1.2	4–7 years
Nitrous oxide	25×10^{-2}	4 years
Nitric oxide	6×10^{-4}	5 days
Ammonia	6×10^{-3}	7 days
Hydrogen sulphide	2×10^{-4}	2 days
Sulphur dioxide	2×10^{-4}	4 days
Hydrogen	5×10^{-1}	?**
Ozone	2×10^{-2}	~ 60 days

* Single values for concentrations, instead of ranges of concentrations, are given to indicate order of magnitude, not specific and universally accepted concentrations.

**Little is known about the residence time.

Finally, the fourth concentration scale is the mass-volume concentration, ρ_p, defined as the ratio of the mass of the pollutant to the volume of air and pollutant combined:

$$\rho_p = \frac{m_p}{v_a + v_p} \tag{2.4}$$

The concentration of a pollutant whether it be gaseous or particulate is at present commonly expressed in terms of ρ_p, usually as micrograms of pollutant per cubic metre of air and pollutant combined, ($\mu g/m^3$). A number of relationships can be developed among these concentrations, but the conversion between y_{ppm} and ρ_p in $\mu g/m^3$ for gaseous pollutants is of special interest. Combining Eqs. (2.2), (2.3) and (2.4) we get,

$$\rho_p = \frac{m_p}{v_p} y_{ppm}\, 10^{-6} \tag{2.5}$$

Assuming that the perfect gas law holds for pollutants that are usually formed in the atmosphere, we can express the mass density of the pure component pollutant, m_p/v_p, in units of $\mu g/m^3$ as,

$$\frac{m_p}{v_p} = \frac{P\hat{M}_p\,(10^9)}{(8.314\times 10^{-2})\,T} \tag{2.6}$$

where P is the total pressure in bars, $\hat{M}_p$ is the molecular weight of the pollutant, T is the absolute temperature in K, and 8.314×10^{-2} is the value of the universal gas constant expressed in units of m^3-bar/kg mol-K. At the standard temperature (25°C) and pressure (1.0133 bars), Eq. (2.6) reduces to,

$$\frac{m_p}{v_p} = \frac{\hat{M}_p\,(10^9)}{24.45} \tag{2.7}$$

Substituting Eq. (2.7) into Eq. (2.5), we get the relation between ρ_p ($\mu g/m^3$) and y_{ppm}:

$$\rho_p = \frac{\hat{M}_p\, y_{ppm}\,(10^3)}{24.45} \tag{2.8}$$

The constant in the denominator becomes 22.4 at 0°C and 1 atm pressure.

Example

Carbon monoxide (CO) is present in standard atmospheric air at a concentration of 50 ppm. Compute y_p, ρ_p, and ω_p values for the CO concentration in the atmosphere.

Solution: From the statement of the problem, $y_{ppm} = 50$. The air density at standard condition is 1.185 kg/m³, and the molecular weight $\hat{M}_p$ is 28. From Eq. (2.3),

$$y_p = y_{ppm}\ 10^{-6} = 0.00005$$

From Eq. (2.8), the mass volume concentration, ρ_p, is

$$\rho_p = \frac{\hat{M}_p\ y_{ppm}\,(10^3)}{24.45} = \frac{28\ (50)\ 10^3}{24.45}$$

$$= 57.26\ (10^3)\ \mu\text{g/m}^3 = 57.26\ (10^{-6})\ \text{kg/m}^3$$

The mass concentration, ω_p, can be calculated as follows: Dividing Eq. (2.1) by (2.2), we have

$$\frac{\omega_p}{y_p} = \frac{m_p}{v_p}\left(\frac{v_a + v_p}{m_a + m_p}\right)$$

But the term $(m_a + m_p/v_a + v_p)$ can be written as

$$\frac{m_a + m_p}{v_a + v_p} = \frac{m_a}{v_a}(1 - y_a) + \frac{m_p}{v_p}\cdot y_p$$

Taking the density of air, ρ_a, as equal to m_a/v_a, we get

$$\omega_p = \frac{\left(\dfrac{m_p}{v_p}\right) y_p}{\rho_a\,(1 - y_p) + \dfrac{m_p}{v_p}\cdot y_p}$$

Substituting the relation $\rho_p = \left(\dfrac{m_p}{v_p}\right) y_p$ in the above equation, we get

$$\omega_p = \frac{\rho_p}{\rho_a\,(1 - y_p) + \rho_p}$$

Substituting the values of ρ_p, ρ_a and y_p into the above equation, we get the mass concentration,

$$\omega_p = \frac{57.26\ (10^{-6})}{1.185\,(1 - 0.00005) + 57.26\ (10^{-6})}$$

$$= 0.000048$$

2.2 CLASSIFICATION AND PROPERTIES OF AIR POLLUTANTS

Classification

The variety of matter emitted into the atmosphere by natural and anthropogenic sources is so diverse that it is difficult to classify air pollutants neatly. However, usually they are divided into two categories of primary pollutants and secondary pollutants. The primary pollutants are those that are emitted directly from the sources. Typical pollutants included under this category are particulate matter such as ash, smoke, dust, fumes, mist and spray; inorganic gases such as sulphur dioxide, hydrogen sulphide, nitric oxide, ammonia, carbon monoxide, carbon dioxide, and hydrogen fluoride; olefinic and aromatic hydrocarbons; and radioactive compounds. The secondary pollutants are those that are formed in the atmosphere by chemical interactions among primary pollutants and normal atmospheric constituents. Pollutants such as sulphur trioxide, nitrogen dioxide, PAN (peroxyacetyl nitrate), ozone, aldehydes, ketones, and various sulphate and nitrate salts are included in this category.

Of the large number of primary pollutants emitted into the atmosphere, only a few are present in sufficient concentrations to be of immediate concern. These are the five major types—particulate matter, sulphur oxides, oxides of nitrogen, carbon monoxide, and hydrocarbons. Carbon dioxide is generally not considered an air pollutant but, because of its increased global background concentration, its influence on global climatic patterns is of great concern. The radioactive pollutants are of specialized nature and they are beyond the scope of this book. The interested reader may refer to Hodges[2].

Secondary pollutants are formed from chemical and photochemical reactions in the atmosphere. The reaction mechanisms and various steps involved in the process are influenced by many factors such as concentration of reactants, the amount of moisture present in the atmosphere, degree of photoactivation, meteorological forces, and local topography.

Properties of Air Pollutants

Particulate Matter

In general the term "particulate" refers to all atmospheric substances that are not gases. They can be suspended droplets or solid particles or mixtures of the two. Particulates can be composed of inert or extremely reactive materials ranging in size from 100 μm down to 0.1 μm and less. The inert materials do not react readily with the environment nor do they exhibit any morphological changes as a result of combustion or any other process, whereas the reactive materials could be further oxidized or may react chemically with the environment.

The classification of various particulates may be made as follows:

Dust—It contains particles of the size ranging from 1 to 200 μm. These are formed by natural disintegration of rock and soil or by the mechanical processes of grinding and spraying. They have large settling velocities and are removed from the air by gravity and other inertial processes. Fine dust particles act as centres of catalysis for many of the chemical reactions taking place in the atmosphere.

Smoke—It contains fine particles of the size ranging from 0.01 to 1 μm which can be liquid or solid, and are formed by combustion or other chemical processes. Smoke may have different colours depending on the nature of material burnt.

Fumes—There are solid particles of the size ranging from 0.1 to 1 μm and are normally released from chemical or metallurgical processes.

Mist—It is made up of liquid droplets generally smaller than 10 μm which are formed by condensation in the atmosphere or are released from industrial operations.

Fog—It is the mist in which the liquid is water and is sufficiently dense to obscure vision.

Aerosol—Under this category are included all air-borne suspensions either solid or liquid; these are generally smaller than 1 μm.

Particles in the size range 1—10 μm have measurable settling velocities but are readily stirred by air movements, whereas particles of size 0.1—1 μm have small settling velocities. Those below 0.1 μm, a submicroscopic size found in urban air, undergo random Brownian motion resulting from collisions among individual molecules. Fig. 2.1 compares the sizes of atmospheric particulates from various sources. Most particulates in urban air have sizes in the range 0.1—10 μm. The finest and the smallest particles are the ones which cause significant damage to health.

The chemical composition of particulate pollutants varies over a wide range. The actual composition is very much dependent upon the origin of the particulate. Particles from soils and minerals primarily contain calcium, aluminium and silicon compounds. Smoke from combustion of coal, oil, wood, and solid waste contains many organic compounds. Insecticide dusts and certain fumes released from chemical plants also contain organic compounds. Hydrocarbons themselves can coalesce into aerosol droplets that constitute one kind of particulate matter. The most harmful components of incomplete combustion are generally grouped as particulate polycyclic organic matter (or PPOM). These materials are derivatives of benz-α-pyrene, a potent carcinogen.

Of all the different types of particulates in the atmosphere, the presence of trace elements such as cadmium, lead, nickel and mercury

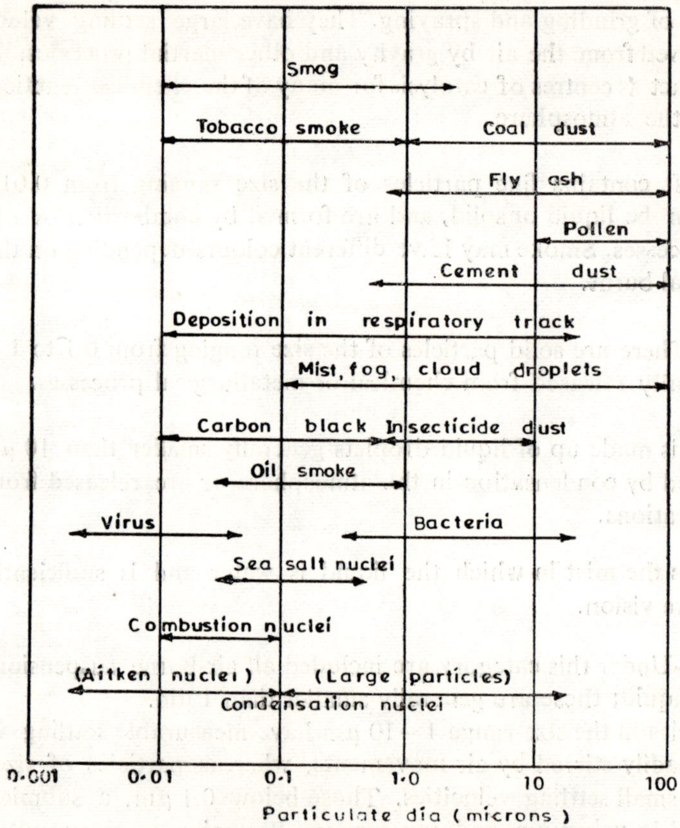

Fig. 2.1 Sizes of atmospheric particulate matter.

may constitute the greatest health hazard. Many of the trace metals are toxic and are concentrated in the finest of particulate matter in a variety of combined forms such as oxides, hydroxides, sulphates, and nitrates.

Oxides of Sulphur

The most important oxide emitted by pollution sources is sulphur dioxide (SO_2). SO_2 is a colourless gas with a characteristic, sharp, pungent odour. It is moderately soluble in water (11.3 g/100 ml) forming weakly acidic sulphurous acid (H_2SO_3). It is oxidized slowly in clean air to sulphur trioxide. In a polluted atmosphere, SO_2 reacts photochemically or catalytically with other pollutants or normal atmospheric constituents to form sulphur trioxide, sulphuric acid and salts of sulphuric acid.

Sulphur trioxide (SO_3) is generally emitted along with SO_2, at about 1–5 per cent of the SO_2 concentration. SO_3 rapidly combines with moisture in the atmosphere to form sulphuric acid which has a low dew point. Both SO_2 and SO_3 are relatively quickly washed out of the atmosphere by rain or settle out as aerosols. This is the reason why SO_2 mass in clean dry air is so small compared to annual emissions from anthropogenic sources (Table 2.2).

Table 2.2 Comparison of the amounts of anthropogenic pollutants with the amounts naturally present in dry, clean air.

Pollutant	Amount (millions of tonnes)	
	Anthropogenic (per year)	Dry, Clean air
Particulates	269	—
SO_2	132	2
NO_2	48	8
CO	400	500

Nitrogen Oxides

Of the six or seven oxides of nitrogen, only three—nitrous oxide (N_2O), nitric oxide (NO), and nitrogen dioxide (NO_2)—are formed in any appreciable quantities in the atmosphere. Often NO and NO_2 are analyzed together in air and are referred to as NO_x.

Nitrous oxide is a colourless, odourless nontoxic gas present in the natural atmosphere in relatively large concentrations (0.25 ppm). The major source of N_2O in the atmosphere is the biological activity of the soil and there are no significant anthropogenic sources. It has a low reactivity in the lower atmosphere and is generally not considered an air pollutant.

Nitric oxide is a colourless, odourless gas produced largely by fuel combustion. It is oxidized to NO_2 in a polluted atmosphere through photochemical secondary reactions. Nitrogen dioxide is a brown pungent gas with an irritating odour which can be detected at concentrations of about 0.12 ppm. It absorbs sunlight and initiates a series of photochemical reactions. Small concentrations of NO_2 have been detected in the lower stratosphere; NO_2 is probably produced by the oxidation of NO by ozone. Nitrogen dioxide is of major concern as a pollutant; it is emitted by fuel combustion and nitric acid plants.

Carbon Monoxide

It constitutes the single largest pollutant in the urban atmosphere. CO is colourless, odourless, and tasteless, and has a boiling point of − 192°C. It has a strong affinity towards the hemoglobin of the bloodstream and is a dangerous asphyxiant. The rate of oxidation of carbon monoxide to carbon dioxide in the atmosphere seems to be very slow; mixtures of CO and O_2 exposed to sunlight for several years have been found to remain almost unchanged. Carbon monoxide is present in small concentrations (0.1 ppm) in the natural atmosphere and has a residence time of about six months. The main sources of CO in the urban air are smoke and exhaust fumes of many devices burning coal, gas or oil.

Hydrocarbons

The gaseous and volatile liquid hydrocarbons are of particular interest as air pollutants. Hydrocarbons can be saturated or unsaturated, branched or straight-chain, or can have a ring structure as in the case of aromatics and other cyclic compounds. In the saturated class, methane is by far the most abundant hydrocarbon constituting about 40 to 80 per cent of the total hydrocarbons present in an urban atmosphere. The unsaturated class includes alkenes (olefins) and acetylenes. Among the alkenes the prominent pollutants are ethylene and propene. The first member of the aromatic class is benzene, but some of its substituted derivatives such as toluene and *m*-xylene are usually present in larger concentrations in the urban atmosphere. Terpenes are a particular class of volatile hydrocarbons emitted largely by natural sources. These are cyclic non-aromatic hydrocarbons found in pine tar and in other wood sources.

The hydrocarbons in air by themselves alone cause no harmful effects. They are of concern because the hydrocarbons undergo chemical reactions in the presence of sunlight and nitrogen oxides forming photochemical oxidants of which the predominant one is ozone. Methane has very low photochemical activity as compared to that of other hydrocarbons. For this reason, it is the non-methane hydrocarbon concentration that is of interest while considering air pollution.

2.3 EMISSION SOURCES

The sources of air pollutants are numerous. They can be grouped according to a variety of methods, including type of source, number and spatial distribution of sources and type of emissions[3].

Classification According to Source Types

Source type refers to natural and anthropogenic sources, as well as to additional subclassifications within each group. Natural sources include wind-blown dust, pollen, sea salt nuclei, volcanic ash and gases, smoke and trace gases from forest fires, and terpenes from forests. Anthropogenic sources cover a wide spectrum of types. Table 2.3 includes a list of major anthropogenic air pollution sources, and their characteristic emissions. The most important of these with regard to quantity are the products of combustion. The combustion of fossil fuels results in the emission of a variety of pollutants into the atmosphere of which the major ones are SO_2, NO_x, CO, and particulate matter such as fly ash. Also emitted are small quantities of water vapour and trace metal oxides, e.g. oxides of mercury, lead and cadmium. Incineration of solid combustible wastes and refuse gives off a wide variety of pollutants depending upon waste composition and combustion conditions. These pollutants include aldehydes, benz-α-pyrene, CO, NO_x, SO_2, hydrocarbons, ammonia and particulates.

Next to combustion systems, the major sources of air pollutants are chemical and metallurgical industries. Thus a major, and often quite concentrated source of sulphur dioxide is the roasting and subsequent treatment of nonferrous sulphide minerals, principally lead, zinc, and copper ores. In addition to SO_2 emission, these operations usually produce dust and fumes of primary metal oxides and small quantities of trace metals.

In the case of iron and steel industry, air pollution originates from large scale high temperature processing of coal and iron ore. The major emission sources include ore sintering, cake ovens, waste water quenching of hot coke and furnace effluents. Some of the significant emissions are dust, fumes, hydrocarbons, tars, H_2S and SO_2. In nonmetallic mineral industries like cement, glass, ceramics and refractories, the operations of mining, crushing, transport and storage of the materials result in the emission of large quantities of hazardous dust.

The chemical process industries are the source of a wide variety of air-pollutants emissions associated with raw material processing, reaction products and their purification. Some of the pollutants emitted from inorganic chemical industries are SO_2 gas, SO_3—H_2SO_4 mist, NH_3, NO_2, hydrogen fluoride, HCl gas and H_2S. Synthetic organic chemical industries emit a variety of hydrocarbon compounds and solvents like C_6H_6, $C_6H_5CH_3$ and CCl_4 into the atmosphere.

Petroleum refining operations are a major source of air pollution, emitting such pollutants as oxides of sulphur and sulphur containing vapours from various stages of the refining process, particulates from catalyst regeneration and recycling, H_2S and mercaptans which are stripped from lighter grades of fuel and hydrocarbons from leaks in valves, pipelines and storage tanks.

Table 2.3 Classification of anthropogenic air pollution sources

Source type	Category	Important sources	Typical pollutants
Combustion	Stationary	Power plants, industrial boilers, diesel generators, municipal or industrial incineration, refuse burning	Oxides of sulphur, NO_x, CO, smoke, fly-ash, trace metal oxides
	Mobile	Motor vehicles, air craft	CO, hydrocarbons, NO_x, SO_2 particulates
Roasting and heating processes	Nonferrous metallurgical	Roasting smelting and refining operations	Dust, smoke, metal fumes (Cu, Pb, and Zn) oxides of sulphur
	Ferrous metallurgical	Materials handling, ore sintering and pelletising, coke ovens blast furnace, steel furnaces	Smoke, fumes, CO, odours, H_2S, organic vapour, fluorides
	Nonmetallic minerals	Crushed stone, gravel, and sand processing, cement, glass, refractories, and ceramics manufacture, coal cleaning	Mineral and organic particulates, SO_2, NO_x dust fumes
Chemicals, petroleum, pulp and paper	Petroleum refining	Boilers, process heaters, catalyst regenerators, flares, reactors, storage tanks, compressor engines	Oxides of sulphur, hydrocarbons, NO_x, particulate matter, CO, aldehydes ammonia, odours
	Inorganic chemicals	Sulphuric acid plants, fertilizer manufacture, nitric acid and ammonia plants, phosphoric acid manufacture	SO_2, HF, H_2S, NO_x, NH_3, particulate matter, H_3PO_4, etc.
	Organic chemicals	Plastics, paint and varnish manufacture, synthetic rubber, rayon, insecticide, soap and detergent manufacture, methanol, phenol, etc.	Particulate matter, odours, SO_2, CO, organic intermediates, product gases and vapours, solvent vapours, etc.
	Pulp and paper (Kraft process)	Digester blow system, pulp washers, recovery furnace, evaporators, oxidation towers	Particulate matter, odourous sulphur compounds (H_2S, methyl mercaptan, dimethyl sulphide) and SO_2 (sulphite process)

Source type	Category	Important source	Typical pollutants
Food and agriculture	Food processing	Drying, preserving, packaging	Vapours, odours, dust.
	Crop spraying and dusting	Pest and weed control	Organic phosphates, chlorinated HC, arsenic, lead
	Field burning	Refuse burning	smoke, flyash and soot

Pulp and paper industry is a notorious source of many air-pollutants emissions into the atmosphere. Sulpur bearing malodourous gases such as H_2S, methyl mercaptan, and methyl sulphides are emitted from blow tanks, evaporators and recovery furnance. In addition, combustion of wood waste results in large particulate emissions into the atmosphere.

Food processing covers a wide range of activities including drying, preservation of food materials and packaging. These activities produce such pollutants as dusts from grinding, milling and handling operations, and odours associated with biological decay of food materials. Use of agricultural chemicals for improvements in crop yield, for control of weeds and insects result in emission of pollutants like nitrates and phosphates, pesticides, arsenic and lead particulates into the atmosphere.

Other Methods of Grouping Air Pollution Sources

Air pollution sources can also be grouped according to number and spatial distribution. These include single or point sources such as steel mills, power plants, oil refineries, and pulp and paper mills, etc., multiple or area sources such as an entire residential area, and line sources which include highways carrying moving vehicles. Another source grouping is by the type of emissions with particulate and gaseous emissions being the two major divisions.

Major Emissions from Global Sources

The estimated magnitudes of major air pollutants emitted from various global sources are given in Table 2.4. As seen from the table the total global emission of particulate matter is estimated to be 2365 million tonnes per year. Of this total emission the natural emissions from both primary and secondary sources account for 2096 million tonnes and anthropogenic sources account for 269 million tonnes. Among the primary sources of natural emissions, sea spray is the largest single contributor, amounting

Table 2.4 Estimated global emissions of major air pollutants (refs. 4, 5, 6, 7, 8)

Pollutant	Emissions (millions of tonnes/year)			
	Natural		Anthropogenic	
Particulate matter				
Primary particle production	Sea salt spray	908	84	
	Soil dust	182		
	Volcanic and forest fires	7		
		1097		
Gas-to-particle conversion	Sulphate from H_2S	182	Sulphate from SO_2	133
	Nitrate from NO and NO_2	390	Nitrate from NO_x	27
	Ammonium	245	Photochemicals from HC	25
	Terpenes	182		
		999		185
Sulphur dioxide	*Emissions expressed as S*		Coal	92
	Biological decay (H_2S)	90	Petroleum	26
	Sea spray (sulphates)	40	Smelting	14
		130		132
Nitrogen oxides	NO : 455[a]		*Emissions expressed as* NO_2[a]	
	N_2O : 537		Coal combustion	24.4
			Petroleum refining	0.6
			Gasoline combustion	6.8
			Other oil combustion	12.8
			Natural gas combustion	1.9
			Other combustion	1.4
				47.9
Carbon monoxide	Oxidation of CH_4 and formaldehyde	3000	400	
	Decay and synthesis of chlorophyll	90		
	Photochemical oxidation of terpene	54		
	Oceans	220		
		3364		
Hydrocarbons	CH_4 : 1450		88	
	Terpenes : 170			

[a]Most NO, whether from natural or anthropogenic sources, is oxidized to NO_2 within a few days; thus, the distinction made here is somewhat arbitrary.

to 908 million tonnes per year with soil dust making up most of the remaining amount. The secondary sources of natural particulate emissions are gas phase chemical reactions. Such reactions involving H_2S, NO_x, NH_3 and terpenes account for the annual production of 999 million tonnes of particulates. In contrast, global particulate emissions of anthropogenic origin by chemical reaction between gaseous pollutants come to around 185 million tonnes per year. This is about 19 per cent of the amount of natural particulates produced by the same process. The direct introduction of (primary) particulates into the atmosphere resulting from the combustion of fuels, cement manufacturing, agricultural burning, and solid-waste disposal operations amounts to 84 million tonnes per year.

Sulphur dioxide is emitted into the atmosphere either directly (e.g. from fuel combustion, petroleum refining and smelting operations, etc.), or through oxidation of H_2S obtained from decomposition of organic matter. The natural sources such as biological decay and sea spray emit about 130 million tonnes of sulphur per year and the anthropogenic sources such as coal combustion, petroleum and smelting operations release an additional 132 million tonnes of SO_2 annually into the atmosphere. The largest single contribution to the total anthropogenic emissions, about 70 per cent, is made by coal combustion. No major natural sources of SO_2 are known, although a small amount is probably present in gases emitted through volcanic activity.

Both natural and anthropogenic sources contribute to the emission of nitrogen oxides to the atmosphere. Global natural emissions, mainly due to bacterial activity in the soil, are estimated to be 455 million tonnes of NO per year. Combustion of coal, oil or natural gas in power plants or the internal combustion engines is the major source of anthropogenic emission of nitrogen oxide. It is estimated that 48 million tonnes of these oxides are emitted per year into the atmosphere from these activities. The amount reported is based on NO_2 since most of the NO is oxidised to NO_2 within a few days. However, it should be noted that it is NO and not NO_2 which enters the atmosphere predominantly.

The major sources of carbon monoxide in the environment are incomplete combustion of carbonaceous fuels, oxidation of methane and formaldehyde derived from biological activity, the decay of chlorophyll in plants, the algae and other biological sources in oceans, and the photochemical oxidation of terpenes. As seen from the Table 2.4 the natural sources far exceed the anthropogenic sources in CO emission. Global emissions from natural sources amount to an estimated 3364 million tonnes per year and these are approximately eight times more than the CO emissions from all human activities. About 88 per cent of this total comes from atmospheric oxidation of CH_4 and formaldehyde and the remaining from other sources.

The major sources of hydrocarbon emissions into the atmosphere are the natural sources. Swamps, marshes and other water bodies generate an estimated 1450 million tonnes of methane per year into the atmosphere. Plants are also natural sources of atmospheric hydrocarbons. They emit an estimated 170 million tonnes of terpene-type hydrocarbons into the atmosphere. The significant anthropogenic sources are incineration operations, evaporation of solvents in industrial coating and cleaning, and incomplete combustion of coal, oil and wood. The largest source of airborne hydrocarbons, however, is the entire cycle consisting of processing and using of petroleum. On a worldwide basis, the annual anthropogenic hydrocarbon emissions amount to an estimated 88 million tonnes.

Importance of Anthropogenic Sources

It is evident from Table 2.4, the pollutant emission from natural sources is much greater than that from anthropogenic sources. Fortunately, the former are distributed throughout the world and eventually they reach a sink such as the ocean or the soil. Thus, the atmosphere can cleanse itself from the pollutants given sufficient time. On the other hand, anthropogenic sources of pollutants are concentrated in specific locations, such as urban areas and reach very high levels in the atmosphere. For example, 95 to 98 per cent of the atmospheric carbon monoxide in an urban area comes from human activities and the resulting levels of CO are commonly 50 to 100 times higher than the characteristic global values of 0.1 to 0.5 ppm. The emission of these pollutants is at such massive rates that they overwhelm the capacity of the environment to cleanse itself.

The extent of man's contribution to global pollution is evident from Table 2.2 where a comparison of the total amount of anthropogenic generation of pollutants is made with their amounts present in dry, clean air. Thus, the influence of anthropogenic sources is quite significant and the air pollutants not only threaten the health and well-being of population in a particular locality, but could also produce adverse effects on a global scale.

Emission Sources in India

In India, air pollution is first and foremost a problem of fuel policy. The estimated nationwide emissions in 1976-77 of the five primary air pollutants from fuels alone are given in Table 2.5. As is seen from the table, a considerable amount of air pollution results from the burning of non-commercial fuels and this far outweighs that resulting from burning of commercial fuels. However, the combustion of fossil fuels and their products is responsible for a sizable amount of air pollution in urban areas. Among commercial fuels, coal is the greatest contributor to air pollution followed by fuel oils and mobile sources.

Table 2.5 Nationwide air pollutant emission from fuel sources in India (ref. 9)

	Particulate matter	SO_2	NO_x	CO	Organics including HC
	(millions of tonnes)				
Commerical (coal, petroleum, natural gas)	3.21	2.15	0.62	1.29	2.80
Non-commercial (firewood, dry cattle dung, vegetable wastes, refuse burning)	7.54	4.67	1.97	2.05	6.19

In addition to burning of fuels, major industries like steel, paper and pulp, textiles, cement, and sulphuric and nitric acid plants contribute relatively small but significant amounts of air pollutants to the nation's atmosphere. Since most of the industries are located in major cities, these significantly add to the pollution burden of the metropolitan areas. Table 2.6 lists the various sources and their contributions to the pollution of a

Table 2.6 Major pollutant emissions in Greater Bombay area (ref. 10)

Source	Particulate matter	SO_2	NO_x	CO	HC
	(Tonnes per day)				
Domestic	19.0	2.7	3.3	7.0	2.0
Commercial & institutional	2.2	1.7	1.0	—	—
Industrial	14.0	236.0	35.7	111.0	41.6
Power generation	28.0	62.0	30.0	2.0	—
Transportation	30.0	40.6	27.7	263.0	37.4
Total	93.2	343.0	97.7	383.0	81.0

typical metropolitan city like Bombay. As seen from the table, industry is the major source of pollution on a tonnage basis, followed by transportation and generation of electrical power. These three sources are responsible for the emission of about 77 per cent of particulate matter, 98 per cent of SO_2 and 94 per cent of NO_x.

The major sources of CO and hydrocarbon emissions are industry and transportation, contributing about 98 per cent of CO and 97 per cent of hydrocarbons. SO_2 and CO are the biggest pollutants of the city environment.

2.4 BEHAVIOUR AND FATE OF AIR POLLUTANTS

Although large amounts of pollutants are discharged annually into the atmosphere, the very fact that their ambient levels have remained very much the same throughout the world suggests that there are certain pathways of exchange from the atmosphere to the Earth, whereby the pollutants are continually removed. These pathways or the scavenging processes, as they are called, may be grouped as follows for both particulates and gases:

Particulates: (a) Wet removal by precipitation.
(b) Dry removal by sedimentation, impaction and diffusion.

Gases: (a) Wet removal by precipitation.
(b) Chemical reaction in the atmosphere to produce aerosols and/or absorption on aerosols with subsequent removal.
(c) Absorption or reaction at land and ocean surfaces.

Wet Precipitation

Wet precipitation has two distinct mechanisms—"rainout" and "washout." The first includes various processes taking place inside clouds, where the contaminants serve as condensation nuclei on which droplets condense. The second mechanism refers to the removal of pollutants below the cloud level by falling rain. Wet precipitation is one of the most effective scavenging processes for both particulate and gaseous pollutants in a global sense. The rainout mechanism is particularly effective for Aitken particles whose size is less than 0.1 μm. These particles are captured by cloud droplets by Brownian diffusion, and the cloud droplets (typically 30 μm) in turn grow in size by coalescence and precipitate. Washout is most effective in removing particles larger than 2 μm. Its scavenging efficiency, however, is influenced by the rain-drop cross-sectional area and the intensity of rain fall. Particles smaller than 2 μm are not usually collected by rain drops (typically 500 μm) because they are brushed aside by the diverging air ahead of the drop.

The scavenging of gaseous pollutants by wet precipitation is more complex and is less well understood. Soluble gas molecules such as SO_2 can migrate to the rain drop by Brownian motion or to the surface of the cloud droplets by diffusion due to the concentration gradient of the gas across the liquid-air interface. However, wet precipitation is very effective in removing the acid droplets and sulphate particles formed after chemical reactions in the atmosphere.

Dry Deposition

Particulate matter smaller than 0.1 μm often coagulates through mutual collisions and forms larger aggregates which are effectively removed by gravitational settling. Brownian motion is the major mechanism of coagulation, although atmospheric turbulence also enhances the diffusive motion of particles. Atmospheric turbulence is particularly effective for coagulating larger particles whose Brownian motion is less pronounced. The rate of settling of the particles depends on their settling velocities according to the Stokes' law:

$$v_t = \frac{g d_p^2}{18\,\mu_a} (\rho_p - \rho_a) \left(1 + \frac{2C}{d_p p}\right) \tag{2.9}$$

where

v_t = terminal settling velocity;

d_p = particle diameter;

ρ_p and ρ_a = density of particle and air, respectively;

μ_a = viscosity of air;

p = air pressure;

and

C = constant [when p is given in millibars and d_p in centimeters $C = 0.0084$ (ref. 11)].

From the above equation, it is seen that the rate of sedimentation is strongly influenced by the particle size. Particles larger than 10 μm have high settling rates and, hence have short residence time in the atmosphere. The terminal velocity is also related to the particle's density, but only by a less sensitive linear dependence. Deviations from the Stokes' law occur due to irregular particle shapes, turbulance in the wake of large particles, and atmospheric vertical velocities; but it is clear that small particles must aggregate to form larger ones if they are to be removed effectively from the atmosphere.

In addition to sedimentation, the mechanisms of inertial impaction and diffusion also contribute to the removal of particulate matter. In inertial impaction, windborne particles strike an obstacle and are deposited; whereas in diffusion small particles migrate to land and ocean surfaces. Dry deposition accounts for about 20 per cent of the total particulate removal from the atmosphere.

Interaction at the Earth's Surface

Gaseous pollutants can be transported to the Earth's surface by atmospheric turbulence where they interact with the ocean surface, vegetation and upper layers of the soil, and are removed by absorption or chemical reaction. Assimilation at these surfaces depends on many factors about which little is known for many pollutants. Two major pollutants, about which some information is available, are sulphur dioxide and carbon monoxide.

At the ocean surface, SO_2 first diffuses through the gas phase, crosses the gas-liquid interface, and finally diffuses into the bulk of the ocean where it is absorbed. Vegetation and upper layers of the soil also act as sinks for SO_2. Here the mode of transfer involved is adsorption. SO_2 first diffuses to the external surface of the solid, penetrates into the pores of the solid, and is subsequently adsorbed on the pore site. An estimated removal of 4×10^7 tonnes of SO_2 per year has been calculated for oceanic absorption[12] and the solid surfaces account for 5.6×10^7 tonnes per year[13]. The importance of these sinks can be seen when their performance is compared with the estimated emission of 14×10^7 tonnes per year SO_2 from the anthropogenic sources.

For carbon monoxide, biological action in soils seems to be an important sink; the role played by soil in the removal of CO has been firmly established[14]. It is theorised that the soil contains certain bacteria which can make use of CO in their metabolism, producing either CO_2 or CH_4:

$$CO + \tfrac{1}{2}\,O_2 \rightarrow CO_2 \tag{2.10}$$

$$CO + 3\,H_2 \rightarrow CH_4 + H_2O \tag{2.11}$$

Nitrogen dioxide also seems to be absorbed by the ocean and other surface waters, but the extent of the contribution of this sink to the overall removal of nitrogen oxides is not clearly established.

Chemical Reactions in the Atmosphere

Many of the gaseous pollutants undergo chemical reactions within the atmosphere and form either new compounds or aerosols. This mode of removal is of great importance for sulphur dioxide. A large part of SO_2 in the atmosphere is oxidized to sulphur trioxide which quickly combines with moisture to form sulphuric acid mist. The overall reaction is represented as:

$$2\,SO_2 + 2\,H_2O + O_2 \rightarrow 2\,H_2SO_4 \tag{2.12}$$

This process has been shown[15] to be catalysed by metal salts such as iron and manganese, commonly formed in the flyash. These particles serve as nucleation sites for droplet formation and the sulphuric acid droplet may in turn react with metal salts (such as NaCl from sea salt particles), metal oxides such as MgO, Fe_2O_3, ZnO, and Mn_2O_3, or ammonia to produce sulphates:

$$2\ NaCl + H_2SO_4 \rightarrow Na_2\ SO_4 + 2\ HCl \tag{2.13a}$$

$$MgO + H_2SO_4 \rightarrow Mg\ SO_4 + H_2O \tag{2.13b}$$

$$2\ NH_3 + H_2SO_4 \rightarrow (NH_4)_2\ SO_4 \tag{2.13c}$$

With these reactions, the sulphuric acid droplet is neutralized and the solubility of SO_2 in the droplets is further increased thereby enchancing the oxidation process. Both the acid droplets and the sulphate particles are rapidly removed from the atmosphere primarily by wet precipitation.

Similarly, a major process of removal of nitrogen oxides appears to be through their conversion to form nitric acid; however, the mechanism of such a conversion has not been clearly determined. The direct conversion of NO_2 to HNO_3 in the presence of moisture appears to be too slow to account for the observed rate of removal:

$$2\ NO_2 + H_2O \rightleftharpoons HNO_3 + HNO_2 \tag{2.14a}$$

$$3\ HNO_2 \rightarrow HNO_3 + 2\ NO + H_2O \tag{2.14b}$$

or

$$3\ NO_2 + H_2O \rightarrow 2\ HNO_3 + NO \tag{2.14c}$$

Since the direct conversion of sulphur dioxide to sulphur trioxide and nitrogen dioxide to nitric acid is slow, other mechanisms have been postulated, based mainly on theoretical conjecture and to a lesser extent on experimental observations. These mechanisms involve the oxidation by reactive species such as radicals, atomic oxygen, ozone, and hydroxyl radicals.

The first step in the oxidation of SO_2 is the photoexcitation of the SO_2 molecule through its absorption of solar radiation:

$$SO_2 + h\nu \rightarrow SO_2{}^* \tag{2.15}$$

where $SO_2{}^*$ represents an excited SO_2 molecule. These molecules in their excited state react more readily with molecular oxygen[16]. Then several reactions follow to complete the oxidation of SO_2 to SO_3:

$$SO_2{}^* + O_2 \rightarrow SO_4^{2-} \tag{2.16a}$$

$$SO_4^{2-} + O_2 \rightarrow SO_3 + O_3 \tag{2.16b}$$

$$SO_2 + O_3 + h\nu \rightarrow SO_3 + O_2 \tag{2.16c}$$

A three-body reaction with atomic oxygen has also been suggested as a possible mechanism:

$$O_2 + h\nu \rightarrow 2\ O \tag{2.17a}$$

$$SO_2 + O + M \rightarrow SO_3 + M \tag{2.17b}$$

The third body, *M*, is required in order to carry off excess energy of reaction.

A proposed rapid mechanism for the conversion of NO_2 to HNO_3 is the reaction of NO_2 with atmospheric ozone to give nitrogen trioxide:

$$NO_2 + O_3 \rightarrow NO_3 + O_2 \tag{2.18}$$

In addition, NO_3 can be formed by reaction with atomic oxygen

$$NO_2 + O + M \rightarrow NO_3 + M \tag{2.19}$$

The NO_3 radical is removed by reaction with NO_2 forming N_2O_5, which in the presence of moisture forms nitric acid:

$$NO_3 + NO_2 \rightarrow N_2O_5 \tag{2.20a}$$

$$N_2O_5 + H_2O \rightarrow 2HNO_3 \tag{2.20b}$$

The HNO_3 so formed is then washed out of the atmosphere in the form of nitrate salts by precipitation.

For carbon monoxide, its reaction with atmospheric oxygen in the presence of sunlight is found to be very slow and accounts for the removal of only 0.1 per cent of available CO for each hour of sunlight. Of major interest as an atmospheric sink for CO is the relatively fast reaction of CO with hydroxyl radicals present in the atmosphere

$$CO + OH\cdot \rightarrow CO_2 + H\cdot \tag{2.21}$$

The above mechanism may account for the removal of a substantial portion of CO from the troposphere depending upon the concentration of OH· radicals. Another possible mechanism is the migration of CO into the stratosphere, where the oxidation to CO_2 may subsequently take place by the hydroxyl radicals. In fact, all these gaseous pollutants, including the hydrocarbons, inter-react by photochemical processes in the phenomenon known as the photochemical smog.

Photochemical Smog

Photochemical smog was first observed in Los Angeles, U.S.A. in the mid-1940's and since then the phenomenon has been detected in most major metropolitan cities of the world. The conditions for the formation of photochemical smog are air stagnation, abundant sunlight, and high concentrations of hydrocarbon and nitrogen oxides in the atmosphere. In India, Bombay and Calcutta are ideal candidates for the formation of photochemical smog, but it may be masked by smoke and sulphur dioxide.

Smog arises from photochemical reactions in the lower atmosphere by the interaction of hydrocarbons and nitrogen oxide released by exhausts of automobiles and some stationary sources. This interaction results in a series of complex reactions producing secondary pollutants such as ozone, aldehydes, ketones, and peroxyacyl nitrates. The reaction mechanisms are complex and are not fully understood.

A broad outline of the principal reactions that occur in a photochemical process are illustrated in Fig. 2.2. The starting mechanism is the absorption of ultraviolet light from the Sun by NO_2. This causes the nitrogen

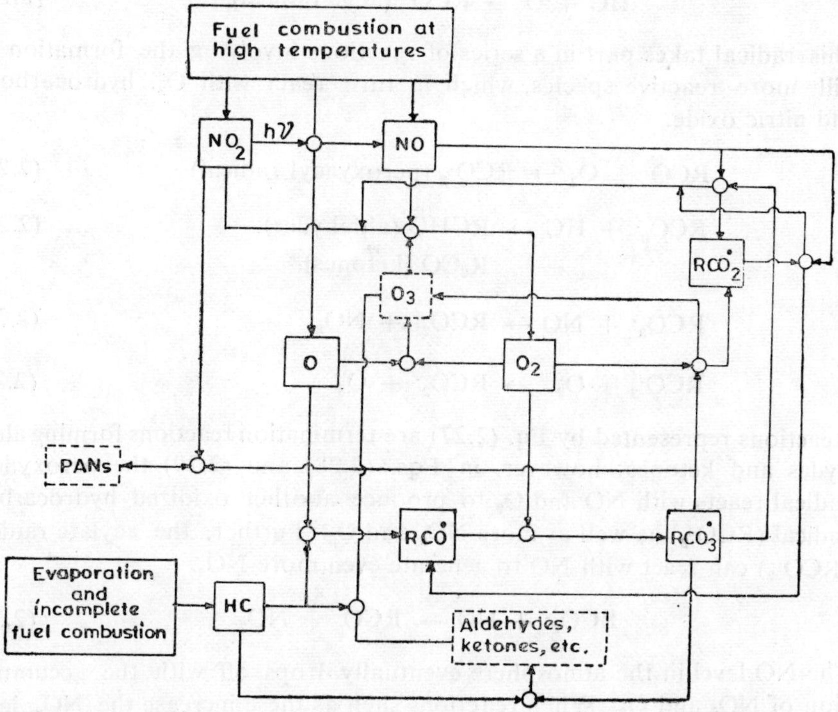

Fig. 2.2 A schematic representation of the formation of photochemical smog.

dioxide to decompose into nitric oxide and highly reactive atomic oxygen

$$NO_2 + h\nu \rightarrow NO + O \tag{2.22}$$

The atomic oxygen initiates oxidizing processes or quickly combines with molecular oxygen to form ozone, which itself is reactive and acts as an oxidant:

$$O + O_2 + M \rightarrow O_3 + M \tag{2.23}$$

$$O_3 + NO \rightarrow NO_2 + O_2 \tag{2.24}$$

In Eq. (2.23) an energy-absorbing molecule or particle (M) is required to stabilize O_3 or else it will rapidly decompose. Under normal conditions, the ozone formed will be quickly removed by reaction with NO to provide NO_2 and O_2 according to Eq (2.24); however, when hydrocarbons are present in the atmosphere this mechanism is partially eliminated as NO reacts with the hydrocarbon radical peroxyacyl ($RCO\cdot_3$) according Eq. (2.28), and as a result ozone concentration builds up to dangerous levels.

Hydrocarbons, indicated by symbol HC, compete for free oxygen released by NO_2 decomposition to form oxygen-bearing free radicals such as the acyl radical.

$$HC + O \rightarrow RCO\cdot \text{ (acyl radical)} \tag{2.25}$$

This radical takes part in a series of reactions involving the formation of still more reactive species, which in turn react with O_2, hydrocarbons and nitric oxide.

$$RCO\cdot + O_2 \rightarrow RCO\cdot_3 \text{ (peroxyacyl radical)} \tag{2.26}$$

$$RCO_3\cdot + HC \rightarrow RCHO \text{ (aldehydes)}, \; R_2CO \text{ (ketones)} \tag{2.27}$$

$$RCO_3\cdot + NO \rightarrow RCO_2\cdot + NO_2 \tag{2.28}$$

$$RCO\cdot_3 + O_2 \rightarrow RCO_2\cdot + O_3 \tag{2.29}$$

Reactions represented by Eq. (2.27) are termination reactions forming aldehydes and ketones; however, in Eqs. (2.28) and (2.29) the peroxyacyl radical reacts with NO and O_2 to produce another oxidized hydrocarbon radical ($RCO\cdot_2$) as well as more NO_2 and O_3. Further, the acylate radical ($RCO\cdot_2$) can react with NO to generate even more NO_2.

$$RCO_2\cdot + NO \rightarrow RCO\cdot + NO_2 \tag{2.30}$$

The NO level in the atmosphere eventually drops off with the accumulation of NO_2 and O_3. When reactions such as these increase the NO_2 level sufficiently, another reaction begins to compete for the peroxyacyl radical.

$$RCO_3\cdot + NO_2 \rightarrow RCO_3\,NO_2 \text{ (PANS)} \tag{2.31}$$

The end products are known as peroxyacyl nitrates or PANS. Numerous PANS could be formed, corresponding to the different possible R groups. Three of the common members of PAN family are:

$$H\overset{\overset{\displaystyle O}{\|}}{C}OO\,NO_2 : \text{ Peroxyformyl nitrate (PFN)}$$

$$CH_3\,\overset{\overset{\displaystyle O}{\|}}{C}OONO_2 : \text{ Peroxyacetyl nitrate (PAN)}$$

$$C_6H_5{-}\overset{\overset{\displaystyle O}{\|}}{C}OONO_2$$: Peroxybenzoyl nitrate (PBzN)

The ozone formed according to Eqs. (2.23) and (2.29) reacts with the hydrocarbons to generate more aldehydes and ketones.

$$HC + O_3 \rightarrow RCO^{\cdot}_2 + RCHO,\ R_2CO \quad (2.32)$$

The above equations represent in a broad sense the nature of the overall photochemical reactions leading to formation of smog and they are by no means the only important mechanisms. It has been observed that carbon monoxide and sulphur dioxide also play a significant part in the process of formation of smog by strongly interacting with many species present in the smog and accelerate the oxidation processes. For example, carbon monoxide does this through a series of reactions whose net effect is to convert CO, NO and O_2 into CO_2 and NO_2 thus accelerating the oxidation of NO. First, CO is oxidized to CO_2 by the $OH^{\cdot}$ radical (Eq. 2.21). In the smogy atmosphere the $OH^{\cdot}$ radical may be produced when aldehydes are attacked by atomic oxygen.

$$O + CH_3\,CHO \rightarrow CH_3CO + OH^{\cdot} \quad (2.33)$$

$$CO + OH^{\cdot} \rightarrow CO_2 + H^{\cdot} \quad (2.21)$$

$$H^{\cdot} + O_2 + M \rightarrow HO_2^{\cdot} + M \quad (2.34)$$

The $H^{\cdot}$ radical reacts with O_2 to form the hydroperoxyl radical $HO^{\cdot}_2$ which is a principal agent for the rapid conversion of NO into NO_2 (ref. 17).

$$HO_2^{\cdot} + NO \rightarrow OH^{\cdot} + NO_2 \quad (2.35)$$

The overall reaction is

$$CO + O_2 + NO \rightarrow CO_2 + NO_2 \quad (2.36)$$

This sequence of reactions provides another route for the oxidation of NO without the participation of O_3 (Eq. 2.24).

Similarly, the reaction of SO_2 with the $HO_2^{\cdot}$ radical may be an important step in the mechanism of the oxidation of SO_2 to SO_3.

$$HO_2^{\cdot} + SO_2 \rightarrow OH^{\cdot} + SO_3 \quad (2.37)$$

In addition, the hydrocarbon radicals may give off an oxygen atom to SO_2 to form SO_3, which in turn is converted to H_2SO_4 droplets resulting in the formation of haze.

Smog behaviour

The formation of photochemical smog is a dynamic process whose nature is illustrated in Fig. 2.3. In the morning the NO and hydrocarbon levels increase followed quickly by increase in NO_2. NO_2 reacts with the sunlight leading to various chain reactions and ultimately to the production of

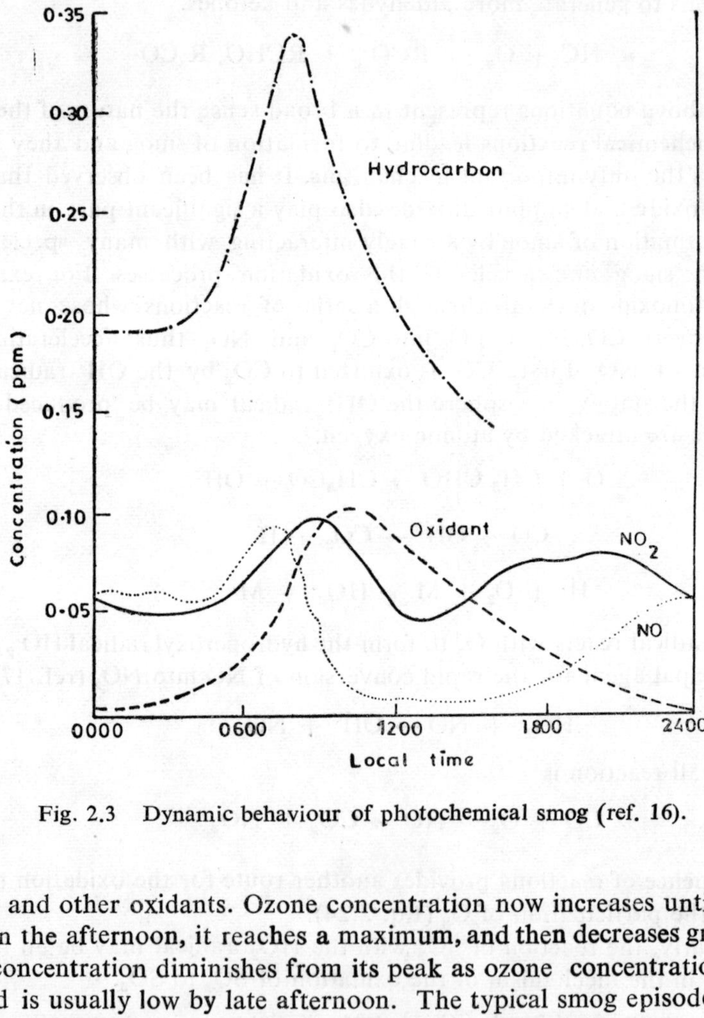

Fig. 2.3 Dynamic behaviour of photochemical smog (ref. 16).

ozone and other oxidants. Ozone concentration now increases until, sometime in the afternoon, it reaches a maximum, and then decreases gradually. NO_2 concentration diminishes from its peak as ozone concentration builds up and is usually low by late afternoon. The typical smog episode occurs in hot, sunny weather under low humidity conditions. The characteristic symptoms of the smog are the brown haze in the atmosphere, reduced visibility, eye irritation, respiratory distress and plant demage.

The control of photochemical smog may require a substantial reduction in NO_x produced in urban areas. At the same time it is necessary to control the release of hydrocarbons from numerous mobile and stationary sources.

2.5 EFFECTS OF AIR POLLUTION

In the previous section the scavenging mechanisms for the major air pollutants were examined. But often the pollutants are discharged into the atmosphere in such high quantities and concentrations as to overwhelm the natural scavenging ability of the atmosphere. As a result, the concentrations of the pollutants persist at levels which are much higher than the allowable background levels. This is particularly true for urban and other industrial regions where the pollutants adversely affect the health of humans and animals, and cause plant and material damage. The global effects of air pollution on climate due to increased levels of carbon dioxide and particulates, and the role of SO_2 in producing acid rain were discussed in Chapter 1. These problems are less apparent than the community-wide problems but they may be of great importance in the long run. In this section the main emphasis will be on the effects of those pollutants which cause community-wide problems and these are presented with particular attention to six ubiquitous pollutants, namely, particulates, sulphur oxides, carbon monoxide, nitrogen oxides, hydrocarbons and photochemical oxidants.

Human Health

Adverse effects of air pollution may be divided into two classes—acute and chronic effects. Acute effects manifest themselves immediately upon short-term exposure to air pollutants at high concentrations, and chronic effects become evident only after continuous exposure to low levels of air pollution. The chronic effects are very difficult to demonstrate and are consequently less obvious. Therefore, much of our knowledge of the effects of air pollution on people comes from the study of acute air pollution episodes. Deaths from these episodes are measured by comparing the number of deaths normally associated with the area and period in question with those that occur during the episode. The difference is referred to as "excess deaths".

In 1930 the Meuse Valley in Belgium was trapped by an inversion in which pollutants accumulated for about five days resulting in the death of about 60 people. Similar stable meteorological conditions occurred in 1948 in Donora, U.S.A., where 20 people died and about 6000 of the Town's 14,000 population became ill. The notorious London episode of 1952 took place under heavy continuous smog conditions and caused an estimated 4000 excess deaths. The episode of 1962, although shorter in duration was very much similar to that of 1952, but the excess deaths were lower. Table 2.7 outlines some of the major air pollution disasters. Most of these were associated with temperature inversions lasting for several days. The general and most wide-spread effects were caused by high concentrations of smoke and SO_2 and in particular by the synergistic action of the two, where the

Table 2.7 Major air pollution disasters (refs. 18, 19, 20)

Location	Conditions and causes	Symptoms and effects
Meuse Valley, Belgium, Dec. 1930	Inversion, smoke, SO_2 (9.6-38.4 ppm.) H_2SO_4 mist	60 excess deaths, thousands ill, eye and nasal irritation, cough
Donora, U.S.A. Oct. 1948	Inversion and fog, SO_2 (0.5-2 ppm.) smoke, zinc particles, H_2SO_4 mist	20 excess deaths, 6000 of town's 14000 population became ill, irritation of eyes, nose and respiratory tract, breathlessness, nausea
Poza Rica, Mexico, Nov. 1952	Shallow inversion, fog and calm conditions, H_2S release due to burner failure	22 excess deaths, 320 hospitalised, irritation of the respiratory tract
London, England Dec., 1952	Low temperature inversion, thick fog, stagnant air, smoke (4500 $\mu g/m^3$) and SO_2 (1.4 ppm) accumulation.	Estimated 4000 deaths, thousands hospitalised for respiratory and heart disease, chronic bronchitis, broncopneumonia
London, England Jan. 1956	Extended fog conditions similar to 1952 episode, particulates (3250 $\mu g/m^3$), SO_2 (0.57 ppm)	1000 excess deaths
London, England Dec. 1962	Shallow inversion, thick fog smoke (2000 $\mu g/m^3$), SO_2 (1.26 ppm)	700 excess deaths, increased illness.
New York, U.S.A. Nov. 1966	SO_2, particles	168 excess deaths
Seveso, Italy July. 1976	Reactor explosion releasing dioxin plume and cloud settled over an area of 1430 hectares	No excess deaths, 187 cases of skin chloraene, 32 official abortions, 15 cases of deformed births and 6 premature offspring, dizziness, diarrhea.
Bhopal, India Dec. 1984	Release of 30 tonnes of deadly methyl isocynate gas from storage tanks due to alleged failure of vent scrubber system	Estimated deaths more than 2500; 1,00,000 people severely affected, vomiting, violent coughing, chemical conjunctivitis, suffocation, cardiac failure.

combined influence of the two pollutants is greater than the sum of their individual effects experienced separately. In this context, it is interesting to compare London episodes of 1952 and 1962. Compared to those in the 1952 episode, the SO_2 levels in 1962 changed relatively little while the particulate concentration decreased by more than 50%. The mortality decrease was also substantial, possibly indicating either a dominating role of particulates or the availability of better medical facilities in 1962. The synergistic damage to health was particularly severe on the aged people who had been suffering from previously existing cardio-respiratory diseases.

The air pollution disasters of Poza Rica, Mexico, Seveso, Italy and the worst ever disaster in Bhopal, India, are examples of so called industrial pollution accidents. Bhopal provides an example of the failure in handling a toxic material in large-scale chemical processing and the consequences that have to be faced by a city in a short time. The disaster struck Bhopal on December 3, 1984 when a pesticide manufacturing plant released a potent toxicant, methyl isocyanate gas, into the atmosphere due to the alleged functional failure of 'vent scrubber' outlet. About 30 tonnes of the gas escaped from underground storage tanks resulting in the death of about 2500 people and severe disability of an additional 1,00,000 of the city's population. The chief causes of deaths and casualities were vomiting, violent coughing, eye infections (chemical conjuctivitis), suffocation, cardiac failure and pulmonary disorders. The full consequences are not yet known.

The greatest impact of air pollutants on human health, however, results from continued exposure to low concentrations under unexceptional conditions. These are chronic effects. It is believed that these effects occur through continued irritation by pollutants which interact with other environmental or biological factors to initiate disease or intensify previously existing diseases. There are two general approaches to studying the chronic effects. One is epidemiological and the other is toxicological.

Epidemiological studies are statistical surveys on the effects of air pollution on human populations under natural conditions. Such studies are extremely important, but due to the multiplicity of unknown factors it is not possible to establish a cause-effect relationship. Toxicological studies are conducted in the laboratory under controlled conditions. The effect of several variables such as pollutant concentration, exposure duration, temperature, humidity, etc., can be experimentally studied. Even though these experimental studies can clearly demonstrate a direct cause-effect relationship between certain pollutants and sickness or death, their relevance to natural setting is sometimes questionable.

Respiratory Effects

Pollutants may enter the body by a number of ways. They can cause eye and skin irritation; certain particulates may be swallowed as a result of internal respiratory cleaning action or certain pollutants could even be ingested. But the primary mode of pollutant transfer into the human body is through the respiratory system.

The respiratory system is composed primarily of two lungs and the air passages which lead to them. The air passages begin at the nose and mouth, and include the windpipe (trachea) and its two branches known as bronchi. The bronchi divide into five main branches (the lobor bronchi) and then subdivide, finally, reaching the terminal bronchi (or respiratory bronchioles) deep within the lungs. At the end of bronchioles are the countless tiny air sacs known as alveoli. Alveoli are the functional units of the lung; it is across their membranes that oxygen diffuses from the air to the pulmonary capillaries and carbon dioxide diffuses in the opposite direction. The total surface area of all the alveoli within the lungs is about 50 square metres.

Particulate matter inhaled may be deposited in various regions of the respiratory system depending on particle size. Particles above 10 μm are almost wholly retained in the nose. Those below 10 μm escape entrapment and generally pass through the upper respiratory system. Fine particles in the size range 0.5 to 5 μm are deposited as far as bronchioles, but few reach the alveoli. The walls of the bronchi and bronchioles are lined with fine hair-like structures called cilia (Fig. 2.4). These are responsible for removing such fine particles along with the mucous by moving them upto the larynx where they may be eliminated by swallowing. The health risk is

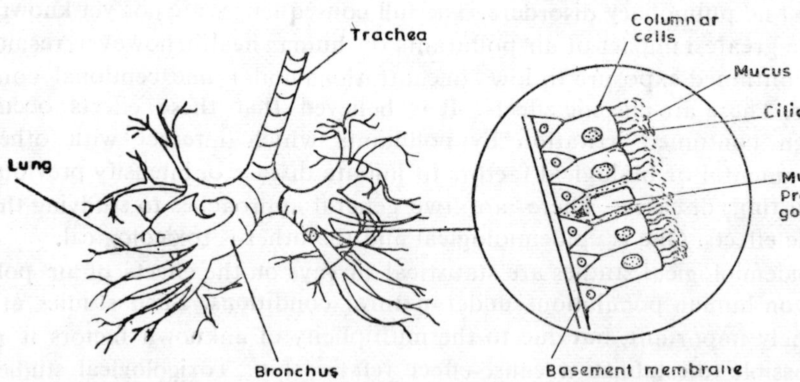

Fig. 2 4 Lower respiratory system showing bronchial lining.

primarily from the deposition of the particles smaller than 0.5 μm in the alveoli where they cause damage to the respiratory organs.

Particulate Matter

The toxic effects of particles can be grouped into three categories:

(a) *Interference of inert particles with the clearing mechanisms of the respiratory tract*: The effect includes a slowing of ciliary beat and mucus flow in the bronchial tree.

(b) *Particle act as carriers of adsorbed toxic gases such as SO_2 and produce synergistic effects*: A summary of some observed relations between particulates and SO_2 levels and the physiological responses is given in Table 2.8. Table 2.9 gives comparative data for SO_2 and particulate levels in major cities of India.

(c) *Particles may be intrinsically toxic because of their physical or chemical characteristics*: Such particles belong to metals which are usually found in the atmosphere in trace quantities but may constitute a great health hazard because of the possibility of their concentrations increasing beyond normal levels (0.01 to 3.0 per cent of all particulate air pollution). Of all the trace metals found in the atmosphere, the one present in largest concentration is lead. In addition to lead, cadmium, nickel and mercury represent known and widespread health hazards. Their sources and the possible effects are summarized in Table 2.10.

LEAD

The poisonous effects of lead on humans have long been recognised. Lead is a neurotoxin whose poisoning results in convulsions, delirium, coma, severe and irreversible brain damage, and death. Inhaled lead is far more

Table 2.8 Observed relations between particulates and SO_2 levels and health effects (ref. 21)

Condition	Effects
1. Particulate level of 750 $\mu g/m^3$ with SO_2 at 715 $\mu g/m^3$ (0.25 ppm), both 24-hr mean	Increased daily death rate; substantial increase in illness of persons with bronchitis
2. Low particulate level with SO_2 levels ranging between 300 and 500 $\mu g/m^3$ (0.11–0.19 ppm), both 24-hr mean	Increased hospital admissions of elderly people respiratory disease cases; increased absenteeism among older workers
3. Particulate level of 300 $\mu g/m^3$ with SO_2 level of 600 $\mu g/m^3$ (0.21 ppm), both 24-hr mean	Worsening of symptoms in persons suffering from chronic bronchitis
4. Particulate level of 185 $\mu g/m^3$ with SO_2 levels ranging between 105 and 265 $\mu g/m^3$ (0.04–0.09 ppm), both annual mean	Increased frequency of respiratory symptoms and lung disease
5. Particulate level of 100 $\mu g/m^3$ with SO_2 level of 120 $\mu g/m^3$ (0.05 ppm), both annual mean	Increased incidence of respiratory disease among school children

Table 2.9 Particulate and SO_2 levels in Indian cities annual averages (ref. 22, 23)

City	Particulate level ($\mu g/m^3$)	SO_2 level ($\mu g/m^3$)
Bombay	240	47
Calcutta	340	33
Delhi	601	41
Madras	100	8
Hyderabad	146	5
Kanpur	543	16
Ahmedabad	306	11

serious than the injested lead. The fine particles emitted by automobile are retained within the lungs and are absorbed by body with an efficiency of about 40 per cent. Extensive measurements of the size of lead particles in automobile exhaust have shown extremely small mean particle size (0.02 μm) besides the highways and much larger size (0.3 – 2.0 μm) in busy urban areas. About 80 per cent of the smallest particles are deposited in the lungs whereas only 20 per cent of the larger particles (0.5 μm) are retained.

Table 2.10 Major toxic metals and their effects

Element	Sources	Health effects
Lead	Auto exhaust (from gasoline), paints, storage batteries, pipes	Neurotoxin, affects blood system, behavioural disorders, death
Cadmium	Coal, zinc mining, incineration of plastic containers, refining of metals, tobacco smoke	Cardiovascular disease and hypertension, interference with zinc and copper metabolism, kidney damage
Nickel	Combustion of coal, diesel and residual oils, tobacco smoke, chemicals and catalysts, steel and non-ferrous alloys manufacture	Respiratory symptoms, lung cancer (as nickel carbonyl)
Mercury	Combustion of fossil fuels, evaporation from ore mining, exhausts from metal smelters, chloralkali cells, paints, pharmaceuticals	Nerve and brain damage, kidney damage

The body maintains about 15–25 μg of lead per 100 g of whole blood. The body responds to any increase in the lead intake by excreting it in the urine as much as possible and the remainder is stored primarily in the bones. At elevated blood lead levels the production of hemoglobin is impaired resulting in oxygen starvation and anemia. High lead levels in the blood may also produce behavioural disorders in children and adults.

CADMIUM

Cadmium is a potential health hazard due to its presence in urban atmosphere and cigarette smoke. Cigarette smoking constitutes a major source for cadmium accumulation in the body. The natural cadmium levels in air vary from 0.002 $\mu g/m^3$ to much higher values of about 0.3 $\mu g/m^3$ found near zinc smelters. Cadmium inhibits the performance of certain enzymes thereby producing hypertension in humans. Exposure to fumes or oxides of cadmium is known to cause cardiovascular diseases. In addition, cadmium can interfere with proper zinc and copper metabolism in the body.

Very little is known about the fate and distribution of cadmium in the environment. It has a very long biological half-life and, therefore, tends to accumulate in the human body. Some of the chronic effects are kidney and liver damage and even death.

NICKEL

The adverse effects of nickel are caused mainly due to nickel carbonyl which is formed when finely divided nickel is emitted into an atmosphere containing carbon monoxide. Nickel carbonyl is also formed in tobacco smoke. Within the lung the carbonyl complex breaks down and deposits finely divided nickel, which may be the main cause of cancer in the lungs. Nickel carbonyl has been shown to cause changes in the alveoli of the lungs, resulting in respiratory damage symptoms.

MERCURY

Mercury in the atmosphere is only a small part of the overall mercury burden of the environment. It is present in gaseous form in the atmosphere because of its relatively high vapour pressure. The gaseous mercury is washed from the air by rain; a portion of it enters the aquatic systems and the remaining is bound to the soil over the land. In both cases the inorganic mercury is generally converted into its methyl or dimethyl compounds by the action of bacteria.

The physiological effects of mercury poisoning include neurological damage, chromosomal aberrations, and even death. Methylmercury can penetrate the membranes separating the bloodstream from the brain, causing injury to cerebellum and the cortex. The effects are blurring of vision and numbness. Another insidious effect of methylmercury is that it can cross the placental barrier in pregnant woman. High levels of mercury can build up in the fetus without the mother showing any signs of illness. The effects are congenital birth defects or the death of the fetus.

Other Toxic Substances

Apart from the four major toxic substances already discussed there are many other substances whose presence in the atmosphere due to industrial activities constitutes a health hazard. Two such substances which may be classified as occupational pollutants are beryllium and asbestos.

Most beryllium emissions are in the form of metallic powder or beryllium oxide particulates. Acute beryllium exposure is a very serious occupational problem affecting the mucus membranes of the eyes and lungs. A chronic condition known as berylliosis is thought to be caused by beryllium concentrations as low as 0.01 to 0.10 $\mu g/m^3$. Berylliosis is a systemic poisoning which starts with progressive shortness of breath, weight loss and cough, and finally affects many organs including the heart. When inhaled as dust, it can also cause lung cancer, but this has not been conclusively proved for humans. Major sources of beryllium emissions into the atmosphere are ceramic industry, processing plants which convert the ore into beryllium powder, rocket motor test facilities, coal combustion, and nuclear power industry.

"Asbestos" is an industrial term for a number of hydrated silicates which separate into strong flexible fibres upon crushing and processing. The most abundant type of asbestos is known as chrysolite, which is the main fibre used in asbestos textiles. Inhalation of asbestos dust or fibres can cause a disabling lung disease known as asbestosis. The disease is characterized by shortness of breath and pleural calcification. Asbestos has also been implicated in lung cancer. The fibres line the membranes of the lungs and abdomen and this can lead to mesothelioma, an incurable and fatal cancer. It has been observed that the frequency of cancer is eight times higher among asbestos workers who smoke compared to that among non-smoking workers. Asbestos in the atmosphere comes from many uses of asbestos products which include insulating materials, asbestos cement and brake linings.

Gaseous Pollutants

The effects of gaseous pollutants on the respiratory system depend on the concentration, period of exposure, and the solubility. Highly soluble gases such as sulphur dioxide are absorbed in the upper part of the respiratory system whereas relatively insoluble gases like CO, NO_2 and ozone penetrate deep and reach the alveoli of the lungs. Some of the adverse effects of these pollutants include pulmonary edema, emphysema, and prevention of oxygen transfer to blood.

(a) Sulphur Dioxide

There is considerable evidence that SO_2 aggravates existing respiratory diseases in humans and contributes to their development. Even healthy individuals experience broncho-constriction when exposed for a few minutes to levels of 1.6 ppm. This condition is accompanied by shallow breathing and an increased respiratory rate. The acute irritant effects of the gas are confined to the upper respiratory tract where more than 95 per cent of inhaled SO_2 is absorbed. The chronic effects resulting from extended exposure to low concentrations include incidence of respiratory infection in children. Table 2.11 summarizes the effects of SO_2 at various concentrations.

Table 2.11 Effects of sulphur dioxide on humans (ref. 8).

Concentration (ppm)	Effects
0.2	Lowest concentration causing a human response
0.3	Threshold for taste recognition
0.5	Threshold for odour recognition
1.6	Threshold for inducing reversible broncho-constriction in healthy individuals
8–12	Immediate throat irritation
10	Eye irritation
20	Immeaiate coughing

The primary threat of SO_2 to urban atmospheres may arise not from SO_2 itself but from the changes it undergoes in the atmosphere, such as the formation of H_2SO_4 and sulphate aerosols. The sulphate particles can be carried deep into the lungs, causing even more severe health problems. SO_2 can also be absorbed on small particulates such as the salts of iron, manganese and vanadium present in the atmosphere and thus enter the alveoli. There, in the presence of moist air, SO_2 is oxidised to H_2SO_4 and the particulates act as catalysts in enhancing the oxidation process.

(b) CARBON MONOXIDE

Carbon monoxide, when inhaled, passes through the lungs and diffuses directly into the blood stream where it combines with the red blood pigment called hemoglobin forming carboxyhemoglobin, COHb. The affinity of carbon monoxide for hemoglobin is 210 times greater than that of oxygen and as a result the amount of hemoglobin available for carrying oxygen for body tissue is considerably reduced. The body tissues are thus deprived of their oxygen supply and death could result by asphyxiation (lack of oxygen). In addition, the presence of COHb in the blood retards the dissociation of remaining oxyhemoglobin, so the tissues are further deprived of oxygen.

Table 2.12 summarizes the effects of various concentrations of COHb in the blood. These are the expected signs and symptoms of exposure to carbon monoxide in healthy individuals and those suffering from heart disease.

The equilibrium level of COHb may be estimated for concentrations of CO below 100 ppm in the inhaled air by using the approximate equation[25]:

$$\% \text{ COHb in blood} = 0.16\,(\text{ppm. CO}) + 0.5 \tag{2.38}$$

Table 2.12 Health effects of COHb blood levels (ref. 24)

COHb blood level (%)	Effects on healthy individuals	Effects on heart patients
1–5	Blood flow to certain vital organs increases to compensate for reduction in oxygen carrying capacity of blood	Heart patients may lack sufficient cardiac reserve to compensate
5–9	Visual light threshold increased	Patients with angina pectoris require less exertion to induce chest pain
16–20	Laboured respiration during exertion, visual evoked response abnormal	May be lethal for patients with severe cardiovascular disease
20–30	Headache, nausea	
30–40	Severe headache; nausea and vomiting, dizziness	
40–50	Slurring of speech; tendency to collapse	
50–60	Convulsions; coma	
60–70	Fatal coma if not treated	

The equilibrium levels are not established instantaneously and it requires at least 8 hrs before the maximum level of COHb in the blood is attained. The time span can decrease considerably with increase in physical activity.

Carbon monoxide levels in cities are usually between 10 and 40 ppm on an annual 8-hr average basis and occasional short term concentrations may exceed 100 ppm. These levels can easily lead to COHb concentration in blood of approximately 2 to 8 per cent. Table 2.13 lists typical concentrations in various urban areas. A concentration of upto 500 ppm in the air, when inhaled for 1 hr produces no observable symptoms but a similar exposure to 1000 ppm can be dangerous. Concentrations of 4000 ppm and above are fatal, usually within 1 hr. Increased COHb in the blood deprives oxygen supply to various vital organs, especially the brain. This leads to impairment of mental performance, visual acuity and other functions.

Table 2.13 Carbon monoxide levels in urban areas.

City	Maximum 1 hour CO values, ppm
London	58
Chicago	46
Los Angeles	43
New York	27
Calcutta	35

The chronic effects of carbon monoxide are not fully known but they may induce heart and respiratory disorders. The long term studies are complicated by the presence of other pollutants such as NO_2 in the atmosphere. While CO itself has not been found to be carcinogenic, there is concern that it may increase the carcinogenic effects of other air pollutants by inhibiting the mucociliary clearance mechanism in the lungs.

(c) Oxides of Nitrogen

The major oxides of nitrogen which affect human health are nitric oxide (NO) and nitrogen dioxide (NO_2). NO is not an irritant and at concentrations that occur in the atmosphere it does not show any adverse health effects. However, its main toxic potential results from its oxidation to NO_2. Table 2.14 gives typical concentration of NO_2 in four Indian cities. These figures, which were reported over 1-2 month periods compare favourably with long-term averages of 0.02–0.05 ppm for London[26] and for several Japanese cities[27].

Table 2.14 NO_2 levels in Indian cities (ref. 29)

City	Range of av. conc., ppm	Range of max. conc., ppm
Bombay	0.008–0.013	0.013–0.05
Delhi	0.008–0.013	0.016–0.035
Calcutta	0.006–0.017	0.023–0.051
Kanpur	0.01 –0.095	0.02 –0.013

NO_2 is relatively insoluble and upon inhalation can reach the moisture-filled alveoli of the lungs. There it is converted to nitrous and nitric acids which are highly irritating and cause damage to the lung tissues. Long term exposure to concentrations of the order of 1 ppm could lead to symptoms resembling emphysema and biochemical alterations in blood. Other observed effects on humans are given in Table 2.15. In addition to the direct effects

Table 2.15 Effects of atmospheric NO_2 (ref. 29).

Effect	NO_2 Conc. (ppm)	Exposure
Increase in acute respiratory disease	0.06–0.1	2–3 years
Increase in acute bronchitis in school children	up to 0.1	6 months
Human olfactory threshold	0.12	<24 hrs
Increase in airway resistance	5	10 min
Pulmonary edema	90	30 min

mentioned therein, NO_2 in combination with hydrocarbons acts as the initiator of photochemical smog leading to the production of secondary pollutants like the oxidants. These oxidants are the ones that cause most damage to human health.

(d) Hydrocarbons and Photochemical Oxidants

At the concentrations usually found in urban air, the hydrocarbons cause no adverse effects on human health. Aliphatic hydrocarbons produce undesirable effects only at concentrations 10^2 to 10^3 times higher than those usually found in the atmosphere. No effects have been observed for levels below 500 ppm. Aromatic hydrocarbons are more reactive than aliphatic ones and cause irritation of the mucous membranes.

The major oxidant produced in photochemical smog is ozone. Contrary to the popular belief, ozone appears to have no effect on the eyes at usual urban concentrations. The respiratory system, however, may respond to very low concentrations. There is a definite increase in airway resistance in some people exposed to concentrations as low as 0.1 ppm. Many other oxidants produced in photochemical smog, particularly the peroxyacyl nitrates, cause eye irritation. Oxidants such as peroxyacetyl nitrate and peroxybenzoyl nitrate irritate the nose and throat, and cause chest constriction. The one with the lowest threshold is peroxybenzyl nitrate, having a value of 0.005 ppm for five-minute exposure. Table 2.16 summarizes the effects of ozone and total photochemical oxidants on humans.

Table 2.16 Effects on health of ozone and total photochemical oxidants (ref. 30)

Effects	Conc. ppm	Exposure
Ozone		
Increased airway resistance	0.1—1.0	1 hour
Extreme fatigue, lack of coordination	1.0—3.0	2 hours
Severe cough	2.0	2 hours
Pulmonary edema	9.0	Unknown
Total oxidants		
Eye irritation	0.1	Instantaneous
Aggravation of asthma	0.05—0.06	1 hour
Impaired performance of athletes	0.03—0.3	1 hour

Table 2.17 Ozone levels in four Indian cities (ref. 28)

City	Range of av. conc., ppm	Range of max. conc., ppm
Bombay	0.003 — 0.017	0.026 — 0.047
Delhi	0.006 — 0.013	0.012 — 0.028
Calcutta	0.013 — 0.018	0.023 — 0.078
Kanpur	0.02 — 0.1	0.028 — 0.135

Typical ozone levels in Indian urban atmosphere are presented in Table 2.17. These values, like those given for NO_2 were reported over 1-2 month periods. These levels, even though of no major consequence to human health, are known to affect vegetation as will be discussed in the next section.

Air Pollution Effects on Vegetation

The most obvious damage caused by air pollutants to vegetation occurs in the leaf structure. The surface of a leaf is covered by a waxy layer known as the cutile. In between the waxy layer is the epidermis, a single layer of cells forming the surface skin of the leaf (Fig. 2.5). Its chief functions are

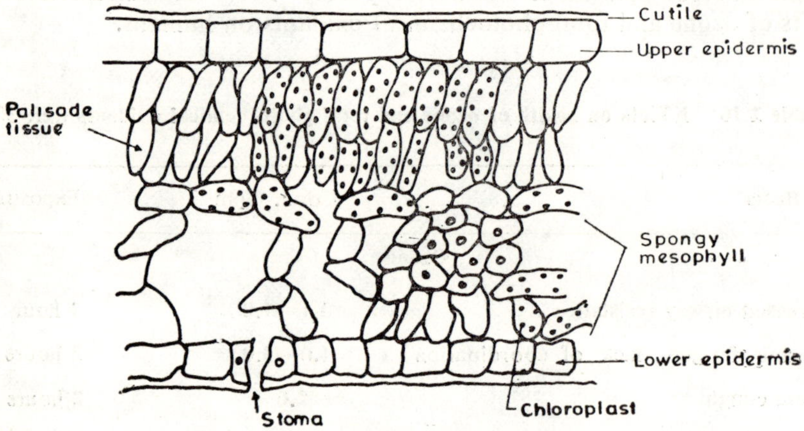

Fig. 2.5 Cross section of a common leaf.

the protection of the inner tissues from excessive moisture loss and the admission of carbon dioxide and oxygen to these internal tissues. The leaf surface is penetrated by a large number of openings called the stomata. Each stoma is protected by a pair of guard cells, which control the opening and closing of the stoma. The internal tissues of the leaf are the palisade, the spongy mesophyll, and the conducting tissues, or veins, which carry water, minerals and food. Of particular interest with regard to air pollution are the stomata through which gases diffuse to the surface of the cells within the leaf.

The damage caused by air pollutants is of several types like necrosis, chlorosis and epinasty. The dead areas on a leaf structure are referred to as necrosis. Chlorosis is the loss or reduction of chlorophyll and leads to the yellowing of the leaf. Epinasty is a downward curvature of the leaf due to higher rate of growth on the upper surface, and the dropping of leaves is called abscission. Few plant species are spared damage on exposure to one or more of the principal air pollutants. In general, the pollutants enter the inner tissues through the stomata, where they destroy the chlorophyll and disrupt photosynthesis. The adverse effects

Table 2.18 Effects of specific air pollutants on vegetation

Pollutant	Level (ppm) and exposure	Effects
SO_2	0.3 to 0.5 for several days	Bleached spots, chlorosis, chronic injury to spinach, and other leafy vegetables
NO_2	0.25 for 8 month	Increased abscission and reduced yield in citrus plants
	0.5 for 10–12 days	Suppressed growth of tomatoes
	3.5 for 21 hours	Spots of mild necrosis on cotton and bean plants
	25 for 1 hour	Acute leaf injury
Ozone	0.03 for 8 hours, time effect reduced if low level SO_2 is also present	Fleck on upper surface; Necrosis and bleaching; damage to tobacco leaves at O_3=0.027 ppm and SO_3=0.24 ppm after 2 hours of exposure
Peroxyacetyl nitrate (PAN)	0.01 to 0.05 for a few hours	Glazing or bronzing of underside of leaf; damage to sensitive plants; young leaves more susceptible to damage
HF	0.001 for 7–21 days can be significant; cumulative effect	Necrosis of leaf tip, grapes are particularly susceptible
Ethylene	0.1 for several hours or 0.05 for several weeks	Epinasty; leaf abscission, flower dropping

range from reduction in growth rate to death of the plant. The effect of particulates on vegetation is not well known. However, some specific dusts have been observed to cause damage. Cement dust deposited on leaves, on combination with mist or light rain, forms incrustations. Plugging of stomata may occur, resulting in plant damage. Chemicals such as arsenic and fluorides when deposited on the leaves can poison animals. The effects of specific pollutants on vegetation are summarized in Table 2.18.

Effects of Air Pollution on Materials

The damage caused by atmospheric pollutants to materials is a well known phenomenon. Particulates such as soot, dust and fumes soil painted surfaces, fabrics and buildings, and because of their abrassive nature, particulates can cause damage to exposed surfaces when they are driven by wind at high velocities. Through their own corrosiveness or in the presence of SO_2 and moisture, they can accelerate the corrosion of steel, copper, zinc and other metals.

The most notorious pollutant responsible for metallic corrosion is sulphur dioxide. It has been reported that corrosion of hard metals such as steel begins at annual mean concentrations of 0.02 ppm (52 $\mu g/m^3$). At levels of 0.09 – 1 ppm, SO_2 affects fabrics, leather and paint. SO_2 is readily absorbed by leather and causes its disintegration. Paper is also discoloured by SO_2 and becomes brittle and fragile. Sulphuric acid mist in the atmosphere causes deterioration of structural materials such as marble and limestone. Many priceless marble sculptures and buildings have suffered damage in the last 30 years as a result of increased SO_2 content in the atmosphere.

Ozone is a very reactive substance. Much of the degradation of materials, such as fabrics and rubber, now attributed to "weathering" is caused primarily by ozone. Ozone causes the cracking of synthetic rubbers at atmospheric levels of 0.01 to 0.02 ppm (20-40 $\mu g/m^3$). It also attacks fabric fibres and the adverse effects increase in the order: fibres made of cotton, acetate, nylon, and polyester. The fading of fibres and the cracking of rubber are attributed to ozone's oxidizing ability. Nitrogen oxides, although less widely publicised than ozone, are known to cause fading in acetate, cotton and rayon fibres at levels of 0.6-2 ppm over 2-3 month period. It has been observed that particulate nitrates attack and damage nickel-brass alloys in the presence of moisture.

Damage to Health, Vegetation and Materials in India

In India no systematic or statistical data are available on the health status of the people exposed to air pollution. Firstly, the data base is extremely poor. According to one estimate, as many as 60 million people are affected in one way or the other by air pollution[31]. Epidemiological data should be collected at the regional and national levels on the effects of air pollutants on human health. The nature of damage caused to tropical and subtropical vegetations by air pollution is almost unknown. Reports of SO_2 damage to plants like banana, grapes, papaya, coconut, etc., are scattered, and no systematic assessment is available. Most of the information available in literature is applicable to countries of moderate climate in Europe and North America, and it is of very little use for assessing the damage to tropical broad-leaved plants.

The information on materials damage due to air pollution is very scanty. However, the information on corrosive effects of acid precipitation on metals is available for a few cities. Due to high concentration of industrial discharges and salinity and humidity in the air, corrosion rates in Bombay are reported to be 3 to 6 times higher than those in other similar coastal areas of the country.

2.6 AIR POLLUTION LAWS AND STANDARDS

The main objective of enacting any pollution law is to control pollutant sources so that ambient pollutant concentrations are reduced to levels considered safe. The oldest antipollution law in India is the "Smoke Nuisance Act" enacted as early as 1920 by the then Bombay Government. The act was originally applicable to Bombay city, but was later extended to cover the entire state. It laid down rules and regulations concerning the size and height of chimneys and boilers in textile mills, besides listing measures to be effected for treatment of smoke. Similarly, a few other states in the country came up with their own smoke nuisance acts. Further, almost all the states have a clause on nuisance control in their municipal corporation acts. However, these acts have either loopholes or they are outdated in context of the development of numerous chemical, metallurgical and power plants. Hence, the need for strict monitoring of air pollutants and their control through detailed regulations is very essential. The Central Air (Prevention and Control of Pollution) Act, 1981 is a significant development in this direction.

This act provides an integrated approach to tackling of problems related to pollution. It empowers the Central Board for the Prevention and Control of Water Pollution, constituted under the Water (Prevention and Control of Pollution) Act, 1974, to exercise the powers and perform the functions of the Central Board for the Prevention and Control of Air Pollution also. It also empowers the state governments to declare air pollution areas and to prohibit the use of any fuel which is likely to cause air pollution. The state governments are also empowered to prohibit the burning of any material which is likely to cause air pollution in any air Pollution control area. The officers of the State Board are empowered to obtain information relating to air pollution and to inspect the concerned premises and to take samples of emissions from sources for analysis.

The Central Air Act will certainly help to maintain performance standards of anti-pollution control measures. In addition, many state governments have their "Environmental Protection Advisory Committees" to take effective measures for pollution control, and these bodies also caution the government about the adverse consequences if such measures are not taken in time.

In many countries the legal standards for ambient air quality are prescribed by the governments. The purpose of these standards is to reduce the pollutants to a certain level which would avoid undesirable effects. The prescribed standards may vary slightly from country to country

Table 2.19 United States air quality standards.

Pollutant	Air quality standard ($\mu g/m^3$)	
	Primary	Secondary
Sulphur dioxide	80, annual arithmetic mean; 365, max. 24-hr value may occur once each year	1300, max. 3-hr value may occur once each year
Oxides of nitrogen	100, annual arithmetic mean	Same as primary standard
Photochemical oxidants (corrected for NO_2)	160, max. 1-hr value may occur once each year	Same as primary standard
Carbon monoxide	10 (mg/m^3), max. 8-hr value may occur once each year; 40 (mg/m^3) max. 1-hr value may occur once each year	Same as primary standard
Particulate matter*	75, annual geometric mean; 260, max. 24-hr value may occur once each year	60 annual geometric mean; 150, max. 24-hr value may occur once each year
Hydrocarbons (corrected for CH_4)	160, max. level that may occur 6–9 A.M. once each year	Same as primary standard

*Standards not applicable to toxic particles such as asbestos, lead or beryllium.

depending upon their meteorological and geographical features, and the population density.

The United States has two types of air quality standards—the primary and the secondary. The primary standards are meant for the protection of human health, whereas the secondary standards pertain to ambient levels considered safe for plants, materials, and personal comforts. Table 2.19 gives a list of air quality standards adopted by the United States Environmental Protection Agency in 1971. These standards are modified from time to time if new data on the effects of air pollution on the health of the people or other conditions necessitate such modification.

REFERENCES

1. J.O'M Bockris, *Environmental chemistry* (Plenum Press, New York), 1977.
2. L. Hodges, *Environmental pollution*. Chapter 15 (Holt, Rinehart and Winston, Inc., New York), 1973.

3. A.T. Rossano, *Air pollution control guidebook for management* (E.R. A. Inc., Stanford, Conn.) 1969.
4. E. Robinson and R.C. Robbins, *Emissions, concentrations, and fate of particulate atmospheric pollutants* (American Petroleum Institute, March SRI Project SCC–8507, Washington D.C. Publication No. 4076), 1971.
5. E. Robinson and R.C. Robbins, "Gaseous sulfur pollutants from urban and natural sources", *J. Air Pollut. Contr. Assoc.*, Vol. 20 (1970), pp. 223-235.
6. E. Robinson and R.C. Robbins, "Gaseous nitrogen compound pollutants from urban and natural sources", *J. Air Pollut. Contr. Assoc.* Vol. 20, (1970), pp. 303-306.
7. L.S. Jaffe, *J. Geophys, Res.*, Vol. 78 (1973) p. 5293.
8. S.H. Stoker and S.L. Seager, *Environmental chemistry: air and water pollution* (Scott Foresman & Co., New York), 1976.
9. K.B. Asthana, S.K. Gadi and S.K. Ghosh, *Water and air pollution laws,* 2nd edition (Law Publishers, Allahabad), 1986.
10. V.B. Shirodkar, "Air pollution in greater Bombay", *Chemical Age of India,* Vol. 28 (1977), p. 461.
11. H.H. Lamb, "Volcanic dust in the atmosphere: chronology and assessment of its meteorological significance", *Phil. Trans. Roy. Soc. (London)*, Vol. A266 (1970), pp. 425-533.
12. D.J. Spedding, "The interaction of gaseous pollutants with materials at the surface of the Earth", in *Environmental chemistry,* edited by J.O'M Bockris (Plenum Press, New York), 1978, p. 218.
13. E. Robinson and R. Robbinson, "Sources, abundance, and fate of gaseous atmospheric pollutants", Final Report, Stanford Res. Inst. Project PR-6755 (Amer. Petrol. Inst., New York), 1968.
14. R.E. Inman, R.B. Ingersoll and E.A. Levy, "Soil: a natural sink for carbon monoxide", *Science,* Vol. 172 (1971), pp. 1229-1931.
15. H.F. Johnstone and D.R. Coughanowr, "Absorption of sulfur dioxide from air", *Ind. Eng. Chem.* Vol. 50 (1958), p. 1169.
16. P.A. Leighton. *Photochemistry of air pollution* (Academic Press, New York), 1961.
17. K. Westberg, N. Cohen and K.W. Wilson, "Carbon monoxide: its role in photochemical smog formation", *Science,* Vol. 171 (1971), pp. 1013-1015.
18. J.R. Goldsmith, "Effects of air pollution on human health", in *Air pollution, Vol. I,* edited by A.C. Stern (Academic Press, New York), 1968.
19. H.M. Dix., *Environmental pollution* (John Wiley & Sons, New York) 1981.
20. *India Today.* Dec. 31, 1984; *Frontline,* Dec. 15, 1984.
21. *Air quality criteria for sulfur oxides and for particulates* (U.S. Public Health Serv. Publ. AP-50), 1969.

22. E. Chauhan. *Current trends in Indian environment* (Today and Tomorrow's Printers and Publishers, New Delhi) 1977, p. 85.
23. S. Kumar, "Air pollution in India", *Chemical Age of India,* July 1976, p. 629.
24. R.D. Stewart, "The effects of carbon monoxide on humans". Annual Review of Pharmacology, Vol. 15 (1975), pp. 409-423.
25. U.S. Environmental Protection Agency, "Air quality criteria for carbon monoxide", AP-62 (USEPA, Washington D.C.), 1970.
26. C.F. Barrett and L.E. Reed *Intern. J. Air and Water Pollution,* Vol. 9 (1965), p. 357.
27. O.J. Tadu, *Sci. Labour* (Tokyo), Vol. 42 (1966), p. 263.
28. P.K. Yennawar, S.N. Dixit, V.I. Pampattiwar, J.N. Dave and S.J. Arceivala, "Short-term air quality surveys in four major cities of India", *Environmental Health,* Vol. 12 (1970). pp. 355-383.
29. U.S. Environmental Protection Agency, "Air quality criteria for nitrogen oxides", AP-84 (USEPA, Washington, D.C.), 1971.
30. *Air quality criteria for photochemical oxidants* (National Air pollution Control Administration, AP-63, Washington, D.C.). March, 1970.
31. M.K. Das, "Air pollution in India and how to tackle it", *Chemical Age of India,* Vol. 28 (1977), pp. 433-435.

SUGGESTIONS FOR FURTHER READING

1. P.V. Hobb, H. Harrison and E. Robinson, "Atmospheric effects of pollutants", *Science,* Vol. 183 (1974), pp. 909-915.
2. V.J. Marchesani, *et al.*, "Minor sources of air pollutants emissions", *J. Air Pollution Control Assoc.*, Vol. 20 (1970), pp. 19-22.
3. F.B. Abeles, *et al.*, "Fate of air pollutants: removal of ethylene, SO_2, NO_2 by soil", *Science,* Vol. 173 (1971), pp. 914-916.
4. A.C. Hill, "Vegetation: a sink for atmospheric pollutants", *J. Air Pollution Control Assoc.*, Vol. 21 (1971), p. 341.
5. John W. Winchester, "Natural and Pollution sources of trace elements in atmospheric particulates", *AIChE Sym. Series,* Vol. 71 (No. 149) (1975), p. 16.
6. S.I. Rasool and S.H. Schneider, "Atmospheric CO_2 and aerosols: effects of large increases in global climate", *Science,* Vol. 173 (1971), p. 138.
7. L.B. Lave and E.P. Seskin, "Air pollution and human health", *Science,* Vol. 169 (1970), p. 723.
8. J. McCanoll and W. Bradley, "Excess mortality as an indicator of health effects of air pollution", *Amer. J. Pub. Health,* Vol. 56 (1966), pp. 1933-42.

9. David L. Coffin, "Air pollution: present and future threat to man and his environment", *Advances in Environ. Sci. and Tech.*, edited by J.N. Pitts and R.C. Metcalf, Vol. 2 (Wiley—Interscience, New York), 1971, pp. 1-38.

PROBLEMS

1. The maximum one-hour carbon monoxide levels in Calcutta reach 35 ppm. Calculate the equivalent concentration in terms of mass fraction (ω_p), and in milligrams per cubic metre at 25°C and 1 atm.
2. The mean concentration of sulphur dioxide in Bombay is 47 $\mu g/m^3$. What is the equivalent concentration in parts per million at 25°C and 1 atm?
3. The exhaust gas from a motor vehicle shows a carbon monoxide concentration of 2 per cent by volume. What is the concentration of CO in mg/m^3 at 0°C and 1 atm? (2.5)
4. A 1000 MW power plant of 35% thermal efficiency is proposed. The plant burns 3 per cent sulphur coal with a heating value of 6000 kcal/kg and emits 64,000 m^3/min of flue gas. What is the concentration of SO_2 in the flue gas in ppm? Assume that the density of SO_2 is 1920 g/m^3. (2800)
5. A captive power plant of 1 MW with 40% thermal efficiency is to be designed which will use residual oil with 1 per cent sulphur content. The emission factors for burning 1 m^3 of oil are: particulates = 1 kg, NO_2 = .10 kg and SO_2 = 16S kg where S is the per cent sulphur content of oil. What are the rates of emission of particulates, NO_2 and SO_2 to the atmosphere if no control devices are provided? The heating value of the oil is 8000 kcal/kg, and its density may be taken as 950 kg/m^3. (0.283 kg/hr; 2.83 kg/hr; 0.045 kg/hr).
6. Indicate the sources of the following trace elements in the atmosphere and their possible effects on human health: manganese, arsenic, zinc and chromium.
7. What are the main sources of fluoride emissions to the atmosphere? Discuss their effects on human health and on plants.
8. What is the difference between epidemiological and toxicological studies? Give specific examples.
9. Define the terms (a) troposphere and tropopause, (b) stratosphere and stratopause, (c) mesosphere and mesopause, and (d) thermosphere. Draw a graph showing the variation of temperature with altitude in each of the above regions, and explain why the curve appears as it does in your plot.

10. Why is the affinity of carbon monoxide to hemoglobin so much greater than that of oxygen?
11. The behaviour of particulate pollution in the atmosphere is influened by their relative sizes. Explain how the size of particulate makes a difference?
12. Explain the mechanisms of agglomeration and coagulation. Particles of what size are most affected by these processes?
13. The dynamic nature of photochemical smog is illustrated in Fig. 2.3 during a sunny day. How do the curves for NO, NO_2, hydrocarbons and oxidants behave on a cloudy day? Explain.
14. Write a summary of the article "Stratospheric pollution: multiple threats to earth's ozone," by A.L. Hammond and T.M. Maugh, published in *Science*, October 25, 1974.

3. Meteorological Aspects of Air Pollutant Dispersion

3.1 TEMPERATURE LAPSE RATES AND STABILITY

Air pollutants emitted from anthropogenic sources must first be transported and diluted in the atmosphere before these undergo various physical and photochemical transformations and ultimately reach their receptors. Otherwise, the pollutant concentrations reach dangerous levels near the source of emission. Hence, it is important that we understand the natural processes that are responsible for their dispersion. Effective dispersion of pollutants in the atmosphere depends primarily on the degree of stability of the atmosphere and on its turbulent structure. In the broadest sense dispersion is controlled by meteorological conditions prevailing in the atmosphere.

The degree of stability of the atmosphere in turn depends on the rate of change of ambient temperature with altitude. The relationship between the temperature and the altitude can be obtained by considering air to be an ideal gas. The change of pressure in the vertical direction may be represented by the relation :

$$\frac{dp}{dz} = -\rho g \tag{3.1}$$

where p is the atmospheric pressure, z is the altitude and ρ is the atmospheric density. Also, the perfect gas relation,

$$p = \rho RT \tag{3.2}$$

holds at any point in the atmosphere where R is the gas constant for air, and T is the absolute temperature. Substitution of Eq. (3.2) into (3.1) gives the general expression for the variation of pressure with altitude:

$$\frac{dp}{dz} = -\frac{pg}{RT} \tag{3.3}$$

If we consider the simple case of isothermal atmosphere, Eq. (3.3) could be integrated directly to give

$$p = p_0 \exp\left(-\frac{g}{RT}z\right) \tag{3.4}$$

where p_0 is the pressure at ground level (1.013 bars). The above equation gives an exponentially decreasing pressure with altitude which is a typical characteristic of compressible fluids such as air. Comparison of this exponential behaviour with observed profile shows a slight quantitative discrepancy and this discrepancy may be attributed to the fact that in the troposphere the temperature decreases with altitude and is not a constant as was assumed in the model. However, this model of the atmosphere may still be used by assuming a series of air layers, each layer having a uniform but different temperature.

A better model for the atmospheric dependence of p and T is the polytropic atmosphere which obeys the relation,

$$T = T_0 \left(\frac{p}{p_0}\right)^{n-1/n} \tag{3.5}$$

Substituting Eq. (3.5) for p in Eq. (3.3) and differentiating, we get,

$$\frac{dT}{dz} = -\frac{n-1}{n}\frac{g}{R} \tag{3.6}$$

Thus, Eq. (3.6) represents the variation of temperature with altitude for a polytropic model where the temperature decreases with altitude linearly with a slope of $-(n-1)\,g/nR$. The decrease in temperature with altitude is known as the lapse rate. Based on meteorological data in the troposphere upto about 10 km, where the temperature decreases linearly with altitude, the environmental lapse rate is found to be about 6.5°C/1000 metres. Putting this value for dT/dz in the equation for polytropic model, we have

$$\left(\frac{dT}{dz}\right)_{env} = -\frac{6.5°\text{C}}{1000 \text{ metres}} = -\frac{n-1}{n}\frac{g}{R} \tag{3.7}$$

From the above equation the value of n comes to be 1.23. The lower atmosphere ends at the tropopause which is at an altitude of about 12 km. Above the tropopause is the stratosphere. In the stratosphere there are two distinct regions of temperature variation. The lower region, extending upto 25 km, has a temperature which is essentially constant, and in the upper region the temperature increases with altitude as a result of ozone formation. In the lower region of the stratosphere the isothermal model based on $n = 1$ is applicable. The temperature profiles based on these models are represented in Fig. 3.1.

The environmental lapse rate, however, is not always constant particularly in the lower troposphere, a zone from the ground level up to an altitude of about 2 km where there is a considerable variation from the normal rate of 6.5°C/1000 metres. This is due to the character of the underlying surface and the rate of radiation emanating from it. On sunny days, because of strong solar heating from the ground, the environmental lapse

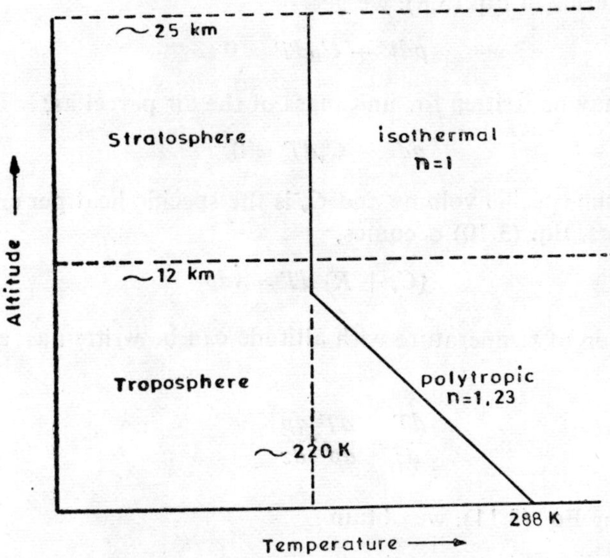

Fig. 3.1 Temperature—Altitude profiles.

rate often exceeds 10°C/1000 metres, although this is usually confined to the lowest 200 metres of the atmosphere. Extreme cases of atmosphere are characterised by a negative lapse rate, where the temperature increases with altitude.

Adiabatic Lapse Rate

The adiabatic decrease in temperature with altitude is especially important in the vertical movement of air pollutants and can be explained by utilizing the concept of an air parcel, which is pictured as a little sphere of air. As the air parcel rises in the atmosphere, it goes through a region of decreasing pressure and expands to accommodate the decreasing pressure. As it expands, it does work on the surroundings. Since the process is usually rapid, there is no heat transfer between the air parcel and the surrounding air. Hence, for the case of an adiabatic process the first law of thermodynamics yields,

$$-dW = dV \tag{3.8}$$

This means that the internal energy decreases thereby decreasing the temperature. The decrease in temperature of a rising, expanding air parcel is an important feature of vertical air motion. Now the work done on the surroundings is equal to pdV and the change in internal energy becomes equal to $C_V dT$, where C_V is the specific heat at constant volume. Substituting these terms in Eq. (3.8), we get

$$pdV + C_V dT = 0 \tag{3.9}$$

Eq. (3.9) may be written for unit mass of the air parcel as,

$$pdv + C_v dT = 0 \tag{3.10}$$

where v is the specific volume and C_v is the specific heat per unit mass. For a perfect gas, Eq. (3.10) becomes,

$$(C_v + R)\, dT = vdp \tag{3.11}$$

The variation of temperature with altitude can be written as a product of two terms:

$$\frac{dT}{dz} = \frac{dT}{dp} \cdot \frac{dp}{dz} \tag{3.12}$$

Rearranging Eq. (3.11), we obtain

$$\frac{dT}{dp} = \frac{v}{C_v + R} \tag{3.13}$$

Substituting Eqs. (3.13) and (3.3) into Eq. (3.12), we get

$$\frac{dT}{dz} = -\frac{g}{C_v + R} \tag{3.14}$$

Noting that $C_v + R = C_p$, the specific heat at constant pressure per unit mass of air parcel, we can write Eq. (3.14) as

$$\frac{dT}{dz} = -\frac{g}{C_p} \tag{3.15}$$

The above expression gives the temperature lapse rate for dry parcel of air moving upwards adiabatically and is known as the dry adiabatic lapse rate. When Eq. (3.15) is evaluated for dry air, the lapse rate is found to be

$$\left(\frac{dT}{dz}\right)_{adia} = -\,9.86\ ^\circ\text{C}/1000 \text{ metres}$$

Thus, the dry adiabatic lapse rate, which is denoted by Γ, is given by

$$\Gamma = \left(-\frac{dT}{dz}\right)_{adia} \simeq 10\ ^\circ\text{C}/1 \text{ km} \tag{3.16}$$

Comparing Eqs. (3.15) and (3.6), we see that for the adiabatic case $n = \gamma = C_p/C_v$, and this becomes equal to 1.4 for dry air.

In the presence of moisture in the atmosphere, a rising parcel of air may cool until the partial pressure of the water vapour equals the saturation vapour pressure of water. In such circumstances, some of the water vapour will condense if sufficient nucleation sites are present and release the latent heat. If λ is the latent heat per gram of water, then the release of this heat is given by,

$$dQ = -\lambda m d\omega \qquad (3.17)$$

where m is the mass of dry air in the parcel and $d\omega$ is the ratio of the amount of water vapour which is condensing to the mass of dry air. Adding this term to the right-hand side of Eq. (3.9) and proceeding further, we get the lapse rate in a saturated condition:

$$\frac{dT}{dz} = -\frac{g}{C_p} - \frac{\lambda}{C_p}\frac{d\omega}{dz} \qquad (3.18)$$

Equation (3.18) is referred to as the wet adiabatic lapse rate equation although the process is not strictly adiabatic. However, the condensing water has negligible effect on temperature. Since $d\omega/dz$ is negative for a rising parcel in which condensation is taking place, the last term in Eq. 3.18 is positive with the result that the temperature decrease upon increase in altitude is less than that for dry air. Also, since the saturation vapour pressure of water increases very rapidly with temperature, the quantity $d\omega/dz$ depends strongly on the temperature. Therefore, the wet adiabatic lapse rate is not a constant, independent of altitude. In warm tropical air the wet adiabatic lapse rate is approximately one-third of the dry adiabatic lapse rate.

Atmospheric Stability

The ability of the atmosphere to disperse the pollutants emitted into it depends to a large extent on the degree of its stability. A comparsion of the adiabatic lapse rate with the environmental lapse rate gives an idea of the stability of the atmosphere. Fig. 3.2 illustrates the temperature profiles for various cases of stability. When the environmental lapse rate and the dry adiabatic lapse rate are exactly the same, a rising parcel of air will have the same pressure, temperature and density of the surroundings and would experience no buoyant force. Such an atmosphere is said to be neutrally stable where a displaced mass of air neither tends to return to its original position nor tends to continue its displacement.

When the environmental lapse rate $(-dT/dz)_{env}$ is greater than the dry adiabatic lapse rate, Γ, the atmosphere is said to be superadiabatic. Hence, a rising parcel of air, cooling at the adiabatic rate, will be warmer and less dense than the surrounding environment. As a result, it becomes more buoyant and tends to continue its upward motion. Since vertical motion is enhanced by buoyancy, such an atmosphere is called unstable. In the unstable atmosphere the air from different altitudes mixes thoroughly. This is very desirable from the point of view of preventing pollution, since the

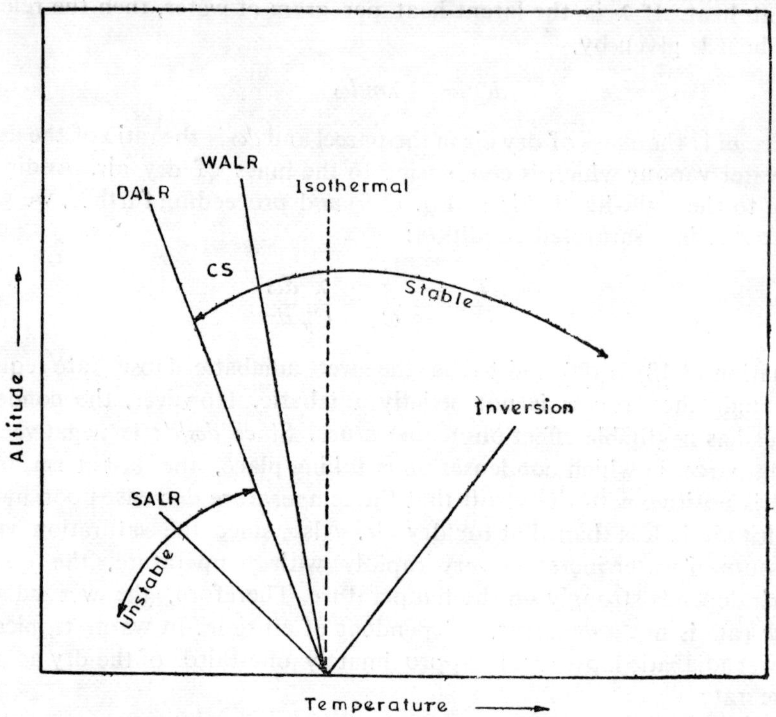

Fig. 3.2 Atmosphere stability regions. (SALR: Super-adiabatic lapse rate; DALR: Dry adiabatic lapse rate; WALR: Wet adiabatic lapse rate; CS: Conditional stability).

effluents will be rapidly dispersed throughout the atmosphere.

On the other hand, when the environmental lapse rate is less than the dry adiabatic lapse rate, Γ, a rising air parcel becomes cooler and more dense than its surroundings and tends to fall back to its original position. Such an atmospheric condition is called stable and the lapse rate is said to be subadiabatic. Under stable conditions there is very little vertical mixing and pollutants can only disperse very slowly. As a result, their levels can build up very rapidly in the environment.

A similar reasoning may be applied to the case of a saturated atmosphere. Thus, instability and stability may be defined with reference to a neutrally stable atmosphere where the environmental lapse rate is equal to the dry adiabatic lapse rate for dry or unsaturated air, and wet adiabatic lapse rate for saturated air. Strictly speaking, however, an absolute instability exists if the environmental lapse rate is greater than the dry adiabatic lapse rate, and an absolute stability exists if the environmental lapse rate is less than the wet adiabatic lapse rate. Conditional stability occurs if the environmental lapse rate is between these two values.

The extreme case of a stable atmosphere, called an inversion, occurs when temperature increases with altitude. Such a lapse rate is known as the negative lapse rate. Under these conditions, the atmosphere is very stable and practically no mixing of pollutants takes place.

Inversions

Atmospheric inversions influence the dispersion of pollutants by restricting vertical mixing. There are several ways by which inversion layers can be formed. One of the most common types is the elevated *subsidence inversion.* This is usually associated with subtropical anticyclone where the air is warmed by compression as it descends in a high pressure system and achieves temperature higher than that of the air underneath. If the temperature increase is sufficient, an inversion will result. Fig. 3.3 illustrates this condition and temperature profile for the subsidence inversion. The subsidence

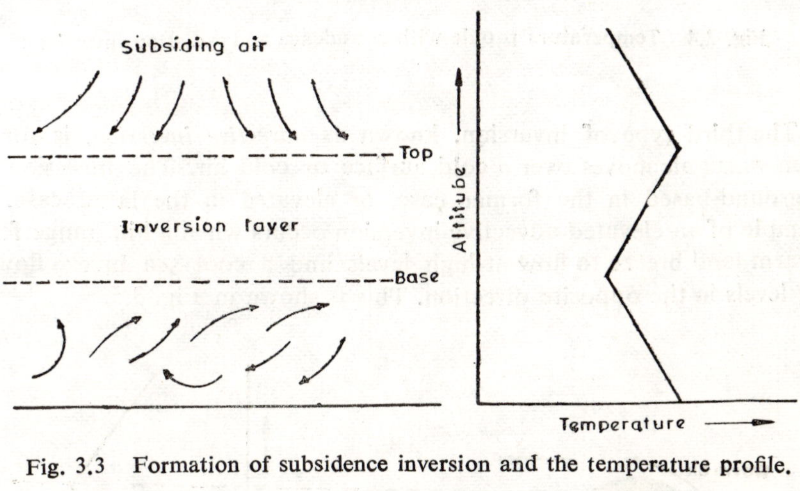

Fig. 3.3 Formation of subsidence inversion and the temperature profile.

is caused by air flowing down to replace air which has flowed out of the high pressure region.

The second common form of inversion is the *radiational inversion.* This results from the normal diurnal cooling cycle. After sunset, the ground cools quickly by radiational heat transfer, and the lowest layer of air in contact with the surface loses sensible heat through conduction and small scale mixing. Consequently, a temperature inversion is set up between the cool low-level air and the warmer air above, in the first few hundred meters above the surface. Valleys and low lying areas are particularly affected by this type of nocturnal inversions because denser, colder air tends to sink down beneath the warmer air. The next day sunlight destroys the inversion as the Earth is warmed and the air previously stratified by inversion is overturned by convective currents. The temperature profile for a combined radiation inversion and subsidence inversion is illustrated in Fig. 3.4. The subsidence inversion is potentially more serious than radiation inversion because the latter usually dissipates quite rapidly after sunrise.

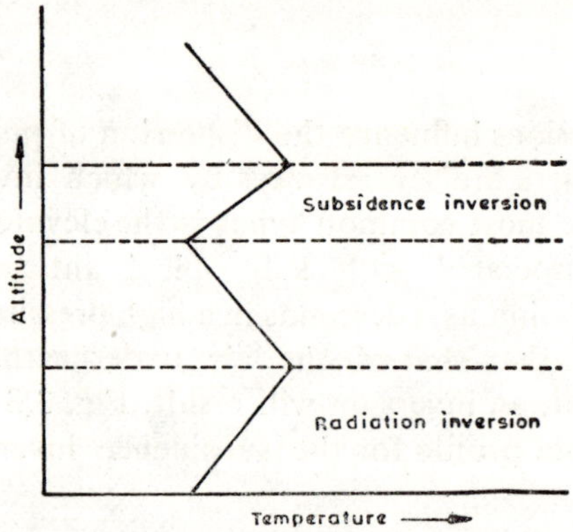

Fig. 3.4 Temperature profile with subsidence and radiation inversions.

The third type of inversion, known as *advective inversion*, is formed when warm air moves over a cold surface or cold air. The inversion can be ground-based in the former case, or elevated in the latter case. An example of an elevated advective inversion occurs when a hill range forces a warm land breeze to flow at high levels and a cool sea breeze flows at low levels in the opposite direction. This is shown in Fig. 3.5.

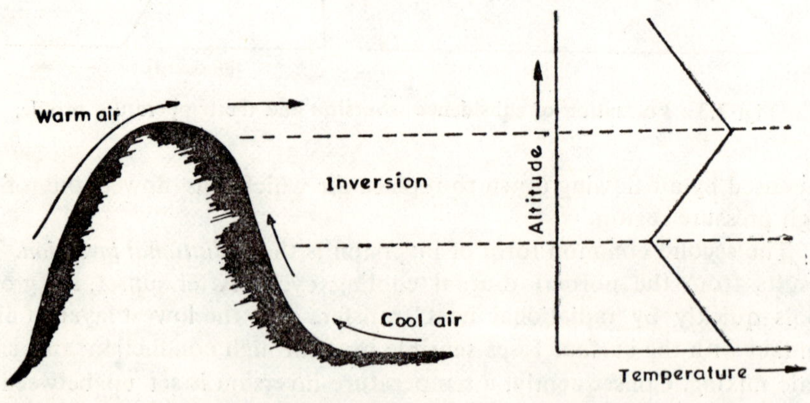

Fig. 3.5 Elevated advective inversion.

As mentioned earlier, during an inversion the vertical mixing of air is practically nil as the atmosphere is very stable, but this condition does not preclude horizontal motion of air currents. It is quite possible that strong winds may be encountered within an inversion.

3.2 WIND VELOCITY AND TURBULENCE

Differential solar heating of the Earth's surface produces pressure and temperature gradients. As a result, the atmosphere is practically in continuous motion with air movements being always turbulent. The motion of air near the surface of the Earth is retarded by friction, which varies with surface roughness. The planetary boundary layer, in which friction is significant, extends to about 1 km above the Earth's surface. The wind velocity profile within the layer is not only influenced by the surface roughness but also by the time of the day. During the day, solar heating causes thermal turbulence or eddies and these eddies set up convective currents so that turbulent mixing is increased. This results in a more flat velocity profile in the day than that at night. Typical profiles during the day and the night are shown in Fig. 3.6. Thermal turbulence also depends on the thermal stability of the atmosphere. It is maximum on a clear sunny day in the afternoon and minimum at night or in the early morning.

The second type of turbulence is the mechanical turbulence which is produced by shearing stress generated by air movement over the Earth's surface—the greater the surface roughness, the greater the turbulence. The effect of terrain on the wind velocity profiles is shown in Fig. 3.7. For smooth surfaces, the air velocity profile becomes very steep near the ground. For rougher surfaces such as those in urban areas more mechanical turbulence is generated and the velocity profile becomes less steep and reaches deeper into the atmosphere.

The mean wind speed variation with altitude in the planetary boundary layer can be represented by a simple empirical power law such as,

$$\frac{u}{u_1} = \left[\frac{z}{z_1}\right]^{\alpha} \tag{3.19}$$

where u is the wind speed at altitude z, u_1 is the wind speed at altitude z_1 and the exponent α varies between 0.14 and 0.40 depending on the roughness of the ground surface (Fig. 3.7) as well as on the temperature stability of the atmosphere. If the environmental lapse rate is dry adiabatic and the terrain is fairly smooth, α is approximately 1/7. The exponent α is observed to increase with increasing stability or with increasing surface roughness. The recommended values of α for a range of conditions are given in Table 3.1. In practice, because of the appreciable change in wind speed with altitude, a wind speed value must be quoted with respect to

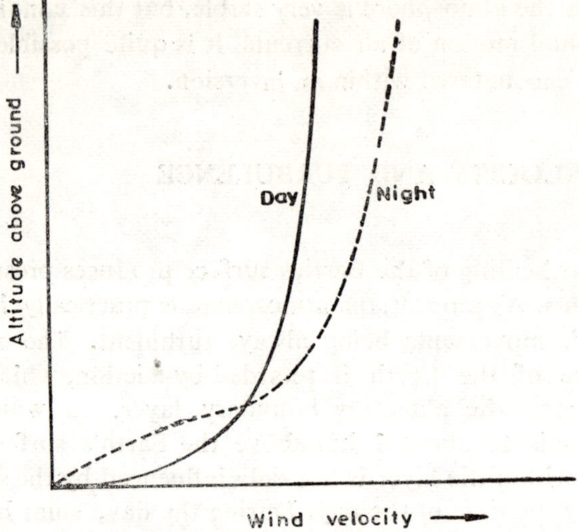

Fig. 3.6 Wind velocity profiles during day and night.

the elevation at which it is measured. This reference height for surface wind measurements is usually 10 meters.

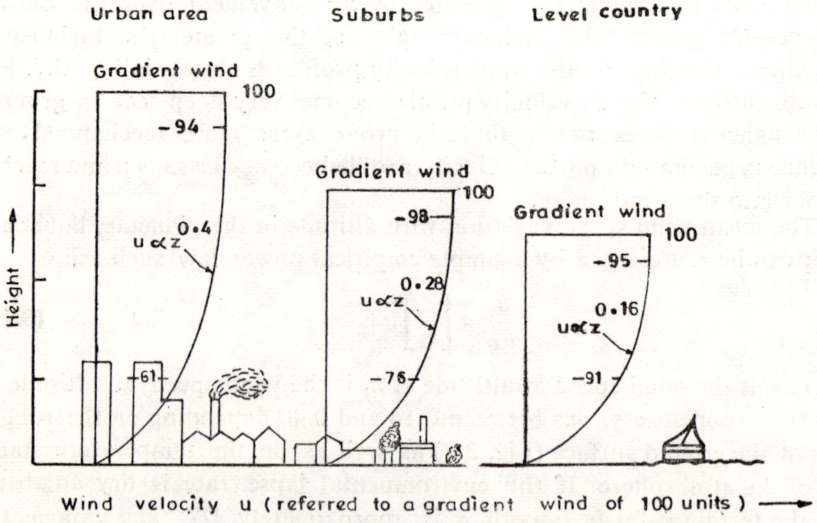

Fig. 3.7 Effect of terrain roughness on the wind velocity profile.

Atmospheric turbulance is characterised by different sizes of eddies. These eddies are primarily responsible for diluting and transporting the

Table 3.1 Values of exponent in Eq. (3.19)*

Surface configuration	Stability	α	Ref.
Smooth open country	Unstable	0.11	(1)
	Neutral	0.14	
	Moderate stability	0.20	
	Large stability	0.33	
Flat open country	Neutral	0.16	(2)
Suburbs	Neutral	0.28	
Urban area	Neutral	0.40	

*From G.H. Strom, in *Air pollution*, edited by A.C. Stern, Vol. 1, 3rd. Edition, (Academic Press New York) 1976.

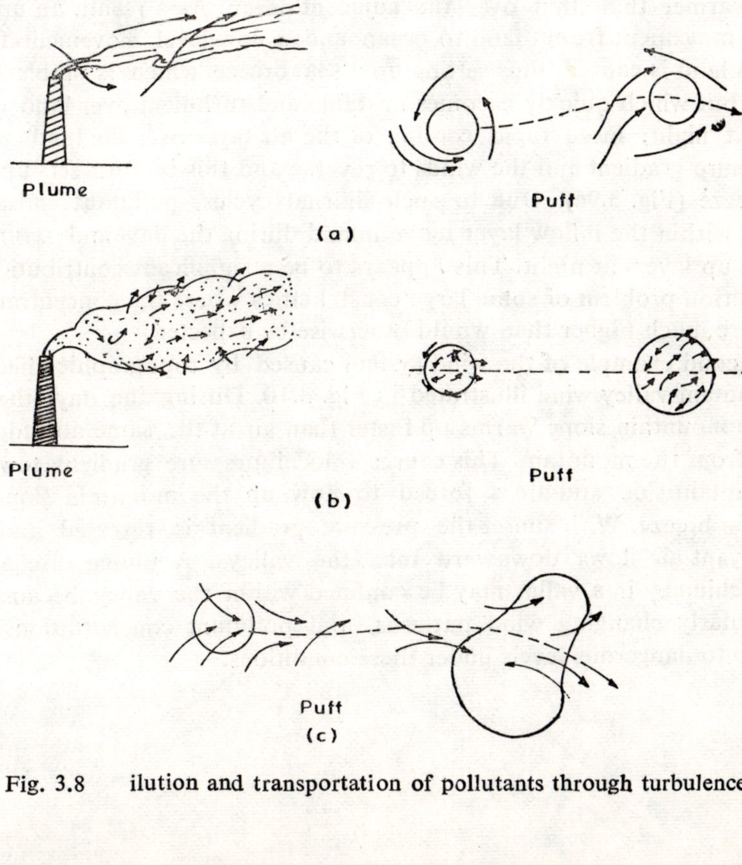

Fig. 3.8 ilution and transportation of pollutants through turbulence.

pollutants injected into the atmosphere. The process of dilution and transportation through turbulence is illustrated in Fig. 3.8. If the size of the eddy is larger than the size of plume or a 'puff', then the plume or the puff will be transported down-wind by the eddy with little dilution as shown in Fig. 3.8a. Molecular diffusion will ultimately dissipate the plume or the puff. If the eddy is smaller than the plume or the puff, the plume or the puff will disperse uniformly as the eddy entrains fresh air at its boundary as shown in Fig. 3.8b. Fig. 3.8c shows a puff where the eddy is of the size comparable to that of the puff. In such cases the eddy will both distort and disperse the puff. The atmosphere usually has a variety of eddy sizes, and these eddies cause the plumes and the puffs to meander and disperse in the wind direction.

Topographical Effects

Surface topography can modify the general pattern of wind speed and direction. One such case is the land and sea breeze (Fig. 3.9). Usually, in the early morning hours the conditions are calm and no pressure gradients exist (Fig. 3.9a), but by afternoon the air over the surface of the coast becomes warmer than that over the adjacent ocean. As a result, an upper-level air movement from land to ocean and a low-level movement from ocean to land is caused, thus setting up a sea breeze which is stable over the sea, but which quickly becomes unstable and turbulent over land (Fig. 3.9b). At night, more rapid cooling of the air layer over the land causes the pressure gradient and the winds to reverse and this in turn sets up the land breeze (Fig. 3.9c). Due to such diurnal cycles, pollutant emissions released within the inflow layer move inland during the day and return to the high up layers at night. This appears to be a significant contribution to the pollution problem of some large coastal cities where the concentrations inland are much higher than would otherwise be expected.

A second example of the wind system caused by topographic effects is the mountain-valley wind illustrated in Fig. 3.10. During the day, the air near the mountain slope warms up faster than air at the same altitude but farther from the mountain. This causes a local pressure gradient towards the mountain side and air is forced to flow up the mountain slope as mountain breeze. With sunset the pressure gradient is reversed and the less byoyant air flows downward into the valleys. A plume discharged from a chimney in a valley may be confined within the valley because of the regularly changing wind patterns, and pollutant concentrations can build-up to dangerous levels under these conditions.

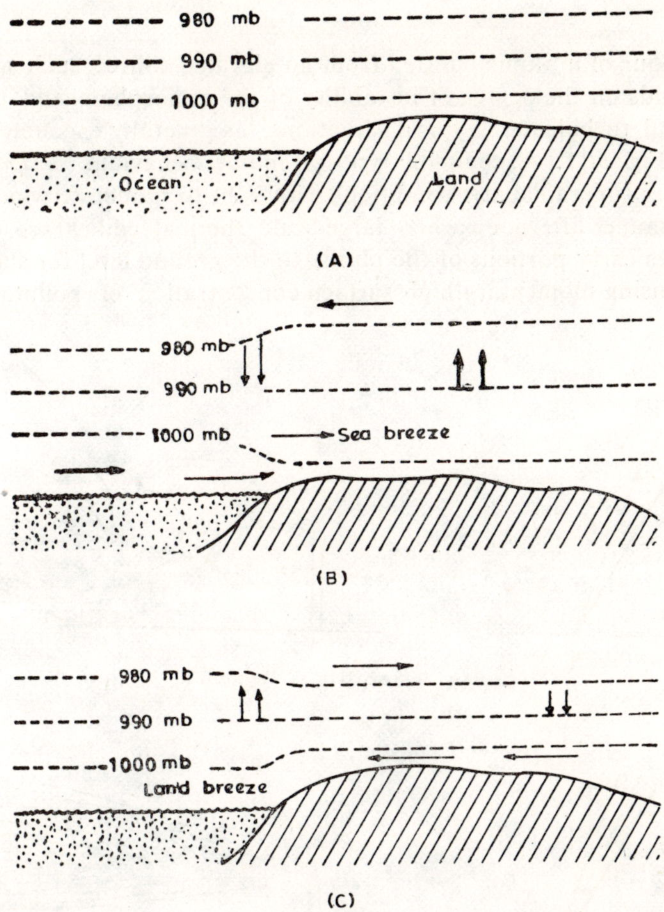

Fig. 3.9 Sea and land breezes. The dashed lines represent isobars.

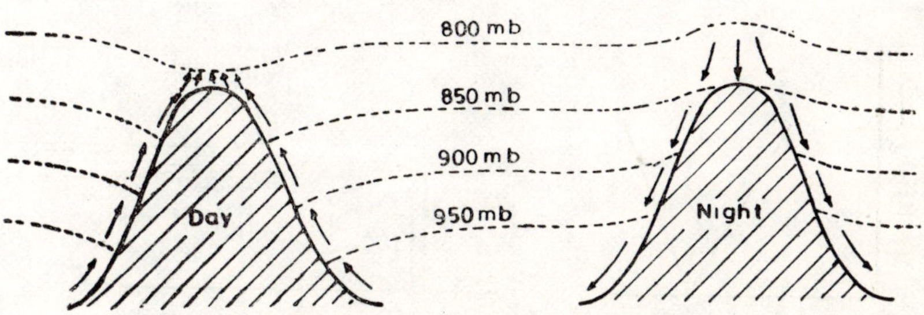

Fig. 3.10 Mountain-Valley wind system.

3.3 PLUME BEHAVIOUR

The behaviour of a plume emitted from an elevated source such as a tall stack depends on the degree of instability of the atmosphere and the prevailing wind turbulence. Typical situations, as generally encountered, are shown in Fig. 3.11. A common type of plume behaviour is *looping*. It occurs under superadiabatic conditions with light to moderate wind speeds on a hot summer afternoon when large-scale thermal eddies are present. These eddies carry portions of the plume to the ground level for short time periods, causing momentary high surface concentration of pollutant near the stack.

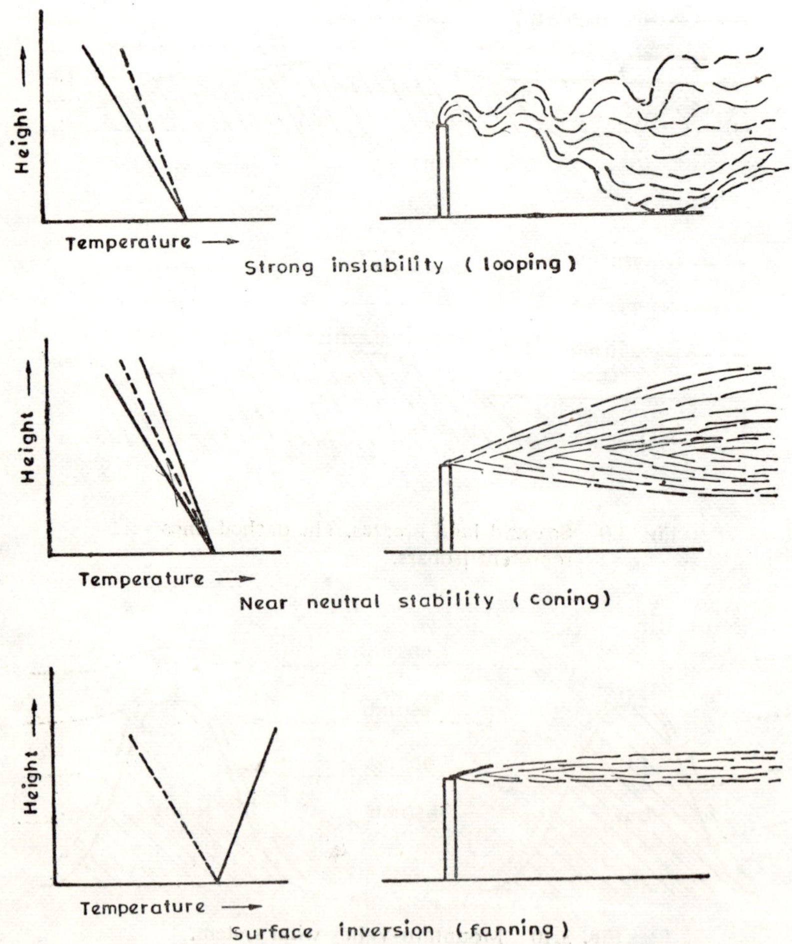

Fig. 3.11 (Contd.)

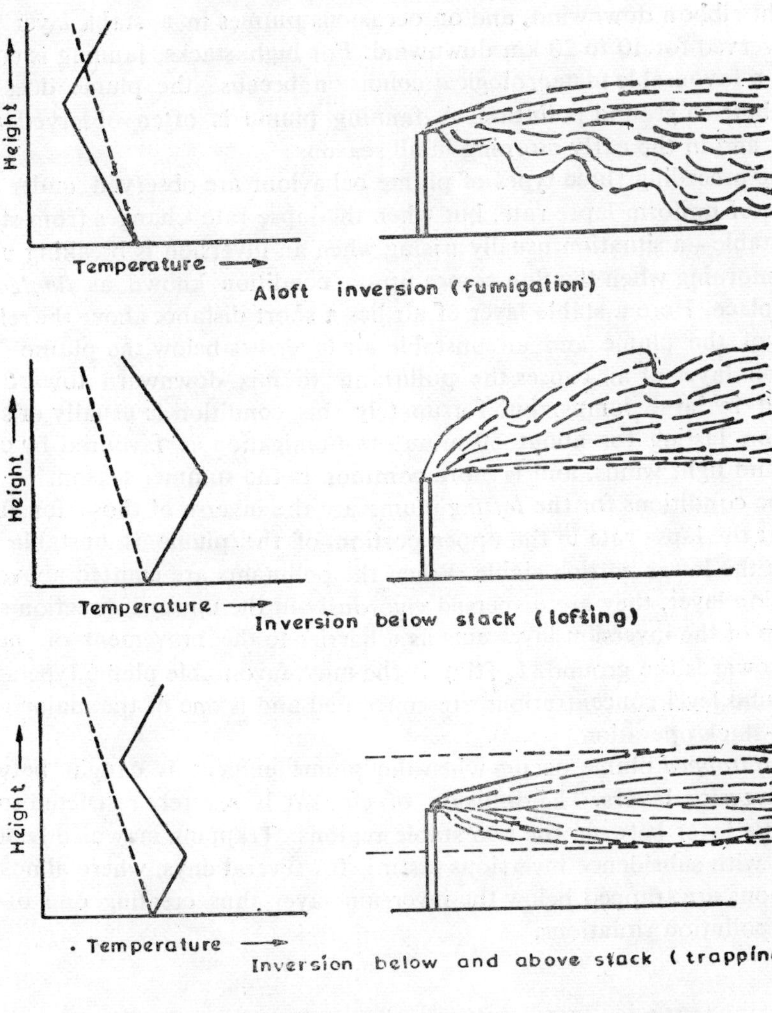

Fig. 3.11 Plume behaviour under various conditions of stability (dry adiabatic lapse rate — — —, ambient lapse rate ———)

Coning plume can occur under cloudy skies both during day and night when the lapse rate is essentially neutral. The plume shape is vertically symmetrical about the *plume line* and a major part of the pollutant concentration is carried downwind fairly far before reaching the ground level. Coning is often associated with moderate to high wind speeds.

Fanning occurs when the plume is dispersed in the presence of very light winds as a result of strong atmospheric inversions. The stable lapse rate suppresses the vertical mixing, but not the horizontal mixing entirely. If the density of the plume is not significantly different from that of the surrounding atmosphere, the plume travels parallel to the ground in a flat, straight ribbon downwind, and on occasions plumes in a stable layer may be observed for 10 to 20 km downwind. For high stacks, fanning is considered a favourable meteorological condition because the plume does not contribute to ground pollution. A fanning plume is often observed at a height and in the early morning in all seasons

The preceding three types of plume behaviour are observed under conditions of uniform lapse rate, but when the lapse rate changes from stable to unstable—a situation usually arising when an inversion is breaking up in early morning when the Sun comes up—a condition known as *fumigation* takes place. Here a stable layer of air lies a short distance above the release point of the plume and an unstable air layer lies below the plume. This unstable layer of air causes the pollutants to mix downward toward the ground in large lumps, but fortunately this condition is usually of short duration lasting for about 30 minutes. Fumigation is favoured by clean skies and light winds, and is more common in the summer season.

The conditions for the *lofting* plume are the inverse of those for fumigation; the lapse rate in the upper portion of the plume is unstable and that in the lower portion stable. When the pollutants are emitted above the inversion layer, they are dispersed vigorously in the upward direction since the top of the inversion layer acts as a barrier to the movement of pollutants towards the ground. Lofting is the most favourable plume type as far as ground level concentrations are concerned and is one of the major goals of tall-stack operation.

The *trapped* plume occurs when the plume effluent is caught between two inversion layers. The diffusion of effluent is severely restricted to the unstable layer between the two stable regions. Trapping may also be associated with subsidence inversions lasting for several days, where almost all emissions are trapped below the inversion layer thus creating one of the worst pollution situations.

Aerodynamic Effects of Structures and Terrain

Deflection and disturbance of the wind field by a structure has a major influence on the distribution of pollutant from a stack on or close to the structure. The flow around an isolated building is characterized by the formation of a cavity behind the building and a wake further downstream. A simplified and idealised representation for a sharp-edged building is shown in Fig. 3.12. When the chimney stack is situated on or adjacent to the building, the plume may be entrained in the characteristic flow pattern. An empirical rule-of-thumb for such stacks is that the height of the stack should be at least two and a half times the height of the surrounding building. If the stack height is much below this value, the plume is trapped in the wake zone of the building (Fig. 3.13). Pollutants remain in the wake because of very poor mixing between the cavity and the mainstream.

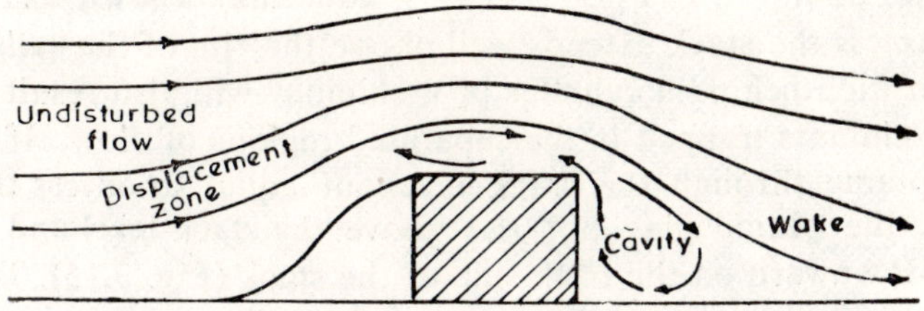

Fig. 3.12 Wind flow field around a sharp-edged building.

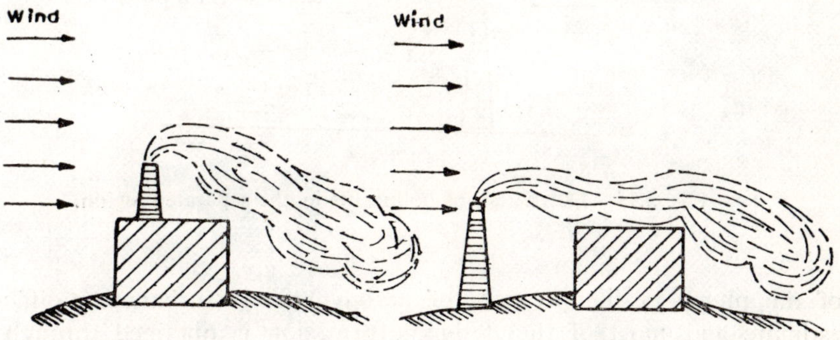

Fig. 3.13 Effect of building on plume behaviour.

Similarly, an irregular terrain surrounding a stack can have a marked influence upon the behaviour of a plume. A plume emitting from a stack located in a deep valley can be carried to the ground when the wind blows

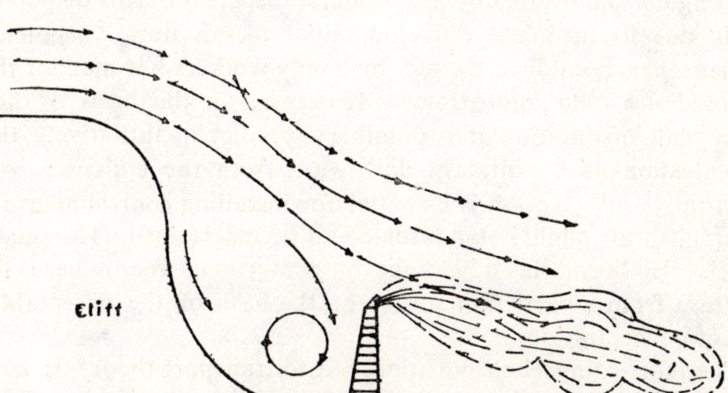

Fig. 3.14 Behaviour of the plume in a valley.

over a cliff as shown in Fig. 3.14. This condition known as "downwash" occurs unless the stack extends well above the rim of the valley.

Separation behind a cylinder or a chimney will also result in downwash of the pollutants trapped in the separated regions of flow. If the velocity of the plume through the stack is about equal to or less than the wind velocity, the plume may not rise above the stack level and may even be carried downward on the back-side of the stack (Fig. 3.15). The behaviour

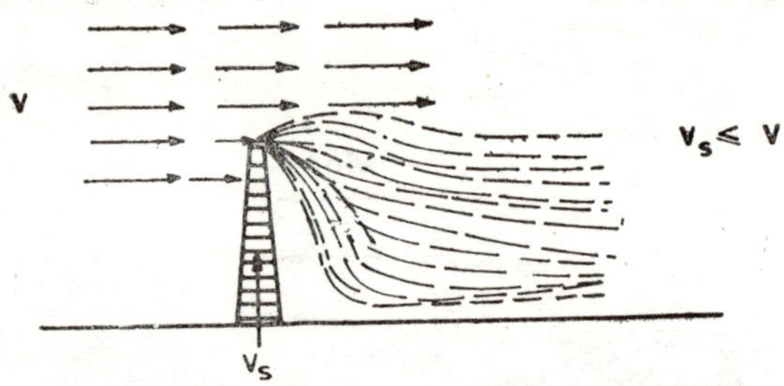

Fig. 3.15 Downwash of pollutants in the separated region.

of the plume in the presence of aerodynamic disturbances is difficult to estimate and most of the design information is obtained through wind tunnel testing.

3.4 DISPERSION OF AIR POLLUTANTS

Pollutants emitted into the atmosphere are mixed thoroughly with the surrounding air and diluted by atmospheric dispersion. This dispersion is primarily due to turbulent diffusion and bulk air flow. Turbulent diffusion problem has been investigated by many workers but most of the theories proposed are only approximate. However, on the basis of the available theoretical information it is possible to predict quantitatively the average concentration of a pollutant downwind from the emission source. The quantitative information is essential for installing control measures so that the ambient air quality standards could be maintained. The mathematical models also have the advantage of prescribing the release limits of a pollutant from a particular source on the basis of the acceptable recipient level of concentration.

The most comprehensive approach to transport theory is based on the eddy diffusion model, known as the 'K-theory' developed by Taylor[3], Richardson[4], and Roberts[5], which in turn involves the use of the concept of mixing length.

The concentration of a species A at different points in space, when motion and chemical reactions are present, can be expressed by the continuity equation:

$$\frac{\partial \rho_A}{\partial t} + \nabla \cdot \rho_A \vec{q}_A = r_A \tag{3.20}$$

where ρ_A is the mass concentration of species A, $\vec{q}_A$ is the absolute velocity of A relative to stationary coordinate axes, and r_A is the rate of production of A by chemical reaction. Eq. (3.20) can be modified by defining the mass flux $\vec{j}_A$ with respect to mass average velocity $\vec{q}$.

$$\vec{j}_A = \rho_A (\vec{q}_A - \vec{q}) \tag{3.21}$$

where $\vec{q}$ for a multicomponent mixture consisting of n species is defined as

$$\vec{q} = \frac{\sum_{i=1}^{n} \rho_i q_i}{\rho} \tag{3.22}$$

In Eq. (3.22), ρ is the total mass concentration of the mixture.

Now

$$\nabla \cdot \vec{j}_A = \nabla \cdot \rho_A \vec{q}_A - \nabla \cdot \rho_A \vec{q} \tag{3.23}$$

Substituting for $\nabla \cdot \rho_A \vec{q}_A$ in Eq. (3.20), we have

$$\frac{\partial \rho_A}{\partial t} + \nabla \cdot \vec{j}_A + \nabla \cdot \rho_A \vec{q} = r_A \tag{3.24}$$

For mass transfer involving molecular diffusion, the mass flux $\vec{j}_A$ is given by Fick's law of diffusion:

$$\vec{j}_A = -\rho D_{AM} \nabla \omega_A \tag{3.25}$$

where ω_A is the mass fraction of species A, and D_{AM} is the mass diffusity of A in the mixture. For a constant density mixture $\nabla \cdot \vec{q} = 0$, and Eq. (3.24) can be simplified to

$$\frac{\partial \rho_A}{\partial t} + \vec{q} \cdot \nabla \rho_A = \nabla \cdot (D_{AM} \nabla \rho_A) + r_A \tag{3.26}$$

Expanding, we get

$$\frac{\partial \rho_A}{\partial t} + u \frac{\partial \rho_A}{\partial x} + v \frac{\partial \rho_A}{\partial y} + w \frac{\partial \rho_A}{\partial z} = \frac{\partial}{\partial x}\left(D_{AM} \frac{\partial \rho_A}{\partial x}\right) + \frac{\partial}{\partial y}\left(D_{AM} \frac{\partial \rho_A}{\partial y}\right) + \frac{\partial}{\partial z}\left(D_{AM} \frac{\partial \rho_A}{\partial z}\right) + r_A \tag{3.27}$$

When the atmospheric flows are turbulent, the instantaneous fluid velocities u, v and w are random variables of space and time. Therefore, we can write $u = \bar{u} + u'$, $v = \bar{v} + v'$, and $w = \bar{w} + w'$. $\bar{u}$, $\bar{v}$, and $\bar{w}$ are the time averaged values, and u', v' and w' are the fluctuating components. These fluctuating components together with molecular diffusion are responsible for mixing of pollutants with air. In contrast to molecular diffusion. the mixing of gases by turbulent fluctuations is called turbulent or eddy diffusion. Since the wind velocities are random fluctuations of space and time, the pollutant concentrations are themselves random variables in space and time.

The determination of ρ_A as a function of space and time is not possible and, hence, we can again define it in terms of some mean $\langle\rho_A\rangle$ and a fluctuating component ρ'_A, where $\langle\rho_A(x, y, z, t)\rangle$ is an ensemble average. Substituting for $\rho_A = \langle\rho_A\rangle + \rho_A'$ together with the velocity components in Eq. (3.27) and taking an average over an infinite ensemble of realisations for ρ_A and time-averaging for velocity components, we have

$$\frac{\partial\langle\rho_A\rangle}{\partial t} + \bar{u}\frac{\partial\langle\rho_A\rangle}{\partial x} + \bar{v}\frac{\partial\langle\rho_A\rangle}{\partial y} + \bar{w}\frac{\partial\langle\rho_A\rangle}{\partial z}$$

$$= \frac{\partial}{\partial x}\left(D_{AM}\frac{\partial\langle\rho_A\rangle}{\partial x}\right) + \frac{\partial}{\partial y}\left(D_{AM}\frac{\partial\langle\rho_A\rangle}{\partial y}\right) + \frac{\partial}{\partial z}\left(D_{AM}\frac{\partial\langle\rho_A\rangle}{\partial z}\right)$$

$$+ \left[\frac{\partial}{\partial x}\left(-\langle u'\rho_A'\rangle\right) + \frac{\partial}{\partial y}\left(-\langle v'\rho_A'\rangle\right) + \frac{\partial}{\partial z}\left(-\langle w'\rho_A'\rangle\right)\right]$$

$$+ \langle R_A\rangle \tag{3.28}$$

where the reaction rate $\langle R_A\rangle$ is a suitably ensemble-averaged value. This term depends on concentration of various species which participate in the formation of a particular component. The correlation terms $-\langle u'\rho_A'\rangle$, $-\langle v'\rho_A'\rangle$, and $-\langle w'\rho_A'\rangle$ are known as the Reynolds dispersion terms. Eq. (3.28) cannot be solved in its present form because it contains the unknown concentration $\langle\rho_A\rangle$ and the three Reynolds dispersion terms which are also unknown. These terms provide an illustration of the closure problem of turbulence. It is because of the closure problem that an Eulerian description of turbulent diffusion will not permit exact solution even for the mean concentration.

Now the problem is to relate $\langle u'\rho_A'\rangle$, $\langle v'\rho_A'\rangle$, and $\langle w'\rho_A'\rangle$ to $\langle\rho_A\rangle$. One of the most common methods to do so is the phenomenological approach involving the mixing-length model. We can express the Reynolds dispersion terms as

$$\left.\begin{aligned} -\langle u'\rho_A'\rangle &= K_{xx}\frac{\partial\langle\rho_A\rangle}{\partial x} \\ -\langle v'\rho_A'\rangle &= K_{yy}\frac{\partial\langle\rho_A\rangle}{\partial y} \\ -\langle w'\rho_A'\rangle &= K_{zz}\frac{\partial\langle\rho_A\rangle}{\partial z} \end{aligned}\right\} \tag{3.29}$$

where K_{xx}, K_{yy}, and K_{zz} are the eddy diffusivities. In addition, Eq. (3.28) can be simplified by comparing the molecular diffusion terms with those of eddy diffusion. Monin and Yaglom[6] suggest that eddy diffusion is $10^5 - 10^6$ times greater than molecular diffusion and that for practical purposes molecular diffusion can be neglected relative to turbulent diffusion. This assumption is implicit in many air pollution models. Thus,

$$D_{AM}\frac{\partial\langle\rho_A\rangle}{\partial x} \ll -\langle u'\rho_A'\rangle$$

$$D_{AM}\frac{\partial\langle\rho_A\rangle}{\partial y} \ll -\langle v'\rho_A'\rangle$$

$$D_{AM}\frac{\partial\langle\rho_A\rangle}{\partial z} \ll -\langle w'\rho_A'\rangle$$

With the above assumptions and substituting Eq. (3.29) into Eq. (3.28), we get

$$\frac{\partial\langle\rho_A\rangle}{\partial t} + \bar{u}\frac{\partial\langle\rho_A\rangle}{\partial x} + \bar{v}\frac{\partial\langle\rho_A\rangle}{\partial y} + \bar{w}\frac{\partial\langle\rho_A\rangle}{\partial z}$$

$$= \frac{\partial}{\partial x}\left(K_{xx}\frac{\partial\langle\rho_A\rangle}{\partial x}\right) + \frac{\partial}{\partial y}\left(K_{yy}\frac{\partial\langle\rho_A\rangle}{\partial y}\right) + \frac{\partial}{\partial z}\left(K_{zz}\frac{\partial\langle\rho_A\rangle}{\partial z}\right)$$

$$+\langle R_A\rangle \tag{3.30}$$

Eq. (3.30) is often referred to as the atmospheric dispersion equation and has been the basis of numerous theoretical and experimental investigations of pollutant dispersion in the atmosphere.

Solutions to the Atmospheric Dispersion Equation

Equation (3.30) can be simplified for certain cases of practical interest. The simplest case is in which $\bar{v} = \bar{w} = \frac{\partial\langle\rho_A\rangle}{\partial t} = 0$. We also assume the pollutants to be chemically inert, and K_{xx}, K_{yy}, and K_{zz} to be constants throughout space (stationary homogeneous turbulence). With the above assumptions, Eq. (3.30) reduces to

$$\bar{u}\frac{\partial\langle\rho_A\rangle}{\partial x} = K_{xx}\frac{\partial^2\langle\rho_A\rangle}{\partial x^2} + K_{yy}\frac{\partial^2\langle\rho_A\rangle}{\partial y^2} + K_{zz}\frac{\partial^2\langle\rho_A\rangle}{\partial z^2} \tag{3.31}$$

where $\bar{u}$ is the mean wind velocity.

A special case of interest in applying Eq. (3.31) to actual atmospheric processes is that of the concentration of pollutants emitted from a continuous point source. Often the transport of pollutants in the x-direction by the wind is dominant over turbulent diffusion in the x-direction. That is,

$$\bar{u}\frac{\partial\langle\rho_A\rangle}{\partial x} \gg K_{xx}\frac{\partial^2\langle\rho_A\rangle}{\partial x^2}$$

By incorporating the above assumption in Eq. (3.31), we have

$$\bar{u}\frac{\partial\langle\rho_A\rangle}{\partial x} = K_{yy}\frac{\partial^2\langle\rho_A\rangle}{\partial y^2} + K_{zz}\frac{\partial^2\langle\rho_A\rangle}{\partial z^2} \tag{3.32}$$

where $K_{yy} \neq K_{zz}$. The solution to Eq. (3.32) must satisfy the following boundary conditions:

(1) $\langle\rho_A\rangle(x, y, z) \rightarrow 0$ as $x, y, z \rightarrow \infty$

(2) $\langle\rho_A\rangle(x, y, z) \rightarrow \infty$ as $x = y = z = 0$

(3) $K_{zz}\frac{\rho\langle\rho_A\rangle}{\partial z} = 0$, for $z = 0$, $x, y > 0$

(4) $\int_0^\infty \int_{-\infty}^\infty \langle\rho_A\rangle \bar{u}\, dydz = Q$ for all $x > 0$

where Q is the source strength.

The third statement asserts that there is no pollutant diffusion through the ground surface. The last boundary condition expresses the conservation of the component at all downwind locations. In other words, the rate at which the pollutant passes through any *yz*-plane downwind from the source must be constant and must equal the emission rate from the source.

For a continuous point source at the origin satisfying boundary conditions (1), (2) and (4), Roberts[5] gives the following solution:

$$\langle\rho_A\rangle(x, y, z) = \frac{Q}{4\pi x (K_{yy} K_{zz})^{1/2}} \exp\left[-\frac{\bar{u}}{4x}\left(\frac{y^2}{K_{yy}} + \frac{z^2}{K_{zz}}\right)\right] \tag{3.33}$$

Eq. (3.33) is applicable for $x \gg \sqrt{y^2 + z^2}$.

At the ground surface ($z = 0$), if there is no diffusion of pollutant into the surface (boundary condition 3) and if all of it is reflected, then the source strength Q is dispersed completely in the upper half space ($z > 0$) rather than in both the upper ($z > 0$) and lower ($z < 0$) half spaces. Then Eq. (3.33) can be modified to give:

$$\langle \rho_A \rangle (x, y, z) = \frac{Q}{2\pi x (K_{yy} K_{zz})^{1/2}} \exp\left[-\frac{\bar{u}}{4x}\left(\frac{y^2}{K_{yy}} + \frac{z^2}{K_{zz}}\right)\right] \tag{3.34}$$

Some of the assumptions made to obtain the above solutions are too restrictive and as a result, Eqs. (3.33) and (3.34) do not permit accurate prediction of downwind concentrations in the atmosphere. One of the assumptions is the homogeneity of turbulence, but it is a known fact that atmospheric turbulence is inhomogeneous, especially in the vertical direction. Also, the wind velocity is not constant but varies with height above the ground.

Equation (3.30) shows that the eddy diffusivities K_{xx}, K_{yy}, and K_{zz} are functions of x, y and z respectively. However, when these coefficients are allowed to vary with position, the analytical solution of Eq. (3.30) becomes difficult.

Now considering Eq. (3.30) and assuming that

$$\bar{v} = \bar{w} = \frac{\partial \langle \rho_A \rangle}{\partial t} = 0$$

and also that

$$\bar{u}\, \frac{\partial \langle \rho_A \rangle}{\partial x} \gg \frac{\partial}{\partial x}\left(K_{xx} \frac{\partial \langle \rho_A \rangle}{\partial x}\right),$$

we get

$$\bar{u}\, \frac{\partial \langle \rho_A \rangle}{\partial x} = \frac{\partial}{\partial y}\left(K_{yy} \frac{\partial \langle \rho_A \rangle}{\partial y}\right) + \frac{\partial}{\partial z}\left(K_{zz} \frac{\partial \langle \rho_A \rangle}{\partial z}\right) \tag{3.35}$$

A further simplification of Eq. (3.35) is necessary to obtain closed-form solutions that incorporate the vertical structure of the atmosphere. This is achieved by assuming an infinite source geometry in the crosswind direction. In this case $\langle \rho_A \rangle$ is constant in the y-direction and $\frac{\partial \langle \rho_A \rangle}{\partial y} = 0$. With this assumption, Eq. (3.35) reduces to,

$$\bar{u}\, \frac{\partial \langle \rho_A \rangle}{\partial x} = \frac{\partial}{\partial z}\left(K_{zz} \frac{\partial \langle \rho_A \rangle}{\partial z}\right) \tag{3.36}$$

Infinite Line Source

The boundary conditions are:

$$\langle \rho_A \rangle (x, z) \to 0 \text{ as } x, y \to \infty$$

$$\langle \rho_A \rangle (x, z) \to \infty \text{ at } x = z = 0$$

$$K_{zz} \frac{\partial\langle\rho_A\rangle}{\partial z} = 0 \text{ for } z = 0, x > 0$$

$$\int_0^\infty \langle\rho_A\rangle(x, z)\, \bar{u}(z)\, dz = Q_l \text{ for all } x > 0$$

where Q_l is the line source strength in g/m-s.

The vertical structure of the atmosphere can be approximated by power law expressions for the vertical profiles of velocity (Eq. 3.19) and turbulent diffusivity K_{zz}[7]:

$$\bar{u}(z) = \bar{u}_1 \left(\frac{z}{z_1}\right)^\alpha, \; K_{zz} = K_1 \left(\frac{z}{z_1}\right)^n \tag{3.37}$$

where $\bar{u}_1$, and K_1 are the values of $\bar{u}$ and K_{zz} at a fixed reference height z_1. The solution valid for $r = \alpha - n + 2 > 0$ is,

$$\langle\rho_A\rangle(x, z) = \frac{Q_l r}{\bar{u}_1\, \Gamma(s)} \left[\frac{\bar{u}_1}{r^2 K_1 x}\right]^s \exp\left[-\frac{\bar{u}_1\, z^r}{r^2\, K_1 x}\right] \tag{3.38}$$

where $s = (\alpha + 1)/r$ and z_1, is taken to be unity. The gamma function $\Gamma(s)$ is given by,

$$\Gamma(s) = \int_0^\infty e^{-x}\, x^{s-1}\, dx \tag{3.39}$$

For a constant shear stress layer the eddy diffusivity K_{zz} may be assumed to vary linearly with height ($n = 1$), and the use of momentum transfer analogy ($K_{zz} = K_M$) gives

$$K_{zz} = k u_* z \tag{3.40}$$

Taking $z_1 = 1$, and comparing Eq. (3.40) with Eq. (3.37), we have

$$K_1 = k u_* \tag{3.41}$$

where k is Von Kármán's constant ($k = 0.40$) and u_* is the friction velocity. The friction velocity u_* may be determined from the equation given by Lettau[8]

$$\frac{u_*}{q_G} = \frac{0.16}{\log_{10} R_0 - 1.8} \tag{3.42}$$

where $R_0 =$ Rossby number $= \frac{q_G}{z_0 f}$, and q_G is the geostrophic wind speed, f is the Coriolis parameter and z_0 is the aerodynamic roughness.

The velocity $\bar{u}_1$ in Eq. (3.38) can be obtained from the logarithmic law:

$$\frac{\bar{u}_1}{u_*} = \frac{1}{k} \ln\left(\frac{z_1}{z_0}\right) \tag{3.43}$$

The values of z_0 derived from wind profiles range from 0.01 to 1 metre respectively for surfaces ranging from open fields to forest or city areas.

For constant wind velocity, $\bar{u} = u_1$ $(\alpha = 0)$ and a linear variation of K_{zz} with height $(n = 1)$, a solution has been given by Calder[9] which is in agreement with the observations :

$$\langle \rho_A \rangle (x, z) = \frac{Q_l}{k u_* x} \exp\left[\frac{-\bar{u}_1 z}{k u_* x}\right] \tag{3.44}$$

Using these equations, we can predict the concentration downwind of a line source situated at ground level in terms of simply estimated meteorological quantities.

The Gaussian Plume Model

The present tendency is to interpret dispersion data in terms of the Gaussian model. The standard deviations are related to the eddy diffusivities through the equations,

$$\sigma_y^2 = \frac{2\, K_{yy} x}{\bar{u}}, \quad \sigma_z^2 = \frac{2\, K_{zz} x}{\bar{u}} \tag{3.45}$$

Substituting the above equations into Eqs. (3.33) and (3.34), we get

$$\langle \rho_A \rangle (x, y, z) = \frac{Q}{2\pi\, \bar{u}\, \sigma_y \sigma_z} \exp\left[-\tfrac{1}{2}\left(\frac{y^2}{\sigma_y^2} + \frac{z^2}{\sigma_z^2}\right)\right] \tag{3.46}$$

and

$$\langle \rho_A \rangle (x, y, z) = \frac{Q}{\pi\, \bar{u}\, \sigma_y \sigma_z} \exp\left[-\tfrac{1}{2}\left(\frac{y^2}{\sigma_y^2} + \frac{z^2}{\sigma_z^2}\right)\right] \tag{3.47}$$

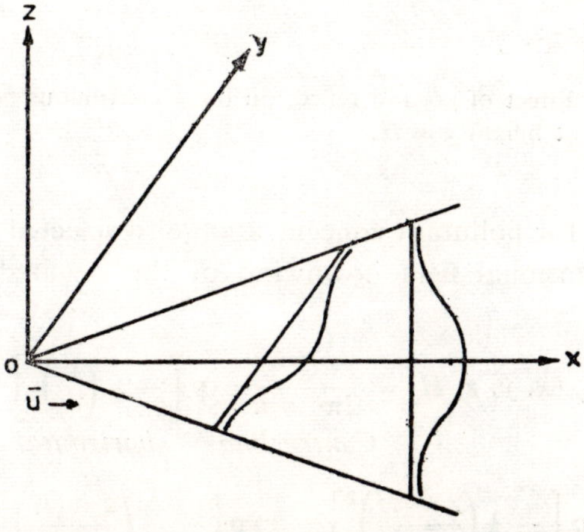

Fig. 3.16 Cross section of a continuous point source showing Gaussian concentration profiles in the y and z directions.

The quantities σ_y and σ_z are referred to as spreading coefficients, since they measure the spreading of the plume in the y and z directions respectively as it blows downwind. From Eqs. (3.46) and (3.47), we can immediately recognise that $\langle \rho_A \rangle \bar{u}/Q$ is the probability density function of two Gaussian-distritbuted variables y and z with standard deviations σ_y and σ_z. This gives rise to the binormal bell-shaped curve whose lateral and vertical distributions arising from Eq. (3.46) are shown in Fig. (3.16).

Extension to Elevated Sources with Reflection

Most pollutants are emitted into the atmosphere from elevated sources such as chimney stacks so that the ground level expression represented by Eq. (3.47) must be modified to account for this. Consider a continuous point source located at $x = y = 0$, and at a height $z = H$. A description of the pollutant concentration $\langle \rho_A \rangle$ could be facilitated by assuming that all the pollutants are reflected when they hit the ground by establishing an image source of equal strength at $x = y = 0$ and $z = -H$. In this method, known as the method of images, the solutions of both the real and image sources are added together as illustrated in Fig. 3.17. The re-

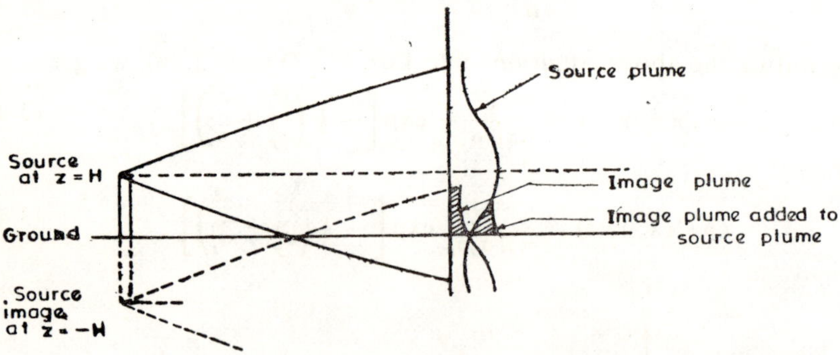

Fig. 3.17 Effect of ground reflection for a continuous point source at height $z = H$.

sultant equation for pollutant concentration at a selected location (x, y, z) in the three-dimensional field downwind of the elevated point source is given by:

$$\langle \rho_A \rangle (x, y, z, H) = \underset{\textit{Centre line}}{\frac{Q}{2\pi\sigma_y\sigma_z\bar{u}}} \; \underset{\textit{horizontal}}{\exp\left[-\tfrac{1}{2}\left(\frac{y}{\sigma_y}\right)^2\right]}$$

$$\underset{\textit{vertical}}{\left\{\exp\left[-\tfrac{1}{2}\left(\frac{z-H}{\sigma_z}\right)^2\right] + \exp\left[-\tfrac{1}{2}\left(\frac{z+H}{\sigma_z}\right)^2\right]\right\}} \tag{3.48}$$

Equation (3.48) consists of three terms representing (1) the centreline concentration as a function of downwind distance, (2) the lateral spread relative to the centreline value, and (3) the vertical spread relative to the centreline value. Fig. 3.18 illustrates the combination of these three terms in the Gaussian plume equation. For $H = 0$, Eq. (3.48) reduces to Eq. (3.47).

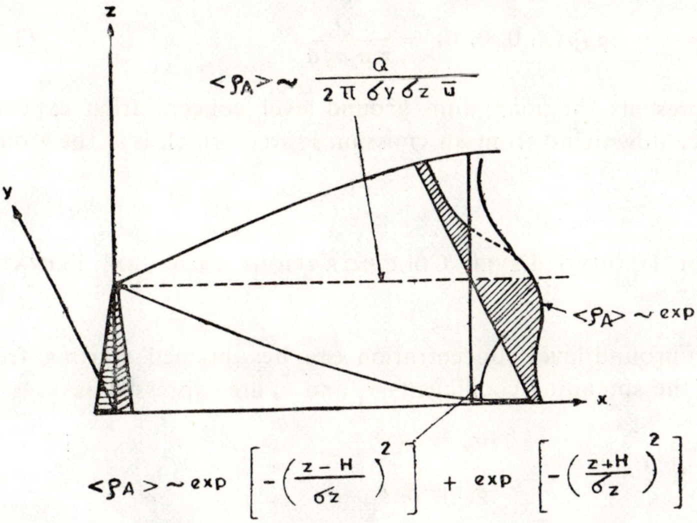

Fig. 3.18 Representation of the centreline concentration term and lateral and vertical terms.

Three common forms obtained from the basic equation are frequently used to analyse specific situations.

(a) Ground Level Concentrations

In this case $z = 0$, and the Eq. (3.48) reduce to

$$\langle \rho_A \rangle (x, y, 0, H) = \frac{Q}{\pi \sigma_y \sigma_z \bar{u}} \exp\left[-\tfrac{1}{2}\left(\frac{y}{\sigma_y}\right)^2\right] \exp\left[-\tfrac{1}{2}\left(\frac{H}{\sigma_z}\right)^2\right] \tag{3.49}$$

In the above equation, the factor z does not appear in the denominator of the coefficient term. It gets cancelled when the two vertical exponential terms are combined.

(b) Ground Level Centreline Concentrations

In this case $z = 0$ and $y = 0$. The first term in the exponential becomes zero, and Eq. (3.49) becomes

$$\langle \rho_A \rangle (x, 0, 0, H) = \frac{Q}{\pi\, \sigma_y \sigma_z\, \bar{u}} \exp\left[-\tfrac{1}{2} \left(\frac{H}{\sigma_z} \right)^2 \right] \tag{3.50}$$

Equation (3.50) is used to estimate the ground level concentrations expected at any distance downwind of an elevated source.

In case when the emission source is also at the ground level, i.e., $H = 0$, Eq. (3.50) further simplifies to

$$\langle \rho_A \rangle (x, 0, 0, 0) = \frac{Q}{\pi\, \sigma_y \sigma_z\, \bar{u}} \tag{3.51}$$

Eq. (3.51) represents the maximum ground level concentration expected at any distance downwind from an emission source, which is at the ground level.

(c) Maximum Ground Level Concentrations from an Elevated Source

The maximum ground level concentration can be obtained starting from Eq. (3.50). If the spreading coefficients σ_y and σ_z are expressed as

$$\sigma_y = Ax^p$$

and

$$\sigma_z = Bx^p$$

where A, B, and p are constants, then the value of $\langle \rho_A \rangle$ in Eq. (3.50) can be maximised by substituting for σ_y and σ_z and setting $\frac{d\langle \rho_A \rangle}{dx} = 0$
Upon simplification of the resultant expression yields

$$\frac{-2p}{x^{2p+1}} + \frac{H^2 p}{B^2\, x^{4p+1}} = 0 \tag{3.52}$$

From Eq. (3.52);

$$Bx^p = \sigma_z = \frac{H}{\sqrt{2}}$$

This gives the downwind position at which the maximum ground level centreline concentration occurs. Substituting for $\sigma_z = H/\sqrt{2}$ in Eq. (3.50) we get

$$\langle \rho_A \rangle_{\max} (x, 0, 0, H) = \frac{2Q}{\pi\, \bar{u} e H^2} \frac{\sigma_z}{\sigma_y} \tag{3.53}$$

Eq. (3.53) clearly shows why the pollutants are emitted into the atmosphere through tall stacks.

Effect of Inversion Layer

When an inversion layer at $z = L$ is imposed, the steady concentration distribution for $H \lhd L$ is given by Bierly and Hewson[10] as:

$$\langle \rho_A \rangle (x, y, z, H, L) = \frac{Q}{2\pi\sigma_y\sigma_z\bar{u}} \exp\left[-\frac{1}{2}\left(\frac{y}{\sigma_y}\right)^2\right]$$

$$\sum_{n=-\infty}^{\infty} \left\{\exp\left[-\frac{1}{2}\left(\frac{z - H + 2nL}{\sigma_z}\right)^2\right] + \exp\left[-\frac{1}{2}\left(\frac{z + H + 2nL}{\sigma_z}\right)^2\right]\right\} \quad (3.54)$$

Since the series coverages rapidly, only the first few pairs of terms ($n = 0, \pm 1, \pm 2$) need be considered in Eq. (3.54).

Infinite Line Source

An infinite line source can often be represented by a series of industries located along a river or by a road along which pollutants are continually emitted by heavy traffic; it is thus spatially extended, and the direction of the wind with respect to the geometry of the source will influence the downwind concentration of pollutants. When the wind direction is normal to the line of emission (y), the ground level concentration downwind (x) of a source emitting at a height H is given by,

$$\langle \rho_A \rangle (x, y, 0, H) = \frac{2Q_l}{(2\pi)^{1/2}\ \sigma_z\ \bar{u}} \exp\left[-\frac{1}{2}\left(\frac{H}{\sigma_z}\right)^2\right] \quad (3.55)$$

where Q_l is the source strength per unit distance. If the source is also at the ground level then Eq. (3.55) can be further simplified:

$$\langle \rho_A \rangle (x, y, 0, 0) = \frac{2Q_l}{(2\pi)^{1/2}\ \sigma_z\ \bar{u}} \quad (3.56)$$

Notice that y does not appear in the above equations since the concentration is uniform in the y-direction at a given downwind distance x.

Estimation of σ_y and σ_z

The values of σ_y and σ_z have been shown to be related to the diffusion coefficients in the y- and z-directions. As might be expected, σ_y and σ_z are functions of downwind distance x from the source as well as the atmospheric stability conditions. Based on experimental observations of the dispersion of real plumes, Pasquill[11] and Gifford[12] have devised a method for calculating σ_y and σ_z of a spreading plume from a knowledge of the atmospheric stability. Six categories of atmospheric stability, A through F, were suggested and these are shown in Table 3.2 as functions of wind and solar radiation.

The values of the wind speed shown in Table 3.2 correspond to measurements made at a height of 10 metres. When the stability category is known from appropriate weather conditions, σ_y and σ_z values for any downwind distance x can be calculated from the equations:

$$\sigma_y = Ax^{0.903} \tag{3.57}$$

$$\sigma_z = Bx^p \tag{3.58}$$

where the values of A, B, and p are given in Table 3.3[14].

Table 3.2 Key to stability categories (ref. 13)

Wind speed (m/s) (at z = 10 m)	Day: Incoming solar radiation			Night	
	Strong	Moderate	Slight	Thin overcast or $\geqslant$4/8 low cloud	$\leqslant$3/8 cloud
<2	A	A–B	B	E	F
2–3	A–B	B	C	E	F
3–5	B	B–C	C	D	E
5–6	C	C–D	D	D	D
>6	C	D	D	D	D

A: extremely unstable, B: moderately unstable, C: slightly unstable, D: Neutral conditions†, E: slightly stable, F: moderately stable

† The neutral class. D, should be assumed for overcast conditions during day or night

Table 3.3 Fitted values for σ_y and σ_z

Class	A	x_1 (metres)	$x \leqslant x_1$: B	$x \leqslant x_1$: p	x_2 (metres)	$x_1 \leqslant x \leqslant x_2$: B	$x_1 \leqslant x \leqslant x_2$: p
A	0.40	250	0.125	1.03	500	0.00883	1.51
B	0.295	1000	0.119	0.986	10000	0.0579	1.09
C	0.20	1000	0.111	0.911	10000	0.111	0.911
D	0.13	1000	0.105	0.827	10000	0.392	0.636
E	0.098	1000	0.100	0.778	10000	0.373	0.587

The above values closely fit the well known curves of Pasquill-Gifford, shown in Figs. 3.19 and 3.20. These correlations are subject to the following limitations:

(1) These concentrations estimated from the use of these charts and equations should correspond to a sampling time of 10 min.
(2) σ_y and σ_z are based on a surface corresponding to an open level country and probably underestimate the plume dispersion in an urban area.
(3) The uncertainties associated with the estimates of σ_y and σ_z will increase with distance from the source.

If the time interval employed in the sampling is other than 10 min, it will be necessary to correct the results predicted by the dispersion model. The following equation gives approximate values of concentration for time intervals less than 2 hours:

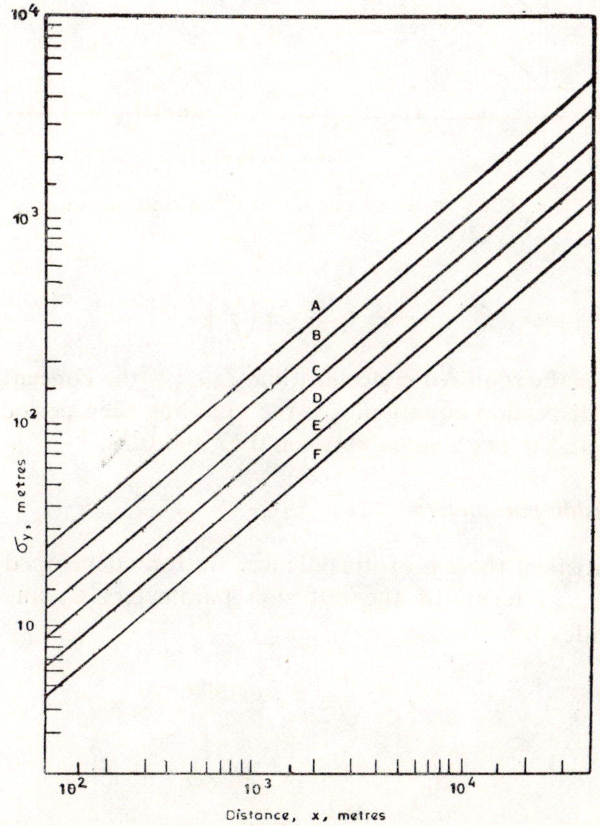

Fig. 3.19 σ_y as a function of downwind distance

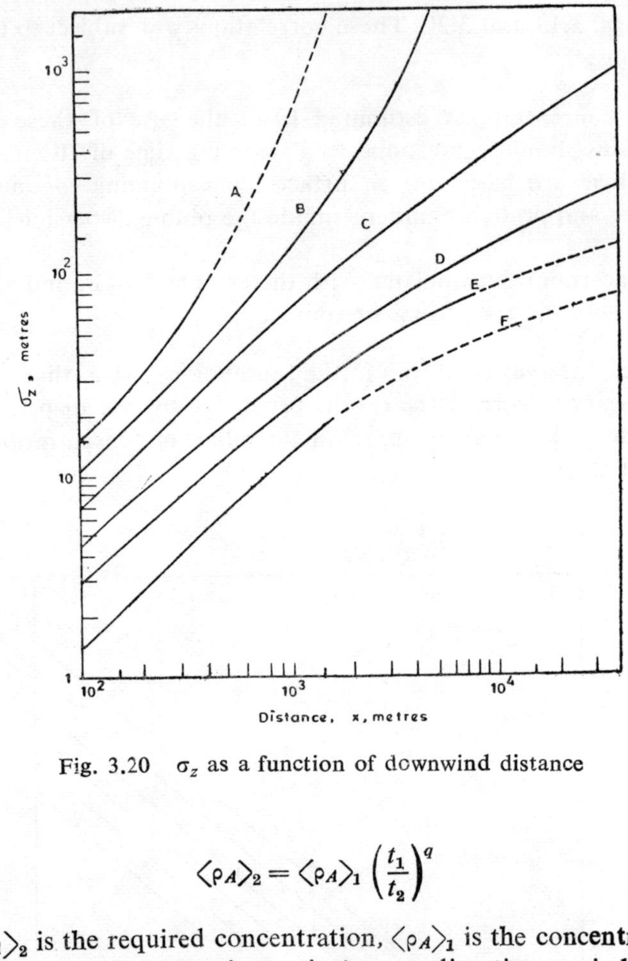

Fig. 3.20 σ_z as a function of downwind distance

$$\langle \rho_A \rangle_2 = \langle \rho_A \rangle_1 \left(\frac{t_1}{t_2}\right)^q \tag{3.59}$$

where $\langle \rho_A \rangle_2$ is the required concentration, $\langle \rho_A \rangle_1$ is the concentration calculated by the dispersion equation, t_2 is the sampling time period in minutes, t_1 is 10 min, and q has a value between 0.17 and 0.20.

Sutton's diffusion parameters

From the statistical theory of turbulence, Sutton[1] developed expressions for σ_y and σ_z in terms of the diffusion parameters C_y and C_z and the turbulence index n':

$$\sigma_y = \frac{1}{\sqrt{2}} C_y \, x^{(2-n')/2} \tag{3.60}$$

$$\sigma_z = \frac{1}{\sqrt{2}} C_z \, x^{(2-n')/2} \tag{3.61}$$

The diffusion parameters C_y and C_z depend on components of turbulence in the y and z directions, respectively, and can be determined in practice by fitting the diffusion equation to concentration profile data. The turbulence index n' depends on the wind velocity profile given by Eq. (3.19), and is related to the exponent α as

$$\alpha = \frac{n'}{(2-n')} \tag{3.62}$$

Thus, knowing the values of C_y, C_z and n', we can calculate spreading coefficients σ_y and σ_z from Eq. (3.60) and (3.61).

Average Wind Speed $\bar{u}$

For elevated sources, $\bar{u}$ is the average wind speed at the stack height H. This can be evaluated by modifying the velocity profile expression given by Eq. (3.19) as

$$\frac{\bar{u}}{\bar{u}_1} = \left(\frac{H}{z_1}\right)^{\alpha} \tag{3.63}$$

where u_1 is usually the meteorological value of the velocity measured at $z_1 = 10$ metres. Smith[15] recommends $\alpha = 0.25$ for unstable and $\alpha = 0.50$ for stable conditions.

Limits to Gaussian Plume Model

The Gaussian plume model provides a simple method by which useful results can be obtained provided its limitations are properly observed. The model is most accurate for downwind distances between approximately 100 and 2000 metres. For distances below 100 metres, the value of $\langle \rho_A \rangle$ has to be corrected by considering the details of the wind flow pattern around the source structure. Similarly, for distances greater than 1000 metres, the local terrain features and meteorological variability have to be taken into consideration.

The dispersion coefficients σ_y and σ_z used in the equation are not very accurate. Also, the model neglects the turning of the wind due to frictional effects. This tends to spread the plume in the cross-wind direction. Also, the consideration of any absorption or deposition of pollutants when the plume reaches the ground is neglected. If the plume is reactive, the model gives inaccurate results. Hence, care is required in using the equations.

Example 1

A chimney with a design stack height of 250 m is emitting sulphur dioxide at a rate of 500 g/s on a sunny day in June with moderate wind speed at the stack altitude. Estimate the concentration of sulphur dioxide downwind for the following situations:

(a) $\langle \rho_{SO_2} \rangle$ (1000, 0, 0, 250)

(b) $\langle \rho_{SO_2} \rangle$ (1000, 50, 0, 250)

(c) $\langle \rho_{SO_2} \rangle$ (1000, 50, 20, 250)

(d) If $\langle \rho_{SO_2} \rangle$ (1000, y, 0, 250) is 100 μg/m^3, what is the value of y in metres?

Solution

On a sunny day in June the incoming solar radiation will be strong. Also, the air will be unstable. A moderate wind speed at the stack altitude will be around 5-7 m/s. Let us take $\bar{u} = 6$ m/s. From Eq. (3.63) the velocity $\bar{u}_1$ at 10 m level can be obtained.

$$\bar{u}_1 = \bar{u}\left(\frac{z_1}{H}\right)^{\alpha}$$

For unstable conditions $\alpha = 0.25$. Therefore,

$$\bar{u}_1 = 6\left(\frac{10}{250}\right)^{0.25} = 2.7 \text{ m/s}$$

This shows that the surface wind speed is between 2 and 3 m/s. Reference to Table 3.2 shows *a* stability class of *A–B*. We choose class *B* as a conservative answer. The values of σ_y and σ_z can be calculated from the information given in Table 3.3. At a distance of 1000 m,

$$\sigma_y = 0.295\,(1000)^{0.903} = 151 \text{ m}$$

using $A = 0.295$ in Eq. (3.57) and

$$\sigma_z = 0.119\,(1000)^{0.986} = 108 \text{ m}$$

using $B = 0.119$ and $p = 0.986$ in Eq. (3.58).

(a) Substituting the values of σ_y and σ_z into Eq. (3.50) and using $Q = 500$ g/s, $\bar{u} = 6$ m/s and $H = 250$ m, we get

$$\langle \rho_{SO_2} \rangle (1000, 0, 0, 250) = \frac{500 \times 10^6}{\pi(151)\,(108)6}$$

$$\exp\left[-\frac{1}{2}\left(\frac{250}{108}\right)^2\right] = 112 \ \mu\text{g/m}^3$$

(b) Using Eq. (3.49), we get

$$\langle \rho_{SO_2} \rangle (1000, 50, 0, 250) = \frac{500 \times 10^6}{\pi(151)\,(108)6}$$

$$\exp\left[-\frac{1}{2}\left(\frac{50}{151}\right)^2\right] \times \exp\left[-\frac{1}{2}\left(\frac{250}{108}\right)^2\right] = 106 \ \mu\text{g/m}^3$$

(c) To find the concentration at 50 m from the x-axis and 20 m above ground, Eq. (3.48) has to be used.

$$\langle \rho_{SO_2} \rangle (1000, 50, 20, 250) = \frac{500 \times 10^6}{2\pi\,(151)\,(108)6} \exp\left[-\frac{1}{2}\left(\frac{50}{151}\right)^2\right]$$

$$\left\{\exp\left[-\frac{1}{2}\left(\frac{20-250}{108}\right)^2\right] + \exp\left[-\frac{1}{2}\left(\frac{20+250}{108}\right)^2\right]\right\}$$

$$= 113\ \mu g/m^3$$

(d) Dividing Eq. (3.49) by Eq. (3.50) we get,

$$\frac{\langle \rho_{SO_2} \rangle (x, y, 0, H)}{\langle \rho_{SO_2} \rangle (x, 0, 0, H)} = \exp\left[-\frac{1}{2}\left(\frac{y}{\sigma_y}\right)^2\right]$$

From the above equation,

$$y = \left[2 \ln \frac{\langle \rho_{SO_2} \rangle (x, 0, 0, H)}{\langle \rho_{SO_2} \rangle (x, y, 0, H)}\right]^{1/2} \sigma_y$$

$$= \left[2 \ln \frac{\langle \rho_{SO_2} \rangle (1000, 0, 0, 250)}{\langle \rho_{SO_2} \rangle (1000, y, 0, H)}\right]^{1/2} (151)$$

$$= \left[2 \ln \frac{112}{100}\right]^{1/2} (151) = 72 \text{ m.}$$

Example 2

The traffic density for a highway is 1200 vehicles/hr and the average vehicle speed is 80 km/hr. The average carbon monoxide emission per vehicle is 40 g/s. Estimate the carbon monoxide concentration 100 m, 250 m, 500 m and 1000 m downwind of the highway if the windspeed normal to the highway is 2 m/s. Assume neutral conditions.

Solution

The source strength per unit length

$$Q_l = 1200 \left(\frac{\text{vehicles}}{\text{hr}}\right) \frac{1}{80} \left(\frac{\text{hr}}{\text{km}}\right) \frac{1}{1000} \left(\frac{\text{km}}{\text{m}}\right) 40 \left(\frac{g}{s} \frac{1}{\text{vehicle}}\right)$$

$$= 0.6 \text{ g/m-s}$$

The values of σ_z may be now calculated for the distances 100 m, 250 m, 500 m and 1000 m using Table 3.3, or alternatively Fig. (3.20). For neutral conditions,

	100 m	250 m	500 m	1000 m
σ_z (m)	4.7	10.1	17.9	31.8

The ground level concentration of carbon monoxide can be calculated from Eq. (3.56).

$$\langle \rho_{CO} \rangle = \frac{2(0.6)}{\sqrt{2\pi}(2)\,\sigma_z}$$

The carbon monoxide concentrations ($\langle \rho_{co} \rangle$) for 100 m, 250 m, 500 m and 1000 m are tabulated below :

Distance (m)	$\langle \rho_{co} \rangle$ (mg/m³)
100	51
250	23.7
500	13.4
1000	7.5

Effective Stack Height

Generally, effluent plumes from chimney stacks are released into the atmosphere at elevated temperatures. The rise of the plume after release to the atmosphere is caused by the buoyancy and the vertical momentum of the effluent. Under windless conditions, the plume rises vertically but more often it is bent as a result of the wind that is usually present. This rise of the plume adds to the stack an additional height ΔH, such that the height H of the virtual origin is obtained by adding the term ΔH, the plume rise, to the actual height of the stack, H_s. The plume cenrteline height $H = H_s + \Delta H$ is known as the effective stack height and it is this height that is used in Gaussian plume calculations. The effective stack height H is illustrated in Fig. 3.21.

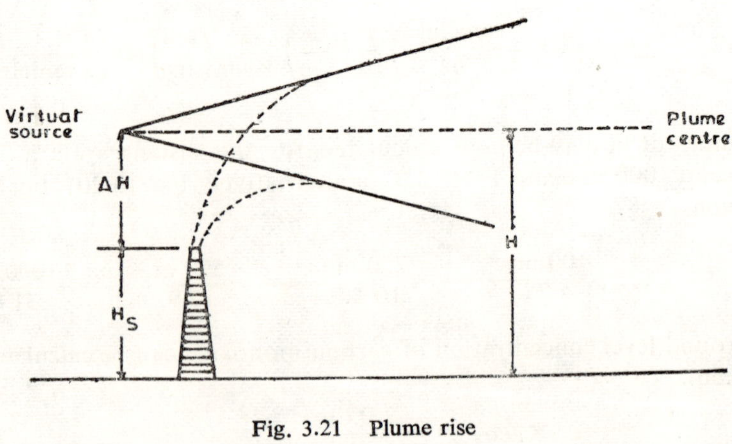

Fig. 3.21 Plume rise

The additional height ΔH due to plume rise is very significant in reducing downwind ground-level concentration and a large number of theoretical and empirical expressions have been proposed for computing ΔH. These are discussed in Section 3.5.

Deposition from Aerosols

Our preceding discussion assumes that the pollutants in the effluent are gaseous. In vigorously turbulent areas, particles of size upto 20 μm do not settle at an appreciable rate since they are more responsive to the eddy motions and effectively follow the surrounding gas at least for a few hours. Thus, the dispersion of such small aerosol particles can also be described by the above equations. When large particles are present, deposition on the ground can be a significant factor in reducing the concentration in wind-borne aerosols.

For a continuous point source located at ground level, the constant source strength Q given in Eq. (3.47) is replaced by $Q(x)$, the effective source strength at the downwind distance x. This can be related to the original source strength Q at the origin by the relation[16]

$$Q(x) = Q \exp\left[-\frac{4v_t x^{n'/2}}{\sqrt{\pi}\, C_z \bar{u}\, n'}\right] \tag{3.64}$$

where v_t is the effective deposition velocity of the particles, C_z is the Sutton's parameter, and n' is the turbulence index. C_z can be replaced by σ_z according to Eq. (3.61), then Eq. (3.64) becomes,

$$Q(x) = Q \exp\left[-\frac{4v_t\, x}{\sqrt{2\pi}\, \bar{u}\, \sigma_z\, n'}\right] \tag{3.65}$$

Substituting $Q(x)$ for Q in Eq. (3.47), we obtain, for the corrected concentration at the point (x, y, z), the expression

$$\langle \rho_A \rangle (x, y, z) = \frac{Q}{\pi \bar{u}\, \sigma_y \sigma_z} \exp\left[-\frac{4v_t\, x}{\sqrt{2\pi}\, \bar{u}\, \sigma_z\, n'}\right] \exp\left[-\tfrac{1}{2}\left(\frac{y^2}{\sigma_y^2} + \frac{z^2}{\sigma_z^2}\right)\right] \tag{3.66}$$

Eq. (3.48) for the concentration from an elevated point source can also be modified to allow for deposition by a similar procedure.

Example 3

A ground level source is emitting smoke particles at a rate of 1 g/s. The wind speed is 5 m/s. Assuming neutral conditions, calculate the ground level concentration on the axis of the smoke cloud at a downwind distance of 100 metres. Take the particle deposition velocity $v_t = 0.04$ m/s.

Solution

At $y = z = 0$, Eq. (3.66) reduces to,

$$\langle\rho\rangle(x, 0, 0) = \frac{Q}{\pi \bar{u}\, \sigma_y \sigma_z} \exp\left[-\frac{4 v_t\, x}{\sqrt{2\pi}\, \bar{u}\, \sigma_z n'}\right]$$

For neutral conditions: $n' = 0.25$, $\sigma_y = 8.3$ m, $\sigma_z = 4.7$ m
Also, $\bar{u} = 5$ m/s, $v_t = 0.04$ m/s, and $Q = 1$ g/s
Substituting these values in the above equation, we get

$$\langle\rho\rangle(100, 0, 0) = \frac{1}{\pi(5)(8.3)(4.7)} \exp\left[-\frac{4\,(0.04)\,100}{\sqrt{2\pi}\,(5)(4.7)\,0.25}\right]$$

$$= 0.551 \text{ mg/m}^3$$

3.5 ESTIMATION OF PLUME RISE

The accurate prediction of the plume rise (ΔH) which is essential for the estimation of the effective stack height (H) remains somewhat of a problem. Many formulae have been proposed in order to describe plume rise under various conditions; however, there is no common agreement among these formulae because of the different assumptions of basic turbulent processes. Some of the formulas which are more commonly quoted are discussed in this section.

Buoyant Plumes

In the case of buoyant plumes, the influence of buoyancy is much greater than the influence of vertical momentum. Such plumes are usually obtained when the release temperatures are more than 50°C greater than ambient atmospheric temperatures.

One of the earliest equations proposed to calculate the plume rise is by Holland[17]. This equation (Eq. 3.67) has been widely used because of its simplicity and has been included in many plume rise studies.

$$\Delta H = \frac{(1.5\, V_S D_S + 4 \times 10^{-5}\, Q_H)}{\bar{u}} \tag{3.67}$$

where V_S = Stack gas exit velocity, m/s; D_S = diameter of stack outlet, m; $\bar{u}$ = wind volocity, m/s, Q_H = heat emission rate, cal/s; and ΔH = plume rise above the stack, m.

Eq. (3.67) may also be expressed as reported by Turner[13],

$$\Delta H = \frac{V_S D_S}{\bar{u}}\left(1.5 + 2.68 \times 10^{-3}\, p_a\, \frac{T_s - T_a}{T_s}\, D_S\right) \tag{3.68}$$

where

p_a = atmospheric pressure, *mb*;

T_s = stack gas temperature, K;

and

T_a = ambient air temperature at stack height, K

Holland suggested[18] that stability can be accounted for and recommended correction factors for use with the equation. The correction factors are 1.15, 1.10, 1.00 and 0.85 for the stability classes A–B, C, D and E–F, respectively.

For stable or "near" neutral conditions Briggs[19] recommends the equation,

$$\Delta H = 2.4 \left(\frac{F}{\bar{u}S}\right)^{1/3} \tag{3.69}$$

where F is the buoyancy flux parameter, given by

$$F = gV_S \left(\frac{D_S}{2}\right)^2 \frac{T_s - T_a}{T_s}, \ \mathrm{m^4/s^3} \tag{3.70}$$

and S, the stability parameter is given by

$$S = \frac{g}{T_a}\left[\left(\frac{dT}{dz}\right)_{env} + \Gamma\right], \ \frac{1}{s^2} \tag{3.71}$$

In Eq. (3.69) "near" neutral conditions mean the stability side of the neutral. For neutral conditions $S = 0$, and hence, the equation is not recommended. For *unstable* or *neutral* conditions, Briggs gives the following equation:

$$\Delta H = \frac{1.6\, F^{1/3}\, (10\, H_S)^{2/3}}{\bar{u}} \ \text{for } x > 10\, H_S \tag{3.72}$$

where x = distance downwind of the stack in meters and H_S = stack height in meters.

Plume Rise Under Stable and Calm Conditions

When there is little or no wind, the bending of the plume is negligibly small and it rises to some height where the buoyancy force is completely dissipated. The recommended equation for such a situation is:

$$\Delta H = 5.0\, F^{1/4}\, S^{-3/8} \tag{3.73}$$

One should calculate ΔH for light-wind case from Eq. (3.73) as well as from Eq. (3.69) and use the smaller value.

Moses and Carson[20] have proposed another formula based on field data from a wide range of stack sizes for different conditions:

Unstable: $(dT/dz < -12.2)$

$$\Delta H = 3.47 \frac{V_S D_S}{\bar{u}} + 10.53 \frac{Q_H^{1/2}}{\bar{u}} \tag{3.74a}$$

Neutral: $(-12.2 \leqslant dT/dz < -1.5)$

$$\Delta H = 0.35 \frac{V_S D_S}{\bar{u}} + 5.41 \frac{Q_H^{1/2}}{\bar{u}} \tag{3.74b}$$

Stable: $(dT/dz \geqslant -1.5)$

$$\Delta H = -1.04 \frac{V_S D_S}{\bar{u}} + 4.58 \frac{Q_H^{1/2}}{\bar{u}} \tag{3.74c}$$

where dT/dz is the environmental lapse rate in °C/km, and Q_H is in kcal/s.

The formulas recommended by the ASME committee[15] are convenient to use for large volume sources, i.e., for volume flow rates greater than 50 m^3/s. For *unstable or neutral* conditions, the recommended equation is

$$\Delta H = 150 \frac{F}{(\bar{u})^3} \tag{3.75}$$

This equation gives reliable estimates of ΔH only when $\bar{u} \geqslant 7$m/s.

Non-buoyant Plumes

For sources at temperatures close to the ambient (or less than 50°C above ambient), and having exit speeds of atleast 10 m/s, the following equation was proposed by the ASME committee:

$$\Delta H = D_S \left(\frac{V_S}{\bar{u}}\right)^{1.4} \tag{3.76}$$

Many of the plume rise formulas discussed above are empirical in nature and thus are accurate only in a statistical sense. They are therefore not designed to estimate ΔH on an hourly basis[21]. This limitation should be recognised when using these formulas along with the plume dispersion equations to predict pollutant concentrations.

Example 4

If in example 1, the stack diameter is 5 m, the sulphur dioxide exit velocity is 13.5 m/s, and the gas temperature of the exit is 145°C, what is the plume rise for an ambient air temperature of 30°C? Calculate the ground level concentration on the plume centreline at the downwind distance of 1 km.

Solution

The volumetric flow rate of the flue gas is 265 m³/s for the above case. Since this is far greater than 50 m³/s, Eq. (3.75) can be applied.*

$$\Delta H = 150\,\frac{F}{(\bar{u})^3}$$

The buoyancy flux parameter F is given by Eq. (3.70).

$$F = 9.8\,(13.5)\left(\frac{5}{2}\right)^2\left(\frac{418-303}{418}\right)$$

$$= 227\ \text{m}^4/\text{s}^3$$

$$\Delta H = 150\,\frac{227}{(6)^3} = 158\ \text{m}$$

This is a substantial addition to the actual stack height of 250 m. The effective stack height $H = H_S + \Delta H$

Therefore, $H = 250 + 158 = 408$ m

Using the value of $H = 408$ m in Eq. (3.50), we get

$$\langle \rho_{SO_2} \rangle\,(1000, 0, 0, 408) = \frac{500\,(10^6)}{\pi(151)\,(108)\,6} \times$$

$$\exp\left[-\frac{1}{2}\left(\frac{408}{108}\right)^2\right] = 1.30\ \mu\text{g/m}^3$$

There is a significant reduction in ground level concentration as compared to the case for a zero plume rise, where

$$\langle \rho_{SO_2} \rangle\,(1000, 0, 0, 250) = 112\ \mu\text{g/m}^3$$

REFERENCES

1. O.G. Sutton, *Quart. J. Royal Meteorol. Soc.*, Vol. 73 (1947), p. 426.
2. A.G. Davenport, "Wind effect on buildings and structures", National Physical Lab.—Symposium (HMSO, London), 1965.
3. G.I. Taylor, "Eddy motion in the atmosphere", *Phil. Trans., A,* Vol. 215 (1915), pp. 1–26.

*With some reservation because $\bar{u} < 7$ m/s

4. L.F. Richardson, "Experiments on the diffusion of smoke emitted from a fixed point", *Phil. Trans. A,* Vol. 221 (1920), p. 1.
5. O.F.T. Roberts, *Proc. Royal Soc., A,* Vol. 104 (1923), p. 640.
6. A.S. Monin and A.M. Yaglom, *Statistical fluid mechanics: mechanisms of turbulence,* Vol. I (English Translation Ed. J.L. Lumley) (MIT Press, Cambridge, Mass.), 1971.
7. O.G. Sutton, *Micrometeorology* (McGraw-Hill, New York), 1953.
8. H.H. Lettau, "Wind profiles, surface stress, and geostrophic drag coefficients", in *Advances in geophysics,* Vol. 6, *Atmospheric diffusion and air pollution* (Academic Press, New York), 1959, pp. 241–256.
9. K.L. Calder, "Some recent British work on the problem of diffusion in the lower atmosphere, air pollution", *Proc. U.S. Tech. Conf. Air Poll.* (McGraw-Hill, New York), 1952, p. 787.
10. E.W. Bierly and E.W. Hewson, *J Appl. Meteorol.,* Vol. 1 (1962), pp. 383–390.
11. F. Pasquill, "The estimation of the dispersion of wind-borne material", *Meteorological Magazine,* Vol. 90 (1961), p. 30.
12. F.A. Gifford, "Use of routine meteorological observations for estimating atmospheric dispersion", *Nuclear Safety,* Vol. 2 (1961), pp. 47-51.
13. D.B. Turner, *Workbook of atmospheric dispersion estimates* (U.S. Public Health Service Publication), 1969, 999-AP-26.
14. R.C. Koch, "Validation and Sensitivity analysis of the Gaussian plume multiple-source urban diffusion model", Geomet. EF-60 APTD 0935 CPA, Geomet. Inc., Rockville, Md. 1971.
15. M. Smith (Ed.), *Recommended guide for the prediction of the dispersion of airborne effluents* (Am. Soc. Mech. Engrs.), 1968.
16. H.L. Green and W.R. Lane, *Particulate clouds: dusts, smokes and mists,* 2nd Edn (E. & F.N. SPON Ltd., Belfast, Northern Ireland), 1964.
17. J.Z. Holland, "Meteorology survey of the Oak Ridge area", ORO-99 (U.S. Atomic Energy Comm. Oak Ridge, Tenn.), 1963.
18. J.R. Mahoney, "Meteorological aspects of air pollution", in *Industrial pollution,* edited by N. Irving Sax (Van Nostrand Reinhold Co., New York), 1974.
19. G.A. Briggs, *Atmos. Environ.* Vol. 6 (1972), p. 507.
20. H. Moses and J.E. Carson, "Stack design parameters influencing plume rise", *J. Air Poll. Control Assoc.,* Vol. 18 (1968), p. 454.
21. T.L. Montgomery and M. Corn, *J. Air Poll. Control Assoc.,* Vol. 17 (1967), p. 512.

SUGGESTIONS FOR FURTHER READING

1. F. Pasquill, *Atmospheric diffusion* (Ellis Harwood Ltd., London), 1974.
2. J.O. Hinze, *Turbulence* (McGraw-Hill, New York), 1959.
3. J.L. Lurnley and H.A. Panofsky, *The structure of atmospheric* turbulence (Wiley, Interscience, New York), 1964.
4. C.H.B. Priestley, *Turbulent transfer in the lower atmosphere* (Univ. of Chicago Press, Chicago), 1959).
5. H.A. Panofsky, "Air Pollution meteorology", *Am. Scientist,* Vol. 52 (1969), pp. 269–285.
6. R.E. Newell, "The global circulation of atmospheric pollutants", *Sci. American,* Vol. 224 (1971), pp. 32–42.
7. F.N. Frenkiel and R.E. Munn (Eds.), *Turbulent diffusion in environmental pollution,* Adv. in Geophysics Series, Vols. 18A and 18B (Academic Press, New York), 1974.

PROBLEMS

1. One model of the atmosphere assumes a constant density. Find the lapse rate for this atmosphere and indicate whether this is a stable or unstable case. Assume that the atmosphere obeys perfect gas law.
2. On the basis of the average temperature gradient in the following situations, classify the degree of stability of the atmosphere: (a) temperature at ground level is 20°C; temperature at 500 m is 25°C; (b) ground-level temperature is 25°C, temperature at 800 m is 15°C; (c) ground-level temperature is 25°C; temperature at 2000 m is 5°C; (d) ground-level temperature is 30°C; temperature at 500 m is 20°C; (e) ground-level temperature is 25°C; temperature at 1000 m is 20°C.
3. Given the following temperature variation with altitude:

Elevation (m):	0	20	40	60	80	100	120
Temp. (°C) :	20	18	16	18	20	18	16

   What type of plume would you expect if the exit temperature of the plume was 20°C and the smoke stack was (a) 20 m tall? (b) 40 m tall? (c) 60 m tall? (d) 80 m tall?
4. Consider a planet whose atmosphere indicates the following composition:

Gas	*Volume Fraction*
H_2	0.866
He	0.132
CH_4	0.0007
NH_3	0.0003
H_2O	0.001

Calculate the adiabatic lapse rate for these atmospheric conditions if $g = 25\ m/s^2$, and compare it with the lapse rate for Earth's atmosphere. (0.2 K/100 m).

5. The stability of the atmosphere is frequently characterized by the potential temperature gradient where the potential temperature θ of dry air is defined as

$$\theta = T\left(\frac{p_0}{p}\right)^{k-1/k}$$

where k is the ratio of specific heats and p_0 is the standard pressure at sea level.

(a) What is the meaning of θ?

(b) Show that the stability parameter S is given by

$$S = \frac{1}{\theta}\frac{d\theta}{dz}$$

(c) What is the significance of $\frac{1}{\theta}\frac{d\theta}{dz}$?

6. An urban region can be thought of as composed of a multitude of point sources that fit into a box. The base of the box is the rectangular boundary of the city and its height is the mixing height above the city. Assume that the sources are uniformly distributed throughout the urban area and the box is a well-mixed vessel. Also assume that wind blows fresh air into one side of the box causing polluted air to be blown out of the opposite side.

Develop an expression for concentration of a pollutant within the city as a function of time. What is the steady state pollution concentration predicted by the model?

7. In the box model of problem 6, if the width of the box is 30 km and its height is 200 m, what will be the steady state carbon monoxide concentration when the emission rate is 10,000 tons/day and the average wind speed is 3 m/s? (6.42 mg/m³)

8. A 1000 MW power plant burns 10,000 metric tons of 1.5% sulphur coal per day. The flue gases are emitted into the atmosphere through a stack whose height is 200 m. The diameter of the stack at the plume exit is 5 m. The velocity and the temperature of the plume at the exit are 10 m/s and 120°C respectively. What is the downwind SO_2 concentration in the plume centreline on the ground at a distance of 5 km on a thin overcast night when the environmental lapse rate is equal to zero? Assume that the ambient air temperature is 15°C and the wind speed at the stack altitude is 6 m/s. (8.7×10^{-3} $\mu g/m^3$)
9. Estimate the total hydrocarbon concentration at a point 200 m downwind from a highway on a overcast night. The wind speed is perpendicular to the highway and has a speed of 5 m/s. The hydrocarbon emission rate is 0.001 g/m-s.
10. Sulphur dioxide is emitted through a stack 50 m high and 2 m diameter at the exit at a temperature of 120°C with a velocity of 10 m/s. The sulphur dioxide emission rate is 200 g/s. Show how the ground level concentration at a downwind distance of 1 km varies with wind speed for stability category *D* when the pressure is 1000 m bar and the ambient temperature is 25°C.

4. Air Pollution Sampling and Measurement

4.1 TYPES OF POLLUTANT SAMPLING AND MEASUREMENT

Sampling and measurement of air pollutants, generally known as air quality monitoring, is an integral component of any air pollution control programme. Through monitoring, the current trends in air quality can be evaluated by comparing the data with the regulated standards. The information so obtained is helpful in implementing control measures for reducing pollutant concentrations to acceptable levels and in assessing the effect of air pollution control strategies. The monitoring programmes of governmental agencies are usually designed for assessing long-term pollution trends, but it is often necessary to obtain short-term sampling to establish the existence of particularly high levels of air pollutants. Care must be taken while expressing the average concentrations of air pollutants since these are quite different for long and short periods.

There are two quite different situations in which the air quality measurement is undertaken. One of these is the ambient air quality measurement where the pollutant levels in the ambient atmosphere are measured. The second type of measurement generally deals with the pollutants emitted from a source such as smoke stack and is known as stack sampling. Stack sampling provides information on the nature and quantities of various pollutants that are emitted into the atmosphere.

4.2 AMBIENT AIR SAMPLING

Probably the most important objective of any air sampling exercise is to obtain a genuine and representative sample. In addition, the sampling should be carried out long enough and at a rate that allows collection of an analytically measurable sample. This is due to the fact that the concentrations of pollutants in the ambient air are likely to be extremely small. There are both gaseous and particulate air pollutants whose concentrations occur at levels below one milligram per cubic metre of air. In some cases the pollutants collected by normal air sampling methods may be reactive and this results in changes after collection. Hence, special measures are necessary to minimize such changes. For example, SO_2 may be oxidized to SO_3 and carrying out the analysis for SO_2 can be misleading in the absence of precautions to prevent the oxidation. Usually, a reducing agent such as isopropanol is used as a medium for collecting SO_2, which inhibits the formation of SO_3. It is, therefore, ideal and desirable to perform the analysis on the unchanged atmosphere, if possible, using continuous automatic analyzers. Such instruments are available for some pollutants and many more are at varying stages of development and evaluation. However, the conventional air sampling methods are still popular and widely used.

A typical air sampling system consists of a sample collector, a flowmeter to measure the airflow through the collector, and a pump to draw the air sample through the system. Fig. 4.1 shows a typical arrangement of air sampling equipment. The flowmeter can be positioned either upstream or

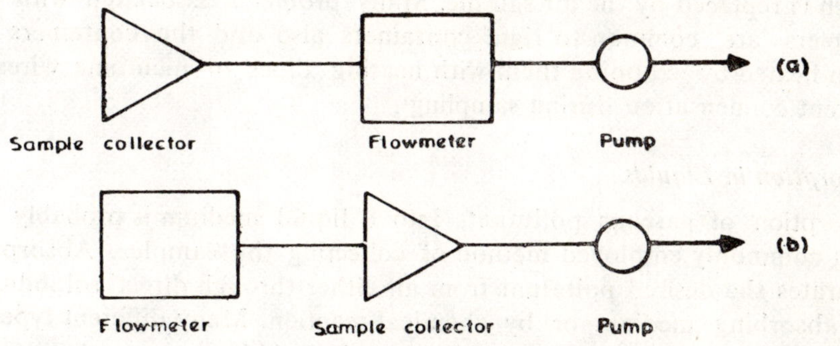

Fig. 4.1 Typical air sampling setup

downstream from the sample collector if the pressure drop across the collector is low, but when this is not the case the flowmeter must be placed upstream of the sample as shown in (b). Otherwise, serious errors can result in the airflow measurement as most of the flowmeters are designed to operate at and are calibrated at atmospheric pressure.

The function of the sample collector is to retain the desired pollutant either alone or as a mixture. The pollutant thus collected is analyzed at a later stage. The choice of collector to be used for collecting the sample is governed by the nature of the pollutant and the method of analysis to be employed. Individual sample collectors are generally grouped into two categories: those that can be used to collect gaseous pollutants and those used for particulate pollutants.

Collection of Gaseous Air Pollutants

Several methods are available for collection of gaseous air pollutants from ambient atmosphere. The common ones are: the grab sampling, absorption in liquids, adsorption on a solid material, and freeze-out sampling.

Grab Sampling

In "grab" sampling the sample is collected by filling an evacuated flask or an inflatable bag. Plastic bags have been widely used for grab sampling and for storage before analysis. Bag sampling is subject to losses caused by moisture condensation or diffusion through the walls of the bag. The losses can be minimized by performing the analysis immediately following collection.

Grab samples may be taken using rigid wall containers made from glass or stainless steel. These containers are first evacuated and then filled by allowing air to enter. Alternatively, a container may be filled with water and then used as a collector simply by draining away the water which is replaced by the air sample. Many problems associated with bag samplers are common to rigid containers also and the containers are often heated by wrapping them with heating tapes or nichrome wires to prevent condensation during sampling[1].

Absorption in Liquids

Absorption of gaseous pollutants into a liquid medium is probably the most commonly employed method of collecting the samples. Absorption separates the desired pollutant from air either through direct solubility in the absorbing medium or by chemical reaction. Many different types of collectors are in use, ranging from simple bubbles to complex devices which provide a high degree of gas-liquid contact. Some of these devices are shown in Fig. 4.2. One such device is the fritted-glass absorber where the gas stream is broken up into extremely small bubbles, thus promoting an intimate contact between the gas and the liquid. Frits designated "coarse" (50-μm pore size) are most often used for air sampling. The glass frit can become blocked and is often difficult to clean after use; therefore, it is often advisable to pre-filter the air prior to sampling.

Probably the most widely used collector is the impinger. In the impinger the gas stream is impinged at high velocity onto a flat surface thus providing good contact between the gas and the liquid. The flat surface can be the bottom of the collector or a specially designed plate. The common impingers are the Greenburg-Smith and the midget types. These devices can handle sample flow rates of abcut 30 and 3 litres per minute respectively. Impingers can collect gases with efficiencies often approaching those of fritted-glass collectors and unlike the latter they are easy to clean and maintain.

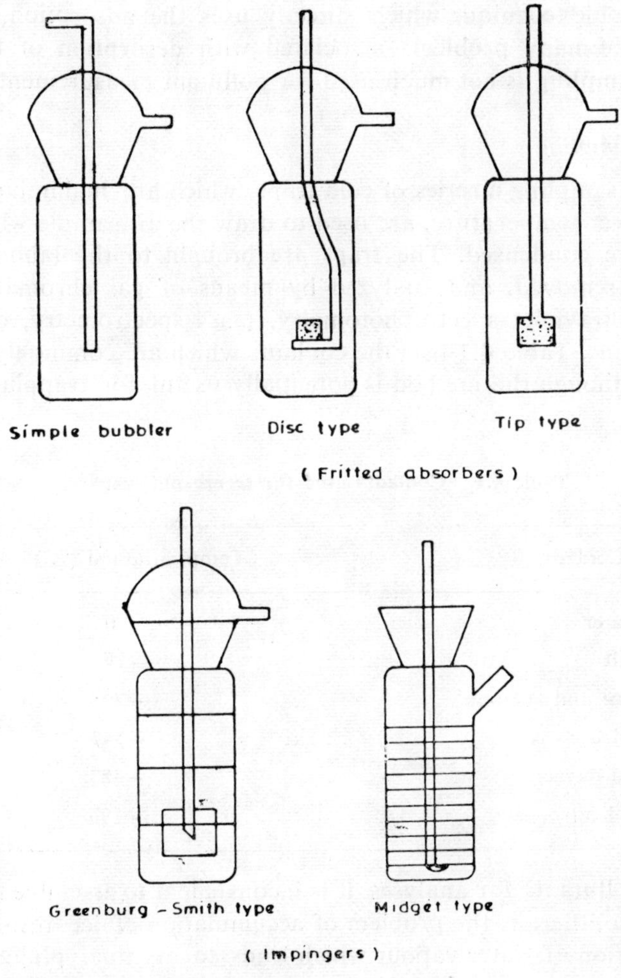

Fig. 4.2 Gas absorbing devices

Adsorption on Solids

This method is based on the tendency of gases to be adsorbed on the surface of solid materials. The sample air is passed through a packed column containing a finely divided solid adsorbent on whose surface the pollutants are retained and concentrated. The most commonly used solid adsorbents are grannular porous solids such as activated charcoal and silica gel with very large surface area.

After adsorption, the sample gases are desorbed for analysis. This may be accomplished by heating the adsorbent to volatilise the trapped material or by washing it with a liquid solvent. Analysis of the gases can be done in a number of ways. Most organic vapours are analyzed by gas chromatographic technique which directly uses the adsorption principle. But because of many problems associated with desorption of the gases, adsorptive sampling is not much used for pollutant measurements.

Freeze-out Sampling

In freeze-out sampling a series of cold traps, which are maintained at progressively lower temperature, are used to draw the air sample whereby the pollutants are condensed. The traps are brought to the laboratory, the samples are removed, and analyzed by means of gas chromatographic, infrared or ultraviolet spectrophotometry, mass spectrometry, or by wet chemical means. Table 4.1 lists the coolants which are commonly used for the traps. Although the method is potentially useful for trapping a large

Table 4.1 Coolants used for freeze-out traps

Coolant	Temp. attained (°C)
Ice-water	0
Ice-salt	–21
Dry ice and acetone	–79
Liquid air	–147
Liquid oxygen	–183
Liquid nitrogen	–196

number of pollutants for analyses, it is inconvenient to assemble the freeze-out traps; also, there is the problem of accumulation of ice resulting from the condensation of water vapour which leads to eventual plugging of the system. For these reasons the technique is not generally used for routine sampling purposes.

Collection of Particulate Pollutants

Particulate pollutants in the atmosphere are grouped generally into those that settle out due to the force of gravity and those that remain suspended as aerosols. The first category, consisting of large particles of size greater than 10 μm diameter, can be collected using sedimentation techniques, while for the second category containing particles of smaller size more sophisticated techniques like filtration, impingement, and electrostatic and thermal precipitation are used. Most instruments used for monitoring either class of particulates are merely collectors whose function is to permit measurement of collected material; the analysis of the weight or number of particles is done separately.

Sedimentation (Dustfall Jar)

The simplest device used for sampling particles larger than 10 μm in diameter is the dustfall collector shown in Fig. 4.3. A typical collector consists of a plastic jar of about 20 to 35 cm height and 10 to 15 cm diameter at the base with a slight inward tapering of the walls from top to bottom.

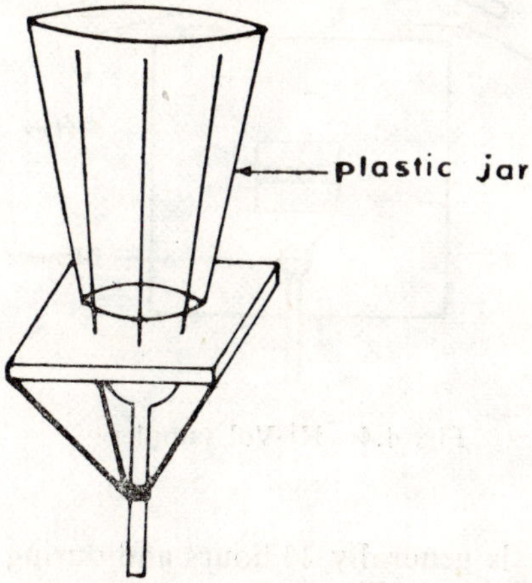

Fig. 4.3 Dustfall collector for particulate measurement

A holder is provided to ensure safe and upright positioning of the collector. The sample is deposited over a period of one month and the material is dried and weighed. Usually, only water-insoluble dustfall is reported in mg/cm^2 or in $tons/km^2$.

Since dust particles larger than 10 μm are seldom carried for distances in excess of 1 km, dustfall stations must be closely spaced for any meaningful data. The method is simple and inexpensive and requires no electrical power or moving parts. However, the method lacks precision and is selective and usually nonrepeatable. Hence, it may only be used to show pollution trends in a specific region over a period of time.

High-volume Filtration (The Hi-vol. Sampler)

The high volume filtration method is popular for measurement of the mass concentration of suspended particulates smaller than 10 μm. The sampler, universally known as "hi-vol," is illustrated in Fig. 4.4. In this method, a known volume of air is sucked by a high speed blower through a fine filter and the increase in weight due to the trapped particles is measured. The filter, usually made of fibrous or granular material, provides a dense porous medium through which an air stream must change direction in a random fashion, allowing the entrained particles to impact on the filter material. A typical glass fibre filter used in the air samplers has an efficiency of over 99 per cent for particles of size 0.3 μm and it is moderately effective for trapping particles of as small size as 0.05 μm.

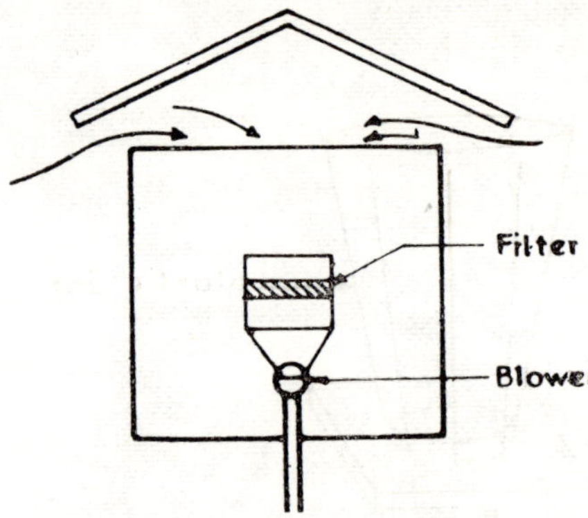

Fig. 4.4 Hi-Vol sampler

The sampling time is generally 24 hours and during this time over 2000 m^3 of air is sucked through the filter. The analysis is gravimetric and the particulate level is generally expressed in terms of $\mu g/m^3$ of air. Based on typical results of such a sampling, the following example would illustrate the calculation:

Air flow through clean filter = 1.7 m^3/min
Air flow through the filter at the end of the test = 1.4 m^3/min
Average air flow = 1.55 m^3/min = 2232 m^3 in 24 hrs
Weight of the clean filter = 5.000 g
Weight of the filter after exposure = 5.348 g
Weight of the particulate dust = 0.348 g = 0.348×10^6 μg

Suspended particulate concentration

$$= \frac{0.348 \times 10^6 \ \mu g}{2232 \ m^3} = 156 \ \mu g/m^3$$

Tape Sampler

Another popular method of sampling particulate level of the atmosphere is the soil-haze particle sampler. In this method a known volume of air is passed through a paper tape which is advanced in discrete time intervals from a supply reel to a take-up reel. During the sampling period, the tape is held stationary. The sampling time could be varied from as little as 10 minutes per sample to 2 hours, depending upon the amount of particulates. The air is passed through a nozzle and the particulates are collected on the paper tape forming a spot. Spots so obtained are evaluated by analyzing the amount of light that can be transmitted through the spot. The measurement is reported as "COH/1000 ft" where COH is known as the coefficient of optical haze, calculated as:

$$\frac{\text{COH}}{1000 \text{ ft}} = \log \frac{T_0}{T} \left(\frac{A}{V}\right) 10^5 \tag{4.1}$$

where $T_0 =$ the transmittance of clean tape (100%), $T =$ the percentage of light transmitted through the spot, $A =$ area of the spot in squart feet, $V =$ volume of the sample in cubic feet or the product of the average sample flow and the sampling time.

The coefficient of haze index is based upon an optical property and this method is quite different from the standard methods of representing the particulate mass concentration. Hence, some guidelines are required for interpreting the COH values in terms of the degree of pollution. Table 4.2 provides a typical rating system for values of COH per 1000 linear feet versus the degree of air pollution.

Table 4.2 Interpretation of COH Index

COH	Degree of Pollution
0–0.9	Light
1–1.9	Medium
2–2.9	Heavy
3–3.9	Very heavy
4–4.9	Extremely heavy

A more realistic analysis of the spot is based upon β-ray attenuation from which the mass concentration of the particulates can be estimated. Here a beam of β-particles of constant source with intensity I_0 is passed through the tape. By measuring the intensities I_c and I_p of the radiation that has passed through a clean tape and through the spot respectively, the mass μ_p of the particulate matter per unit area of the spot can be evaluated according to Eq. (4.2):

$$\mu_p = \frac{1}{k} \ln \left(\frac{I_c}{I_p}\right) \tag{4.2}$$

where k is a constant which depends on the energy of the β-particles; the intensity of radiation can be measured by a Geiger counter. The mass concentration of the particulates in the air ρ_p is then given by,

$$\rho_p = \mu_p \left(\frac{A}{V}\right) \tag{4.3}$$

where A is the area of the spot and V is the volume of the sample in appropriate units. The measurement set up is illustrated in Fig. 4.5.

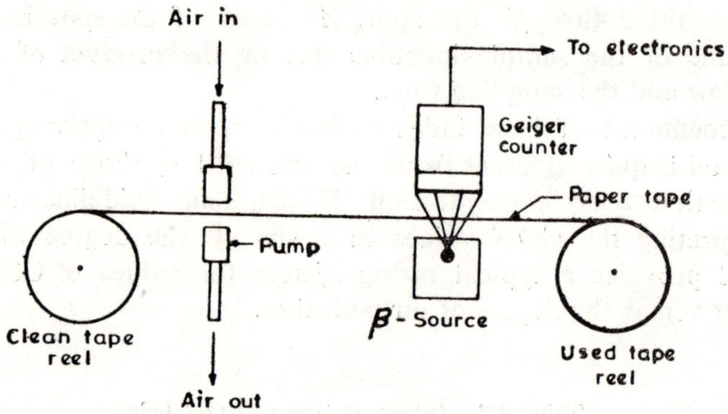

Fig. 4.5 Tape sampler with β-ray attenuation

Impingement

In an impinger the separation of particulates from the airstream takes place by collision against a flat surface. Wet impingement is used for collection of particulates in a liquid phase. As in gas sampling the airstream containing the particulates impinges at a high velocity onto a flat surface immersed in a liquid. The gas breaks up into small bubbles, and the particles are collected by change in the direction of gas flow and promoting their contact with the liquid. The Greenburg-Smith type impinger which is shown in Fig. 4.2 achieves 90 to 99 per cent efficiency for particles of the size equal to or greater than 1 μm.

Dry impingers operate on the principle of impaction on a dry surface for particle collection. These impingers are not as efficient as wet devices for collecting smaller particles (< 2 μm); however, they have the advantage of collecting the sample directly without the need of filtration or evaporation of a liquid. Certain dry impingers, called cascade impactors, are designed for making particle size measurements. A typical cascade impactor consists of 5 to 10 collection stages in which the sample airstream is constrained to pass through a series of jets where pollutant particles are directed against collection surfaces placed normal to the jets. Smaller particles having lower momentum, are collected in later stages, where the jet velocities become progressively higher. In most cascade impactors a filter is placed after the last stage to collect any fine particles that might have passed through the impactor. Fig. 4.6 shows the schematic diagram of a

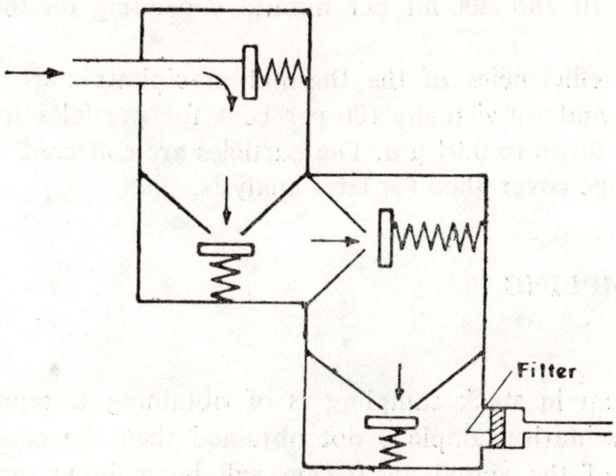

Fig. 4.6 Four-stage cascade impactor

cascade impactor. After sampling, the material collected at each stage is analysed for its weight and chemical composition. The impactors are reported to be capable of sizing particles down to 0.2 μm, and an efficiency of 100 per cent for particles as small as 0.6 μm has been obtained.

Electrostatic Precipitation

Air sampling can be done on the basis of the principle of electrostatic precipitation. Several types of portable commercial precipitators are available in the market. During their operation a negative charge is imparted to a wire placed axially inside a cylinder which is positively charged. When the particle laden airstream passes through the cylinder, the particles acquire a negative charge from a corona discharge occurring on the central wire. The particles migrate towards the inner surface of the cylinder and are removed for subsequent chemical or microscopic analysis.

Small samplers capable of field operation have been designed and these exhibit good separation characteristics for particles in the size range of 0.01 to 20 μm. Nevertheless, they are not as widely used for ambient air sampling as filters, but are more suitable for source sampling where the particle loading is relatively high.

Thermal Precipitation

Thermal precipitators operate on the principle that small particles, under the influence of a strong temperature gradient between two surfaces, have a tendency to move towards the lower temperature and get deposited on the colder of these two surfaces. These temperature gradients are normally of the order of 3000°C per centimetre. To maintain such high gradients the gas velocity through the sampling device must be maintained low; this is usually between 10 and 200 ml per minute depending on the type of device.

The collection efficiencies of the thermal precipitators are quite high for small particles and are virtually 100 per cent for particles in the size range from 10 μm down to 0.01 μm. The particles are collected on a grid or a thin microscope cover glass for later analysis.

4.3 STACK SAMPLING

The major problem in stack sampling is of obtaining a representative sample. If a representative sample is not obtained then the concentration and composition of the actual gas stream will be different, and serious errors would result in the analysis. The important factors in obtaining a representative sample are the selection of the sampling site and the number of sampling points required. The sampling site should be located at least eight stack or duct diameters downstream and two diameters upstream from any source of flow disturbance such as bends, fittings, or constrictions. The gas stream in a stack is normally under turbulent flow conditions and any flow disturbance would cause non-uniform and unstable gas flow profiles and non-uniform particle concentration patterns. These problems can be minimized by providing proper distance so that adequate mixing may occur.

However, in some stacks, it is not always possible to ensure uniform flow, so concentration patterns and, hence, multiple samples are required to obtain a representative sample. That is, the actual sampling must be performed at a number of traverse points in the stack. Two sets of traverse readings, 90° apart, are taken in the same plane for circular stacks. The number of traverse points required on each of the two perpendiculars may be estimated from Table 4.3. According to the earlier sampling point criterion, the number of traverse points required on each of the two perpendiculars is six. If this criterion is not met, then a different number of traverse points is required depending upon the location of downstream and upstream disturbances. It has been suggested that the greater of the two values of the traverse points should be chosen. The location of the traverse points in a circular stack is at the centre of the annular rings of equal area as shown in Fig. 4.7.

For rectangular stacks, the sampling site can be located by calculating the equivalent diameter according to the following equation:

$$D_{eq} = 4 \times \frac{\text{Cross-sectional area of flow}}{\text{Wetted perimeter}} \tag{4.4}$$

Table 4.3 Minimum number of traverse points (ref. 2)

Number of stack diameters		
Location of flow disturbance		Minimum number of Traverse points on each diameter
Upstream	Downstream	
8 +	2 +	6
7.3	1.8	8
6.7	1.7	10
6.0	1.5	12
5.3	1.3	14
4.7	1.2	16
4.0	1.0	18
3.3	0.8	20
2.6	0.6	22
2.0	0.5	24

The sampling points are located at the centre of rectangles of equal area as shown in Fig. 4.7, and a minimum of nine traverse points should be selected for proper sampling.

Other problems associated with stack sampling are the high temperatures and high velocities of the gases encountered inside the stack. These conditions, especially high temperatures, cause difficulties in collecting the sample and require a number of additional measurements such as the moisture content, pressure, temperature, the flow rate of the gases and their compositions. Accurate measurements of these factors are necessary for valid sampling.

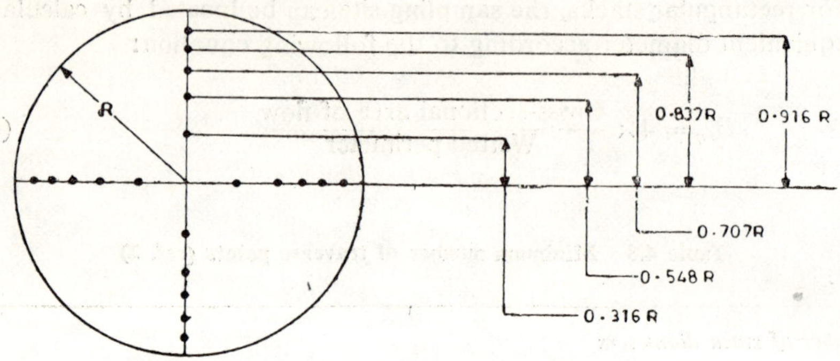

Circular stack (10-point traverse)

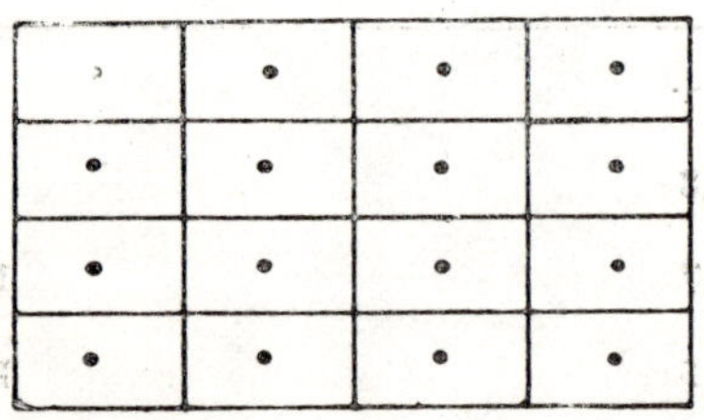

Rectangular stack (measure at centre of at least 9 equal areas).

Fig. 4.7 Traverse point locations for multiple sampling

Sampling System

Stack sampling is carried out by diverting a part of the gas stream through a sampling "train" of which a general arrangement is shown in Fig. 4.8. The train consists of a nozzle placed in the gas stream, a sampling probe through which the sample is drawn at different traverses, particulate and gas collection devices, a flow measuring device, and a prime mover such as a vacuum pump or an ejector.

The nozzle at the end of the probe is normally sharp edged, pointing inward from the outside edge and the traversing probe is usually made of stainless steel with glass or teflon lining. For sampling hot gases whose temperatures are above 400°C, these probes are provided with a circulating coolant system to prevent combustion of particulate materials inside the probe and to prevent the temperatures from exceeding the maximum allowable temperatures of filtration materials.

Similar to the devices for ambient air collection, a number of devices are used for collection of particulates and gases in stack sampling. These include filtration, wet or dry impingement, impaction, electrostatic and thermal precipitation devices for collecting particulates, and absorption, adsorption and freeze-out devices for collecting gases. A summary of the

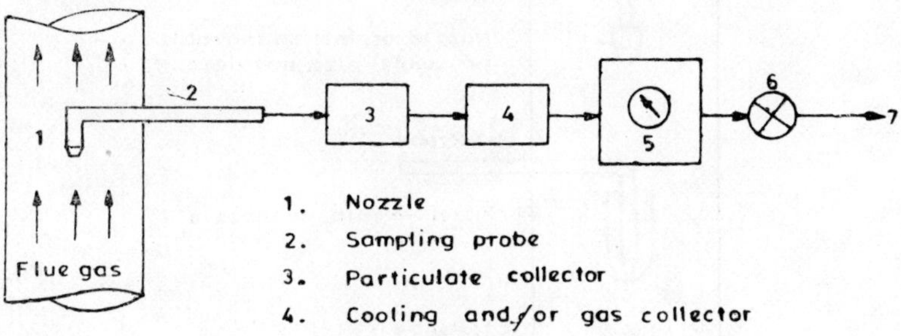

Fig. 4.8 General arrangement of sampling train

uses and properties of various gas and particulate collectors are given by Bloomfield[3].

Flow measurements can be made with devices such as a rotameter or orifice meter, and with a dry gas meter if the information on the total volume of the gas sampled is required. These devices are normally placed after the collection devices and upstream of a vacuum pump so that any leak in the vacuum pump will not cause errors in the gas volume measurements. Fig. 4.9 summarizes the major features of several commonly used components in stack sampling systems.

Particulate Sampling

The common technique employed for sampling particulate-laden gas streams is the so called "isokinetic technique". In this technique a sample is drawn into the probe such that the conditions at the tip of the probe are the same as those in the free gas stream at the sampling point. When these conditions are fulfilled, then the sample extracted through the probe is representative of an isokinetic condition. Under isokinetic conditions, the static pressure at the tip of the probe must be equal to the static pressure in the free stream at the same cross section. This implies that when the two pressures are equalized then the corresponding velocities must be equal.

When a sharp-edged probe is introduced into a single-phase fluid stream, then the flow pattern around the probe may be represented as in

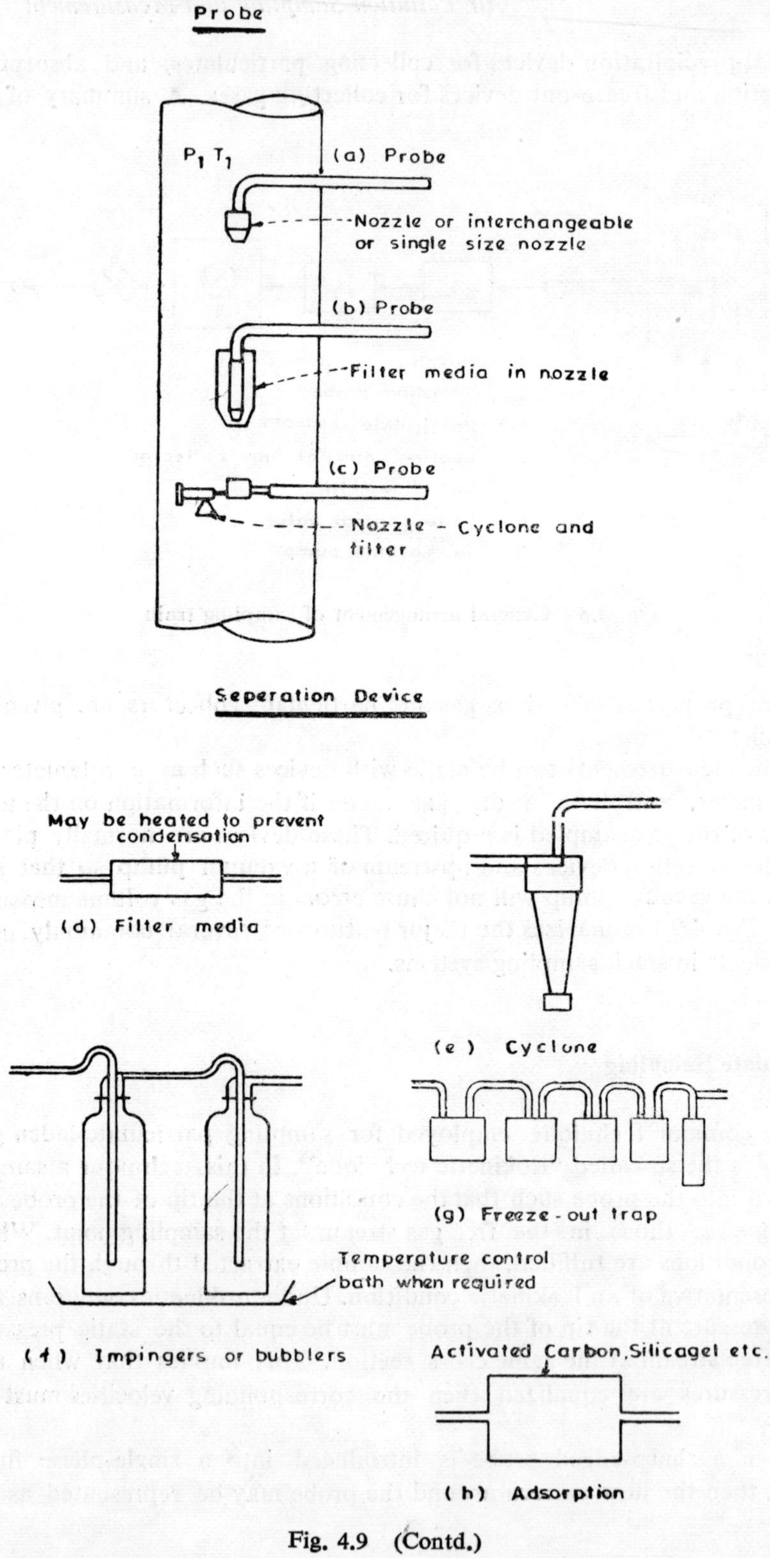
Probe
P1 T1
(a) Probe
Nozzle or interchangeable or single size nozzle
(b) Probe
Filter media in nozzle
(c) Probe
Nozzle - Cyclone and filter
Seperation Device
May be heated to prevent condensation
(d) Filter media
(e) Cyclone
(g) Freeze-out trap
Temperature control bath when required
(f) Impingers or bubblers
Activated Carbon, Silicagel etc.
(h) Adsorption

Fig. 4.9 (Contd.)

the case of a simple pitot tube. The velocity inside the probe is zero and the pressure at the tip of the probe becomes equal to the stagnation pressure.

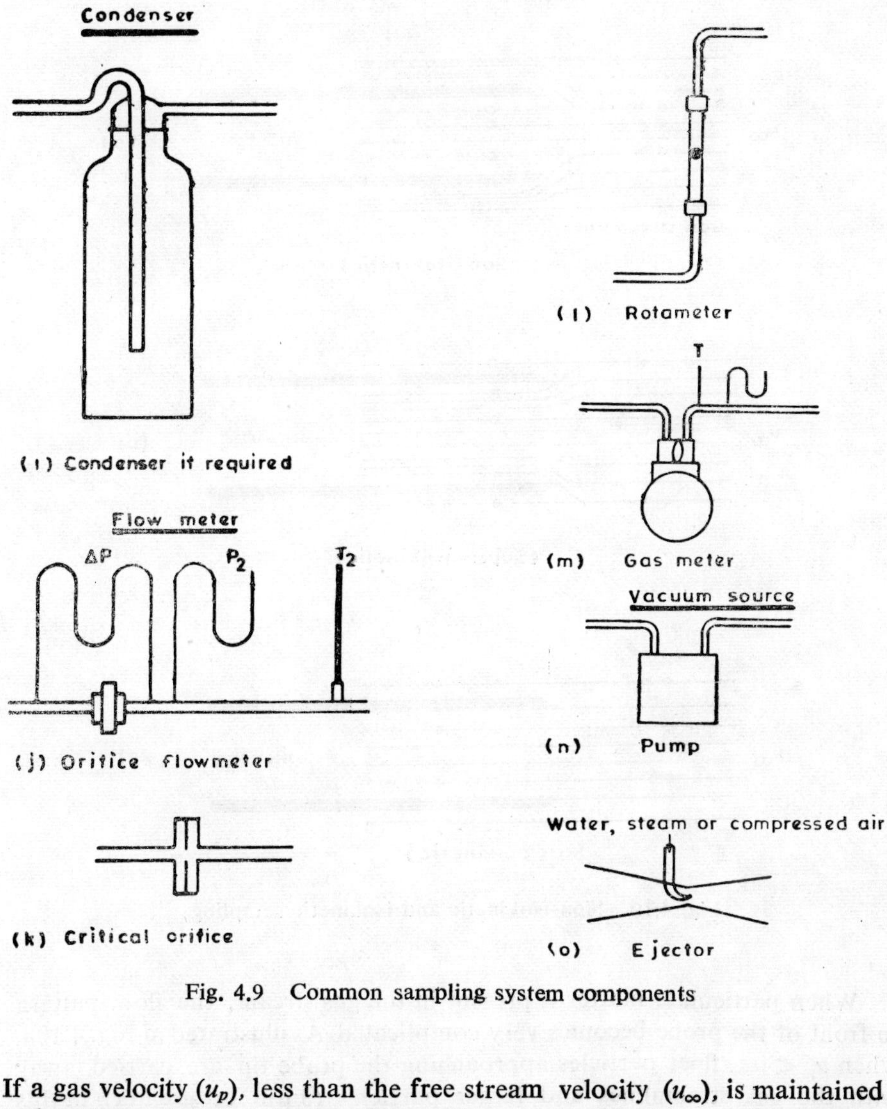

Fig. 4.9 Common sampling system components

If a gas velocity (u_p), less than the free stream velocity (u_∞), is maintained inside the probe, then the gas flow pattern at the tip of the probe may be represented as in Fig. 4.10a. In this case the static pressure at the tip of the probe is greater than the free stream static pressure at the same cross section. On the other hand, if the sampling velocity (u_p) is greater than the free stream velocity (u_∞), then the flow pattern may look somewhat like that represented in Fig. 4.10b. Here the static pressure at the tip of the probe is less than the free stream pressure, and too much gas will be sampled in proportion to the probe area. Only when the sampling velocity is equal to the free stream velocity, the pressures become equal and true isokinetic conditions are attained. Under isokinetic conditions the flow pattern in front of the probe is not disturbed as shown in Fig. 4.10c.

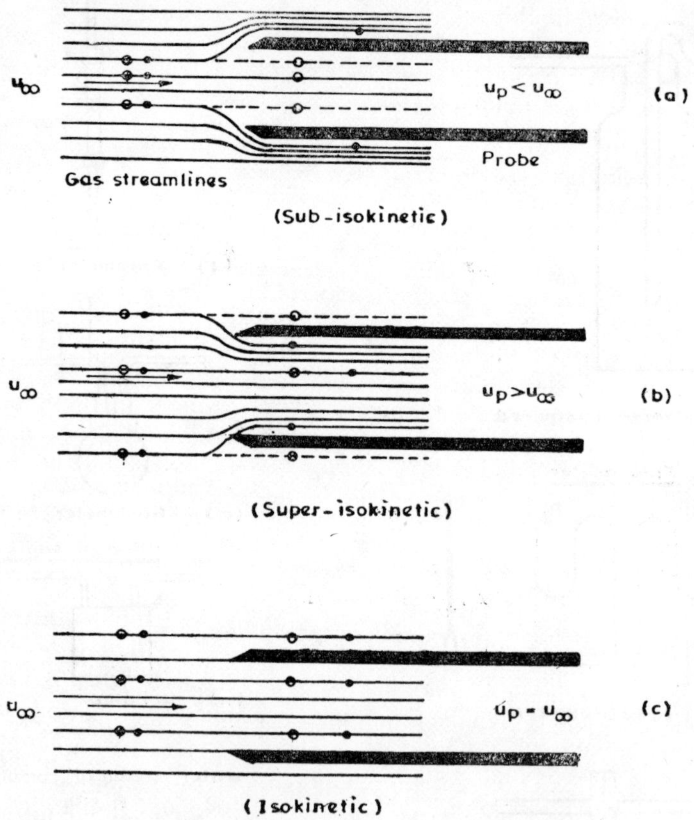

Fig. 4.10 Non-isokinetic and isokinetic sampling

When particulate matter is present in the gas stream, the flow pattern in front of the probe becomes very complicated. As illustrated in Fig. 4.10a, when $u_p < u_\infty$, finer particles approaching the probe tip are carried away with the gas streamlines and larger particles (3 μm or greater) in this portion of the gas travels in straight-line trajectories due to their higher inertia and enter the probe tip and continue into the probe. In this case the particle-capture efficiency* of the probe will be less than one and the

*The capture efficiency is defined as the ratio of actual capture rate of particles to that for straight line trajectories.

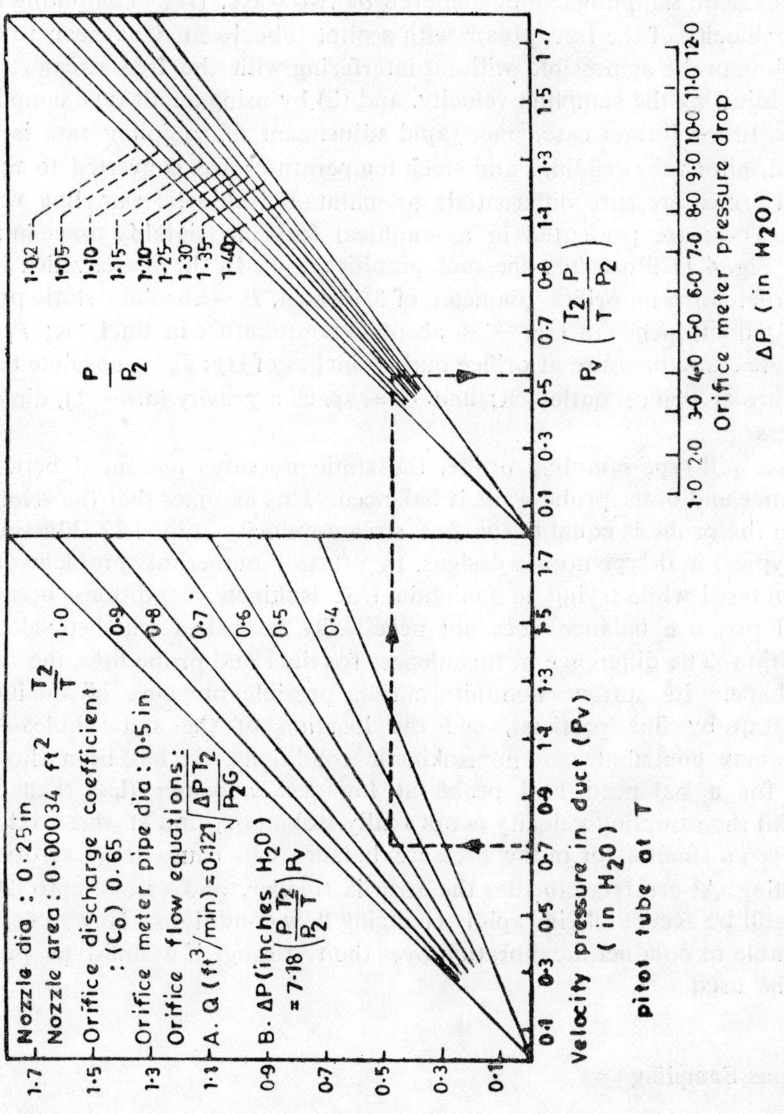

Fig. 4.11 Conversion of pitot tube readings to orifice pressure differentials (ref. 4)

sample will indicate high concentration of coarse particles. Conversely, in Fig. 4.10b the capture efficiency will be greater than one and the sample will contain a large amount of finer particles. Only when the sampling is isokinetic, the capture efficiency will be equal to one and the sample will be a representative sample as shown in Fig. 4.10c.

Isokinetic sampling can be achieved by two ways: (i) by computing the point velocity of the free stream with a pitot tube located as near to the sampling probe as possible without interfering with the free stream, and then adjusting the sampling velocity, and (ii) by using a null-type sampling probe. In the former case, since rapid adjustment of sampling rate is required, pitot tube readings and stack temperatures are converted to equivalent orifice pressure differentials to maintain isokinetic sampling velocities. These are presented in a graphical form to simplify usage in the field. Fig. 4.11 illustrates one such graphical aid[4]. In the figure, which was constructed for an orifice diameter of 3/16 inch, P = absolute static pressure in duct, inches of Hg; T = absolute temperature in duct, °R; P_2 = absolute static pressure at orifice outlet, inches of Hg; T_2 = absolute temperature at orifice outlet °R; and G = specific gravity (air = 1), dimensionless.

In a null-type sampling probe, the static pressures measured between the inner and outer probe walls is balanced. This assumes that the velocity within the probe is equal to the free stream velocity. Fig. 4.12 illustrates two typical null-type nozzle designs. In practice numerous problems are encountered while trying to maintain true isokinetic conditions because a null pressure balance does not necessarily indicate a matched velocity condition. The difference in turbulence for duct and probe flow, the nozzle shape, its surface nonuniformities, possible plugging of the inner static tap by fine particles, and the location of the static holes—all these may contribute to nonisokinetic conditions. It has been shown[7] that for a balanced null probe at low gas velocities (less than 350 m/min) the sampling velocity is not really isokinetic, and at these velocities even a small error in the pressure balance can cause large errors in sampling. At greater velocities the error is smaller, and errors upto 20% may still be acceptable in rapidly changing flow conditions. However, it is advisable to conduct a calibration over the test range if a null-type probe is to be used.

Gaseous Sampling

The basic routine for obtaining a gas sample from stack emissions is similar to that employed in sampling of particulate matter. However, in this case, the sampling is much easier because it is not necessary to sample at isokinetic conditions. The sample can be withdrawn from the stack at a constant rate independent of the flow rate in the stack. Nevertheless, there

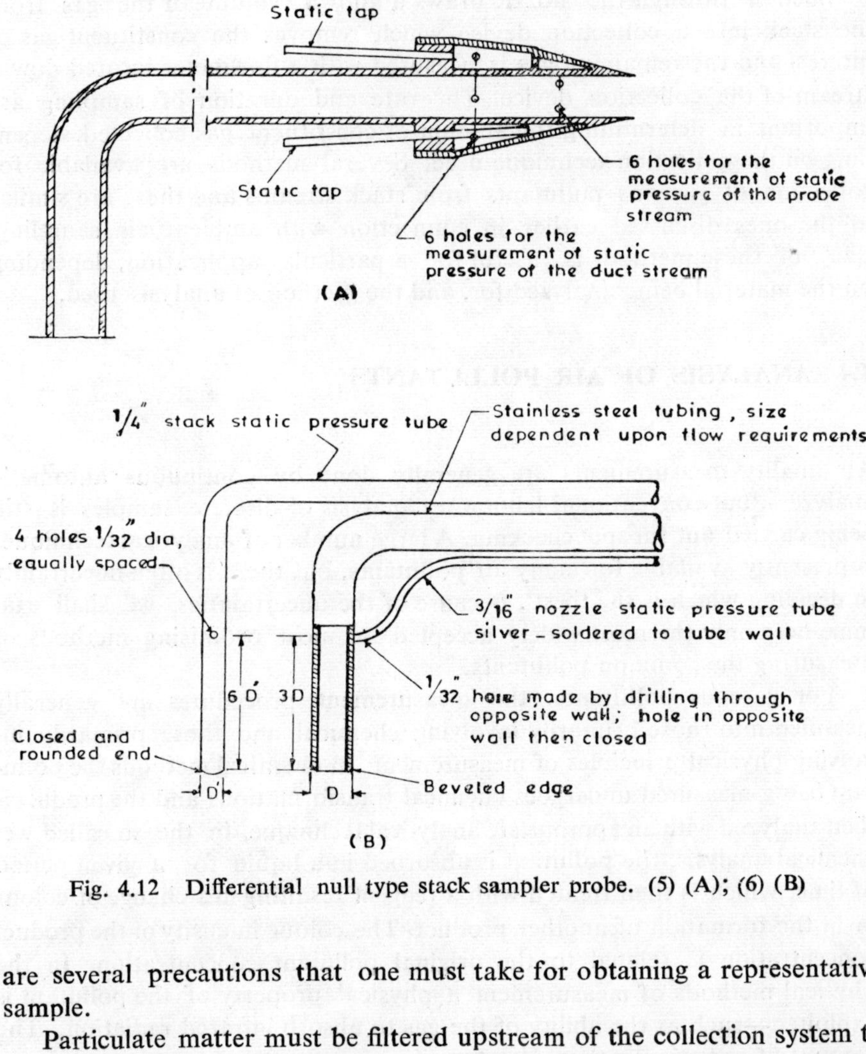

Fig. 4.12 Differential null type stack sampler probe. (5) (A); (6) (B)

are several precautions that one must take for obtaining a representative sample.

Particulate matter must be filtered upstream of the collection system to prevent downstream line plugging as well as to minimize losses of gaseous pollutants due to reaction with particulates on cooling. It is useful to point the probe downstream to minimize the amount of particulates that are pulled into the sampling line. If a straight probe is used, it can be fitted with an integral filter such as glass of pyrex wool.

If moisture is present in stack gases it could get condensed in the sampling line and dissolve the gaseous constituents of interest. To prevent the losses resulting from condensation, the sampling line may be heated. Care must be taken to use the tubing and the sampling probe made up of inert material so as to avoid reaction with the gas to be measured. Stainless steel is usually an acceptable probe material and in some special applications teflon is the preferred material.

Suction through the nozzle draws a known volume of the gas from the stack into a collection device which removes the constituent gas of interest and the remaining gas is measured with a flowmeter located downstream of the collection device. The rate and duration of sampling are important in determining the amount of constituent gas collected, depending on the collection technique used. Several methods are available for collection of gaseous pollutants from stack streams and these are similar to the ones discussed earlier in connection with ambient air sampling. Each of these methods is useful for a particular application, depending on the material being analyzed for, and the method of analysis used.

4.4 ANALYSIS OF AIR POLLUTANTS

Air quality measurements are generally done by continuous automatic analyzers, but conventional laboratory analysis of discrete samples is still being carried out for spot checking. A large number of analytical techniques is presently available for many air pollutants, but there is often uncertainty in deciding which is the 'best'. Because of the uncertainties, we shall examine here only the most widely accepted or most promising methods of measuring the common pollutants.

For gaseous pollutants, the measurement procedures are generally classified into those primarily involving chemical and those primarily involving physical principles of measurement. In chemical methods the pollutant being measured undergoes chemical transformations and the product is then analyzed with an appropriate analytical technique. In the so called wet chemical analysis, the pollutant is absorbed in a liquid for a given period of time, which is then treated with a reagent resulting in a change of colour or in the formation of another product. The colour intensity or the product concentration is related to the original pollutant concentration. In the physical methods of measurement a physical property of the pollutant is exploited—such as the ability of the gas to absorb infrared radiation. The amount of radiation is then detected.

Sulphur Dioxide

Ambient Air Measurements

The most common manual methods for measuring atmospheric SO_2 are based on colourimetry, iodimetry, or turbidimetry. Of these, the modified West—Gaeke[8] colourimetric procedure has gained wide popularity and is considered the standard or reference method. In this method, SO_2 from a measured quantity of air is absorbed in a solution of sodium tetrachloromercurate to form a stable and nonvolatile dichlorosulphitomercurate complex. This is then reacted with formaldehyde and bleached pararosaniline, yielding a magenta-coloured pararosaniline sulphonic acid product. The colour intensity of this acid, which can be detected by photometric methods, is proportional to the concentration of SO_2 over a range of 0.002 to 5 ppm.

Iron and other heavy metal salts cause interference by lowering the apparent SO_2 concentration. Ethylenediaminetetraacetic acid (EDTA) is often added to sequester such metal interferences. Nitrogen dioxide interference can be eliminated by adding a small amount of sulphamic acid which destroys the nitrate ion prior to colour formation.

There are several automatic instruments for monitoring sulphur dioxide; the most common of these are based upon conductometric, colourimetric, and flame photometric principles. In the conductometric technique, the sampled air containing SO_2 is passed through a dilute solution of hydrogen peroxide in dilute sulphuric acid. SO_2 is oxidised to H_2SO_4, thereby increasing the electrical conductivity of the solution. The increase in conductivity, which is measured, is proportional to the concentration of SO_2 in the sample. With this technique, as low SO_2 concentrations as 0.01 ppm can

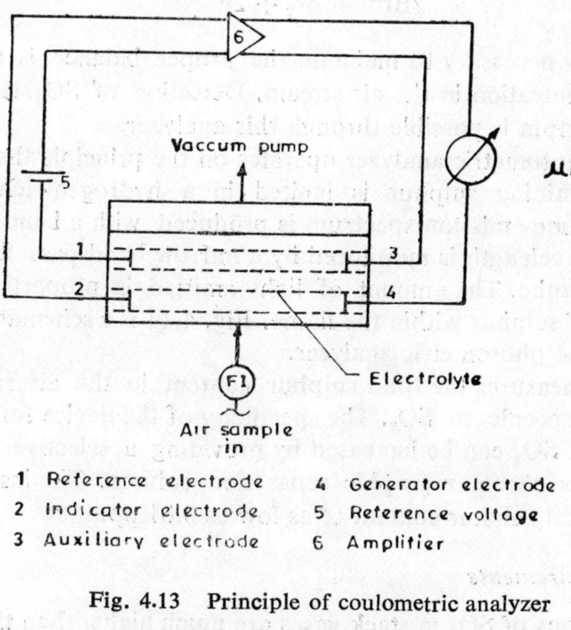

Fig. 4.13 Principle of coulometric analyzer

be detected. However, the conductometric analyzers suffer from a wide variety of interferences. Acidic gases such as hydrogen chloride give positive errors, while ammonia interferes negatively.

One of the first methods used for the automatic monitoring of SO_2, coulometry, is still one of the best methods. In the bromocoulometric analyzer, air containing SO_2 is drawn continuously through an electrolytic cell which contains acidified bromine solution and two sets of electrodes as shown in Fig. 4.13. The indicator reference set of electrodes is used to detect the bromine concentration while the other set comprising, the generating electrodes and the auxiliary electrode, is used to generate any bromine necessary to maintain the proper balance. SO_2 in the air sample is oxidised by bromine, causing a reduction in bromine concentration according to the reaction:

$$2H_2O + SO_2 + Br_2 \rightarrow SO_4^{2-} + 4H^+ + 2Br^- \quad (4.5)$$

This reduction in bromine concentration changes the oxidation-reduction potential of the reagent. A voltage is developed between the indicator electrode and the reference electrode, and this voltage is compared to a reference voltage. The difference between the two voltages is sensed by the other electrode system causing an electric current to flow between them and generating sufficient bromine to maintain the original concentration according to the reaction:

$$2Br^- \rightarrow Br_2 + 2e \quad (4.6)$$

This current flow necessary to maintain the proper balance is a measure of the SO_2 concentration in the air stream. Detection of SO_2 level in the range of 0.1–10 ppm is possible through this analyzer.

The flame photometric analyzer operates on the principle that when an air stream containing sulphur is ignited in a hydrogen-rich flame, a characteristic flame emission spectrum is produced with a band centred at 394 mμ. This wavelength is monitored by a narrow band-pass filter and a photomultiplier tube. The amount of light emitted is proportional to the concentration of sulphur within the flame. Fig. 4.14 is a schematic diagram of a typical flame photometric analyzer.

The device measures the total sulphur content in the air sample and as such is not specific to SO_2. The specificity of the device for the actual concentration of SO_2 can be increased by providing a selective filter or by the use of a gas chromatographic separation column. The instrument is sensitive to a total sulphur content of as low as 0.01 ppm.

Stack Gas Measurements

The concentrations of SO_2 in stack gases are much higher than those found in the atmospheric air. Some of the methods discussed for measuring SO_2 concentrations in ambient air are applicable to stack gas analysis also after diluting the sample. Even though the colourimetric methods are used for measuring SO_2 concentrations in the stack gases, they are not generally

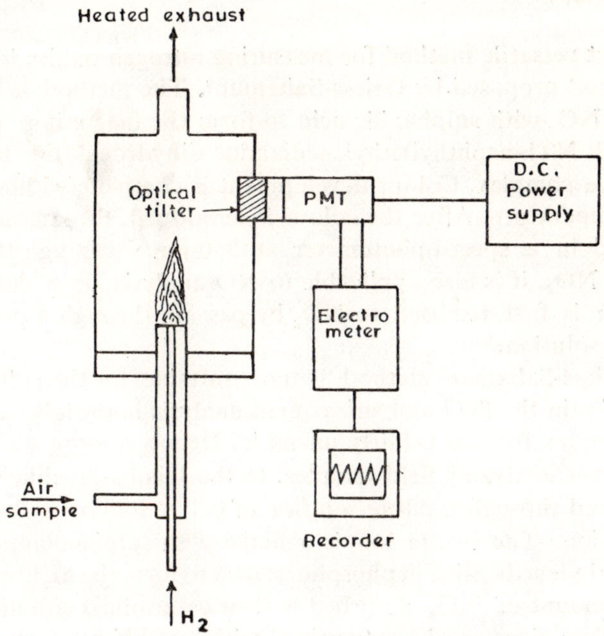

Fig. 4.14 Flame photometric analyzer.

recommended because of the interference by nitrogen oxides and other gases.

A technique which is relatively new and which is directly applicable to the analysis of SO_2 stack emissions is the electrochemical method. In the devices based on this method, SO_2 gas diffuses through a semipermeable membrane and a thin electrolyte layer to get absorbed at the sensing electrode where it undergoes an electrochemical reaction. The resulting current is proportional to the SO_2 concentration. Other preferred techniques for source monitoring are the infrared and the ultraviolet spectrophotometry. These techniques are based on the selective absorption of light at a given wavelength by SO_2. The degree of absorption is proportional to the SO_2 concentration. These spectrophotometric analyzers require efficient small-particle filtration, water removal, and sulphuric acid mist removal; achieving these without also removing some SO_2 is difficult. In addition, these instruments are bulky and difficult to transport from one location to another. Hence, they have proved to be least advantageous for field application. On the other hand, the electrochemical analyzers are reasonably specific to SO_2 and appear to have adequate sensitivity and stability (inspite of calibration drift) for both source and ambient SO_2 monitoring. Moreover, they offer the advantages of simplicity, portability, lower costs, and immediate results.

Nitrogen Oxides

The most versatile method for measuring nitrogen oxides is the colourimetric method proposed by Griess-Saltzman[9]. The method is based on the reaction of NO_2 with sulphanilic acid to form the diazonium salt. This is coupled with N-(1-naphthyl)ethylenediamine dihydrochloride to get a pink-coloured dye complex. Colour development is complete within 15 minutes at room temperature. After the colour is developed, the amount absorbed is measured in a spectrophotometer at 550 nm. Although the method is specific for NO_2, it is also applicable to NO analysis or a combined analysis if NO is first oxidised to NO_2 by passing through a potassium permanganate solution.

The Griess-Saltzman method is not suitable for the collection of 24-hour samples in the field and subsequent analysis in the laboratory because the dye complex formed is fairly unstable. Hence, a more stable technique is required for analyzing field samples. In the Jacobs-Hochheiser method[10], NO_2 is passed through a dilute solution of 0.1 *N* sodium hydroxide to form the nitrite ion. The nitrite ion is reacted with sulphanilamide and N-(1-naphthyl)ethylenediamine in phosphoric acid to form the highly coloured azo dye. The amount of NO_2 absorbed is then determined colourimetrically.

When the oxides can be determined without differentiation as NO, the method most commonly used is the phenoldisulphonoic acid method[2]. This method can be used for NO_x concentrations between 15 and 1500 ppm by volume and as such is suitable for stack gas emissions. The stack gas is drawn into a collection flask, evacuated previously to the vapour pressure of the absorbing solution. The absorbing solution consists of 3% hydrogen peroxide and 0.1 *N* sulphuric acid. The presence of hydrogen peroxide causes the oxides of nitrogen (except nitrous oxide) to get oxidized to nitric acid. Analysis is carried out by reacting the liquid with phenoldisulphonic acid, which is subsequently treated with ammonium hydroxide to obtain a yellow compound. The intensity of the colour is read using a spectrophotometer.

The method is specific and accurate for the oxides of nitrogen but the analytical procedure is complicated and time-consuming. The method can be used for ambient air sampling of industrial areas where the NO concentrations often reach 25 ppm or higher. The method is inefficient at lower levels as the absorption of the gas in the reagent solution is very slow and incomplete.

Several continuous analyzers are available for the measurement of NO and NO_2. These include colourimetric, nondispersive infrared and chemiluminescent type devices. All these instruments measure either nitric oxide or nitrogen dioxide concentration. These can be measured separately and their concentrations summed-up to get the total nitrogen oxide concentration, or one oxide can be converted to the other before the measurement is made. Out of the several analyzers available, the chemiluminescent type device is gaining wide popularity because of its low cost compared to that of the spectrophotometric devices and also because of its accuracy, sensitivity ($\sim 10^{-3}$ ppm NO), specificity, and applicability for a wide concentration range (0.004 ppm to more than 1000 ppm).

In the chemiluminescence technique[11], NO reacts rapidly with ozone to form NO_2 and oxygen. About 7% of the NO_2 produced by this reaction is in an excited state and reverts to the ground-state with the emission of radiant energy:

$$NO + O_3 \rightarrow NO_2^* + O_2 \tag{4.7}$$

$$NO_2^* \rightarrow NO_2 + h\nu \tag{4.8}$$

The emitted radiation is received by a photomultiplier tube whose output is amplified and fed to a recorder. The intensity of radiation is proportional to the amount of nitric oxide. The chemiluminescence reaction with ozone is specific for NO and if there is any NO_2 present in the sample gas, it has to be dissociated first into NO before the gas enters the reaction chamber. A thermal converter made of stainless steel coil is provided with the instru-

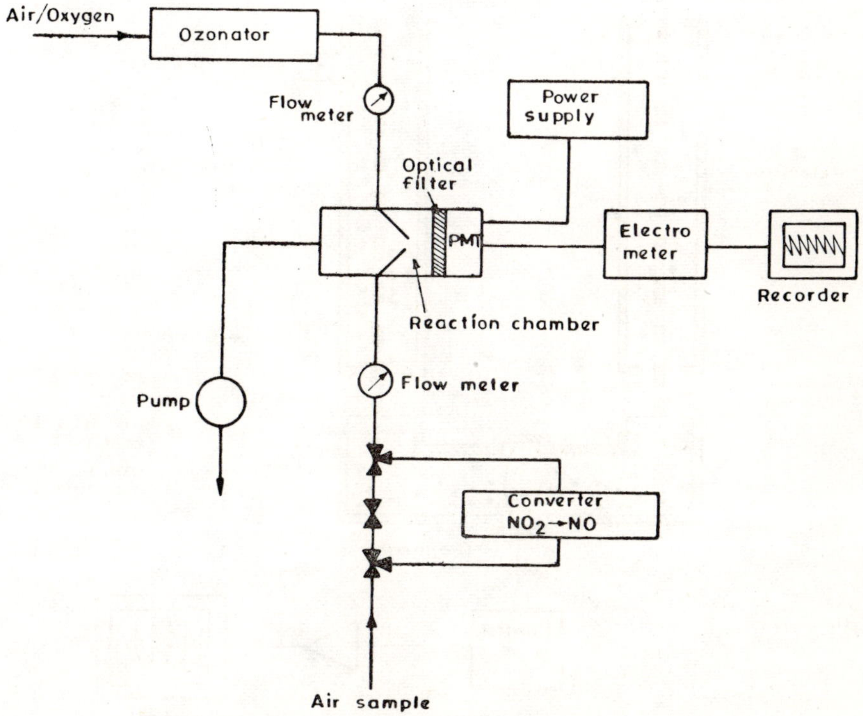

Fig. 4.15 Chemiluminescent analyzer.

ment and the NO_2 content of the sample can be measured by taking the difference in readings when the gas flows through the converter and when it is bypassed. Fig. 4.15 is a schematic diagram of a typical NO—NO_2 chemiluminescent analyzer.

Carbon Monoxide

The standard method for measuring atmospheric carbon monoxide levels is the nondispersive infrared spectrophotometry (NDIR). This method allows continuous analysis based on the capacity of carbon monoxide to absorb infrared radiation. A typical NDIR analyzer (Fig. 4.16) consists of

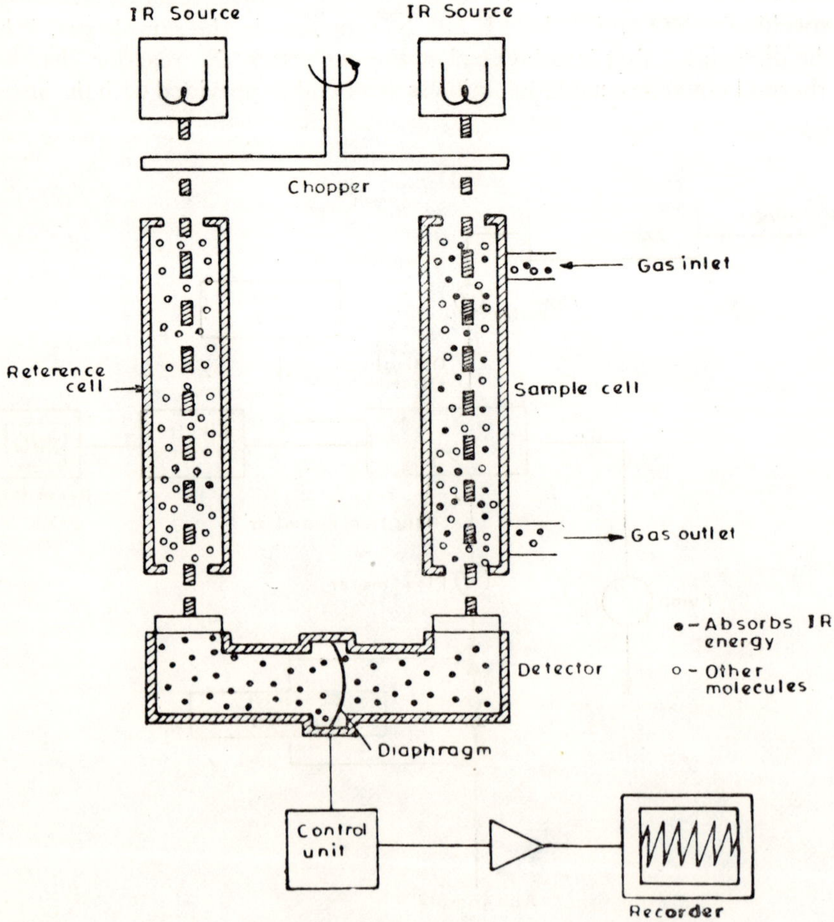

Fig. 4.16 Nondispersive infrared analyzer.

a sample cell and a reference cell, two infrared sources, and a detector cell. The reference cell is filled with a non-absorbing gas, such as nitrogen, and the sample cell is continuously flushed with the sample containing carbon monoxide which absorbs radiation at 4.6 μm. The detector consists of two compartments separated by a thin metal diaphragm and filled with carbon monoxide.

In operation, an optical chopper alternately exposes the reference and sample cells to the infrared sources. The reference cell passes almost all of the infrared energy on to one compartment of the detector cell, while a varying amount of infrared energy, inversely proportional to CO concentration in the sample cell, reaches the other detector compartment. Since more radiation enters the reference cell side of the detector compartment, the CO contained therein expands causing a pressure differential across the diaphragm which in turn causes the diaphragm to pulse back and forth in phase with the chopper action. The magnitude of diaphragm distention is a function of the absorbing CO concentration in the sample cell. The mechanical movement of the diaphragm is converted into an electrical signal which can be amplified and recorded.

Commercial instruments can detect CO concentrations from 1.0 ppm upto 25 to 100 ppm depending on path length, operating pressures and electronic amplification. Water vapour is the main cause of interference and it can be removed by providing a drying stage before measurements are made. NDIR analyzers have rapid response and good sensitivity over a wide range of concentrations but suffer from the problems of drift.

Another instrumental technique that has recently become available as a commercial analyzer is that involving gas chromatographic analysis. CO is separated from methane and converted to methane subsequently by hydrogenation. The methane thus formed is detected by a flame ionization unit. This technique is quite suitable for measuring ambient CO concentrations and is specific to CO. The limit of detection ranges from a minimum of 0.05 ppm to the highest urban concentration levels. The analyzer is relatively expensive and requires hydrogen for the conversion of CO into CH_4 and also for the flame in the flame ionization unit.

Other techniques for measuring CO levels include the electrochemical analysis, the so called ultraviolet mercury replacement method, and the catalytic analysis. In the electrochemical analysis[12] the gas sample is passed over hot iodine pentoxide at 150°C when iodine gas is liberated:

$$I_2O_5 + 5CO \rightarrow 5CO_2 + I_2 \tag{4.9}$$

The liberated iodine is absorbed by an electrolyte and is reduced at the cathode of an amperometric cell by the electrode reaction:

$$I_2 + 2e \rightarrow 2I^- \tag{4.10}$$

The current produced is a quantitative measure of the CO content of the sample. Minimum detectable concentration is 0.1 ppm with commercial instruments, but the response time of the analyzer is relatively slow.

The ultraviolet mercury replacement method is based on the reduction of mercuric oxide by CO at 210°C to produce CO_2 and mercury vapour.

$$CO + HgO(s) \xrightarrow{210°C} CO_2 + Hg(g) \qquad (4.11)$$

The mercury vapour formed is determined through its absorption of radiation from a mercury lamp. The range of applicability of this method is from 0.025 to 10 ppm of CO.

The catalytic analysis involves the oxidation of CO by a catalyst known as Hopcalite which contains a mixture of a metal oxides (CuO and MnO_2). The resultant temperature rise can be used to measure the carbon monoxide concentration.

The techniques used for stack gas analysis depend on the expected concentration and desired degree of accuracy. The common methods are NDIR absorption and a wet chemical technique known as the Orsat method. The Orsat method of analysis is useful for CO concentrations greater than 0.2 per cent by volume, or 2000 ppm. In the Orsat analyzer carbon monoxide is absorbed in a solution of cuprous chloride and the difference in volume between readings taken before and after the absorption step gives the amount of CO present in the sample.

Oxidants and Ozone

Oxidants are those substances that oxidize a reference material or solution under certain conditions. These include ozonides, peroxides of hydrocarbons and substances such as peroxyacyl nitrates. The standard method for the net determination of total oxidants is the neutral phosphate-buffered KI colourimetric technique of Saltzman[14]. In this method, a sample of air at a known flow rate is passed through a solution of 1% neutral buffered potassium iodide and the liberated iodine is measured colourimetrically at 352 nm.

$$2KI + O_3 + H_2O \rightarrow O_2 + 2KOH + I_2 \qquad (4.12)$$

Since one mole of ozone releases one mole of iodine, the I_2 concentration measured is directly related to the total oxidant concentration in the air sample. The method is applicable to oxidant concentrations in the range of 0.01 to 10 ppm (as ozone).

The standard KI method has been automated using a continuous flow colourimeter and 10–20% neutal buffered KI solution in order to minimize the effects of reagent depletion by the reaction. SO_2 interference can be eliminated by incorporating in the sampling train a glass-fibre paper filter impregnated with chromic acid. The chromic acid absorber will also oxidize

NO to NO_2 necessitating an additional correction; otherwise, NO does not cause any interference. Therefore, when SO_2 is less than 20% of the NO concentration, the use of chromic acid filter is not recommended.

The coulometric or Mast and Saunders analysis[15] is also widely used in continuous analyzers for measuring total oxidants. Here, the air sample is passed through a sensing cell containing a thin film of the KI solution. The liberation of iodine at the cathode of the cell causes an external current flow which is proportional to the iodine (or oxidant) concentration and is measured with a microammeter. Sensitivity of the method is less than 0.01 ppm, while its best working range lies between 0.01 and 0.20 ppm (ozone). Interferences include those from NO_2 and SO_2 and are roughly the same as those for colourimetric method.

One highly specific technique, which is currently being used for the automatic monitoring of ozone, is based on the chemiluminescence of the ozone-ethylene reaction[16]. In this method ozone is exposed to a large excess of ethylene in a reaction vessel. The chemiluminescent light released during the reaction is then measured using a photomultiplier tube, and related to the ozone concentration. There are no interferences from other atmospheric components and the device is capable of measuring concentrations less than 0.001 ppm. However, regular calibration is required to allow for changes in such system parameters as the flow rates of air and ethylene, and the supply voltage to the photomultiplier tube.

Hydrocarbons

Hydrocarbon pollutants can be determined as 'total hydrocarbons' or as individual chemical species. The most abundant hydrocarbon methane, is not a hazardous air pollutant and as such the total hydrocarbon content is generally reported on a non-methane basis. The total hydrocarbon content is most commonly determined by the flame ionization technique.

In the flame ionization detector (FID) the sample gas is injected into the flame created by burning hydrogen in either air or oxygen. When hydrogen alone burns, relatively few ions are generated but when hydrocarbons are present, the flame produces a complex ionization in which a large number of ions are present. An electric field is set up in the vicinity of the flame (Fig. 4.17) by making the burner jet positive with respect to a wire loop. The electric field induces ion migration in such a manner that a small ionization current is established between the electrodes, and this current is proportional to the concentrations of ions in the flame. The current is amplified and displayed on an output meter. Since the response of the detector is approximately proportional to the number of carbon atoms present in the hydrocarbon being consumed in the flame, the data are usually expressed in units of the gas used in calibration, e.g., "ppm of *C* as methane".

The FID is uniquely suitable for hydrocarbon analysis because it does not respond to other gases such as CO, CO_2, H_2O, SO_2 and nitrogen oxides. The recording range of these analyzers varies from approximately 2 to 3000 ppm hydrocarbons as methane. Ordinarily, the 0 to 20 ppm range is most useful for ambient-air analysis.

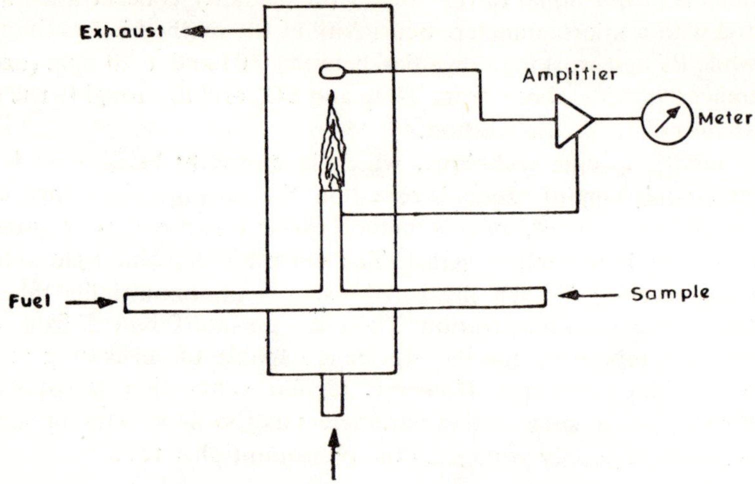

Fig. 4.17 Flame ionization detector.

The non-methane hydrocarbon analysis can be accomplished by a differential measurement. The sample is passed through a charcoal column which is saturated with methane but is still capable of retaining other hydrocarbons. This column can then be used with FID as an analyzer for the analysis of methane only. By differences between the results obtained, the non-methane hydrocarbons can be assessed.

The FID is sensitive only to the total quantity of hydrocarbons present and does not differentiate between various individual components. When the measurement of a specific component is of interest, gas chromatography may be employed to separate the components. Spectroscopic methods may also be used for individual hydrocarbons, but these devices are complex and expensive; hence, they are not competitive unless already in use for other reasons.

Particulate Matter: Stack

Methods of measurement of particulate emissions in flue gases generally fall into two categories: opacity measurements and mass measurements. Opacity measurements are necessary because the particulate matter emitted from a stack may produce a visible plume which may be itself objectionable and which can cut out sunlight and produce a haze. Plume visibility has been measured using the Ringelmann scale which consists of a black rectangular pattern superimposed on a white card. The lines are of definite width and spacing such that in different sections of the card the ratio of the black to white changes. This is indicated in Table 4.4. In practice, the chart is held at eye level, at a distance of several metres, such that the patterns merge to form a range of greys which can be compared with the plume greyness. The observer notes the number of the chart most nearly corresponding to the shade or opacity of the plume. The Ringelmann measurements are subjective, depending on the background effects, time of day, distance, and other factors.

Table 4.4 Spacing of lines on Ringelmann scale (ref. 17)

Ringelmann No.	Width of black lines [mm]	Width of white spaces (mm)	Percentage black
0	All white	—	0
1	1	9	20
2	2.3	7.7	40
3.	3.7	6.3	60
4	5.5	4.5	80
5	All black	—	100

A number of photometers have been developed to quantify the Ringelmann measurement. The simplest comprises a light source placed on one side of the stack, with a beam of light directed across to the opposite side where a detector photocell is mounted. Particulate matter in the flue gas obscures part of the light beam, and this effect is translated into a reading of per cent opacity to indicate the optical properties of the material in the stack. When the particulate concentration in the flue gas is zero, the photocell gives its maximum reading. This is indicated as zero. The conventional method of setting the zero is to wait for an off-load period when the stack is clear. However, in some continuous operations, recalibration is not possible. In such situations, the newer and more sophisticated instruments are used that incorporate some automatic calibration devices through which the 100% transmittance point (zero setting) can be checked as often as desired.

In-stack opacity measurement of solid particulate matter cannot be made when water droplets are present, such as those following a scrubber, because condensed water vapour prevents transmission of the light beam.

In such cases it is necessary to locate the detection cell outside the stack and draw flue gas through the unit. The sample stream must be heated above its dew point to prevent condensation so that the particulate matter present may be detected.

Many techniques are available for monitoring the mass concentration of particulate matter, including some continuous and semicontinuous ones; but the most common technique by far is that of intermittent sampling and weighing. One available instrument for continuous monitoring of mass concentration is the piezoelectric monitor. In this device, particles in the sample stream are electrostatically deposited onto an osciallating quartz crystal. The resonant frequency of the crystal decreases linearly with the added weight of the particulate matter and this shift is detected and used to provide a signal which is proportional to the mass present. Commercially available quartz crystals show a frequency shift of 2,000 Hz/μg of material deposited.

High particulate loadings such as those found in many stacks cannot be handled by piezoelectric monitors without a complicated sampling and dilution step[18]. Hence, their applicability to stack emissions is limited.

A promising method which could be used for monitoring of stack particulates is the β-attenuation technique, the principle of which was discussed earlier in Section 4.2. The main advantage of this technique is that the analysis depends only on the mass of the collected material and not on its chemical or physical properties.

REFERENCES

1. R.O. Blosser and H.B.H. Cooper (Ed.), "Analytical equipment and monitoring devices for gases and particulates", *Proc. of the west coast region air quality control workshop, atmospheric pollution technical bulletin No.* 35 (National Council for Air and Stream Improvement, New York), March 6, 1968.
2. N.L. Marrow, R.S. Brief and R.R. Bertrand. "Sampling and analysis of air pollution sources", *Chem. Engg.*, Vol. 79 (1972), pp. 84–98.
3. B.D. Bloomfield, in *Air pollution*, Vol. II, edited by A.C. Stern (Academic Press, New York), 1968, Chapter 28, pp. 487–536.
4. C.A. Gallaer, "How to measure dust in stacks and ducts", *Power,* Vol. 101 (1957), pp. 88-91.
5. *Methods for determination of velocity, volume, dust and mist content of gases,* WP 50 (Western Precipitation Corp., Los Angeles, Calif,), 1968.
6. W.C.L. Hemeon and G.F. Haines, *J. Air Pollut. Contr. Assoc,,* Vol. 4 (1954), p. 159.

7. R. Dennis, W.R. Samples, D.H. Anderson and L. Silverman, *Ind. Engg. Chem.*, Vol. 49 (1957), p. 294.
8. P.W. West and G.C. Gaeke, "Fixation of Sulphur dioxide as sulfitomercurate (III) and subsequent colourimetric determination", *Anal. Chem.*, Vol. 28 (1956), p. 1816.
9. M. Katz, Nitrogen compounds and oxidants", in *Air Pollution*, Vol. III, edited by A.C. Stern (Academic Press, New York), 3rd Edition, Chap. 7, 1976, pp. 259-305.
10. M.B. Jacobs and S. Hochheiser, "Continuous sampling and ultramicro determination of nitrogen dioxide in air", *Anal. Chem.*, Vol. 30 (1958), 426–428.
11. A. Fontijin, A. Sabadell and R.J. Ronco, "Homogeneous chemiluminescent measurement of nitric oxide with ozone", *Anal. Chem.*, Vol. 42 (1970), pp. 575-579.
12. D.J. Kroon, "Instrumental methods for automatic air monitoring systems", in *Air Pollution Control*, Part—III, edited by W. Strauss (John Wiley and Sons, New York). 1978, pp. 317–386.
13. R.C. Robbins, K.M. Borg, E. Robinson, "CO in atmosphere". *J, Air Poll. Control Assoc.*, Vol. 18 (1968), p. 106.
14. *Air quality criteria for photochemical oxidants,* Publ. No. AP–63, U.S. National Air Pollution Control Administration, Dept. of Health, Education and Welfare, 1970.
15. G.M. Mast and H.E. Saunders, "Research and development of instrumentation of ozone testing", *Instr. Soc. Am. Trans.*, Vol. 1 (1962), pp. 325–328.
16. G.I. Warren and G. Bobcock, *Rev. Sci. Instr.*, Vol. 41 (1970), p. 280.
17. L.R. Burdick, *Ringelmann Smoke Charts* (revised), U.S. Bureau of Mines, Information Circular, 7718, 1955.
18. G.J. Sem, J.A. Borgos and J.G. Olin, "Automatic monitors of particulate mass emissions from stationary fossil-fuel combustion sources" paper presented at the AIChE 68th National Meeting, Houston, Texas, March 3, 1971.

SUGGESTIONS FOR FURTHER READING

1. P.O. Warner, *Analysis of air pollutants* (John Wiley & Sons, New York), 1976.
2. H.B.H. Cooper and A.T. Rossano, *Source testing for air pollution control* (McGraw-Hill Book Co., New York), 1974.
3. J.D. Butler, *Air pollution chemistry*, Chapters 3 and 4 (Academic Press, London), 1979.

4. T.H. Maugh, "Air pollution instrumentation: a trend toward physical methods", *Science*, Vol. 177 (1972), pp. 685–687.
5. M.B. Jacobs, *The chemical analysis of air pollutants* (Interscience, New York), 1960.
6. B.G. Liptak (Ed.), "Air pollution control instrumentation and air quality detectors", in *Air pollution*, Vol. 2, *Environmental engineers handbook* (Chilton Book Co., Pennsylvania), 1974.
7. W. Liethe, "*The analysis of air pollutants* (Ann Arbor Science Publishers, Inc., Ann Arbor, Michigan), 1970.

5. Air Pollution Control Methods and Equipment

5.1 CONTROL METHODS

The most effective means of dealing with the problem of air pollution is to prevent the formation of the pollutants or minimize their emission at the source itself. In the case of industrial pollutants, this can often be achieved by investigating various approaches at an early stage of process design and development, and selecting those methods which do not contribute to air pollution or have the minimum air pollution potential. These are known as source correction methods. Application of these methods to existing plants is difficult, but still some of these correction methods could be applied without severely upsetting the economy of the operation. Control of the pollutant at the source can be accomplished in several ways through raw material changes, operational changes, modification or replacement of process equipment, and by more effective operation of existing equipment.

When source correction methods cannot achieve the desired goal of air pollution control, use is made of effluent gas cleaning techniques. These involve many of the chemical engineering unit operations and at present form the main part of pollution control technology.

5.2 SOURCE CORRECTION METHODS

Raw Material Changes

If a particular raw matarial is responsible for causing air pollution, use of a purer grade of raw material is often beneficial and may reduce the formation of undesirable impurities and byproducts or may even eliminate the troublesome effluent. A typical example of this approach is the use of low-sulphur fuel in place of high-sulphur ones. Burning of natural gas produces less pollution than that of coal, but a major drawback of such low-sulphur fuels is their limited availability for wholesale use.

Fuel desulphurization is an attractive alternative, but removal of sulphur from fuels such as coal poses formidable technical problems. The most promising way of using coal in combustion processes with minimum air pollution appears to be through coal gasification because sulphur and some other unwanted materials can be removed from the gas much more readily than from solid coal. Coal gasification can be carried out in two possible steps:

1. Gas can be produced by destructive distillation of coal. The greater part of the volatile matter is driven off and coke is left as a by-product.
2. Gasification of the coke residues of the carbonization process can be done with steam, CO_2, O_2, air, or a mixture of these gases. The main products are H_2, CO, CO_2, and undecomposed steam. Hydrogen can also be used as a gasifying medium. In the case of hydrogasification methane is produced; however, high pressures are required to obtain considerable yields.

In all these processes, the sulphur is recovered by passing the gases through an absorbing medium. Surveys of some industrial desulphurization processes are given by Strauss[1].

Ore handling operations usually result in the emission of large quantities of dust into the atmosphere. In steel industry, replacement of raw ore with pelleted sintered ore has gradually reduced dust emissions and also helped to reduce the blast furnace "slips."

Process Changes

Process changes involving new or modified techniques offer important ways of lowering atmospheric pollutant emissions. Radical changes in chemical and petroleum refining industries have resulted in minimizing of the release of materials to the atmosphere. The volatile substances are recovered by condensation and the non-condensable gases are recycled for additional reactions. Hydrogen sulphide, which was once flared in refineries, is now recycled and used in Claus process to recover elemental sulphur.

Rotary kilns are a major source of dust generation in cement plants. Some degree of dust control may be achieved in the kiln operation by adjusting operating conditions. Reduction of the gas velocities within the kiln, modification of the rate and location of feed introduction and employment of a dense curtain of light-weight chain at the discharge end of the kiln can lead to dust control.

Smelting and paper industries are notorious for emitting highly objectionable sulphurous materials. These emissions are being curtailed by major process changes such as hydro-metallurgical separations of ores and use of no sulphides in paper making.

In the steel industry, a radically different process has been proposed to lower sulphurous emissions during combustion. In this process the sulphur bearing fuel, limestone and air are injected into a molten iron bath. The combustibles in the fuel are partially oxidized to carbon monoxide within the molten iron bath; the gaseous CO comes off at the top of the molten iron and is burnt efficiently in a conventional manner. The sulphur is retained in the iron bath and forms a slag with the limestone, which is removed.

Other examples involving process changes include: (1) reduction of the formation of nitric oxides in combustion chambers by low excess air combustion in two stages, flue gas recirculation and water injection, (2) washing the coal before pulverization to reduce the fly ash emissions, and (3) substitution of bauxite flux for fluorine-containing fluorspar in the open hearth method.

Equipment Modification or Replacement

Air pollutant emissions can be minimized by suitable modification or replacement of process equipment. For example, the unburnt carbon monoxide and hydrocarbons in the cylinders of an automobile engine, which are otherwise emitted into the atmosphere through the tail pipe, can be burnt by injecting air into the hot exhaust manifold of the engine. Similar results can be obtained by suitable modifications in the carburation and ignition systems.

In petroleum refineries, hydrocarbon vapours are released into the atmosphere from storage tanks due to temperature changes, direct evaporation, and displacement during filling. These losses can be minimized by designing the tanks with floating roof covers or by pressurising the tanks.

Replacement of the open hearth furnace by oxygen furnace in steel industry and development of an alternative power source for automobiles in place of the internal combustion engine are additional examples of equipment alteration.

In addition to the above mentioned source correction methods, air pollutant emissions from industrial operations can be reduced by proper equipment maintenance, housekeeping and cleanliness in the facilities and premises. Often changes in the design of local exhaust hood and proper installation can minimize the emission of pollutants to the atmosphere. Chemical process plants often have excessive leakage around ducts, piping, valves and pumps. Many such leaks can be prevented by checking the seals and gaskets routinely. Floors, decks, storage bins and silos, loading areas, and material transfer conveyors must be kept clean to reduce dust pollution.

5.3 CLEANING OF GASEOUS EFFLUENTS

The technology for the removal of gaseous pollutant emissions after their formation has probably received the maximum attention. The cleaning techniques are applied to those cases where emissions of pollutants cannot be prevented and pollution control equipment is necessary to remove them from the main gas stream. Normally, it is more economical to install the control equipment at the source where the pollutants are present in the smallest possible volume in relatively high concentrations rather than at some point away from the source where the pollutants are diluted by other process gases or air. The size of the equipment is directly related to the volume being treated, and equipment costs can be drastically reduced by decreasing the exhaust volume. At the same time, the equipment is more efficient for the handling of higher concentrations of pollutants. Gas cleaning techniques used alone, or in conjunction with source correction methods form the basis of present-day air resource management concepts. The methods are often integrated into chemical processes, which eliminate pollutant discharges and conserve materials in the same unit operation.

Emission control equipment may be classified into two general types: particulate control type, and gases and odours control type. For gaseous pollutants, essentially two alternative classes or methods are available. The first class comprises the wet absorption and dry adsorption methods, and the second class includes methods depending on the chemical alteration of the pollutant, usually through combustion or catalytic incineration. The basic mechanism in all these is the diffusion of the particular gas either to the surface of an absorbing liquid or adsorbing solid or catalyst, or to the reaction zone of a chemical reaction. On the other hand, the removal of particles relies only partly on their diffusion from the gas stream to a collecting surface or space, and other mechanisms which take special advantage of the nature and properties of particulate matter play a more important role.

5.4 PARTICULATE EMISSION CONTROL

Particulate matter emitted in gaseous streams consists of discrete and minute suspended particles. Particle sizes range generally from 100 μm down to 0.1 μm and even less. The choice of collection devices depends upon a number of factors: the physical and chemical characteristics of particulates, the particulate size and concentration in the gas, volume of particulates to be handled, and the temperature and humidity of gaseous medium. In particular, factors like toxicity and inflammability must be taken into consideration when evaluating operating efficiency.

Collection Efficiency

The operating or collection efficiency measures the system performance. In existing systems, it represents the actual performance. The most common way of expressing the efficiency is in terms of the weight of the material caught in or retained by the collector compared to the total amount entering the collector. This is known as the gravimetric efficiency and is usually expressed as a percentage.

$$\eta_T = 100 \times \frac{\text{Weight of material collected}}{\text{Total amount entering collector}} \tag{5.1}$$

In particulate collection systems, the efficiency of collection varies with particle size. This variation of efficiency is often expressed in the form of fractional efficiency. It is the efficiency with which particles of a specified size-range are collected under specific conditions of gas flow rate, temperature, and particle composition. From a knowledge of the particle size distribution, a fractional efficiency curve can be constructed by considering the efficiency in a series of narrow ranges of the

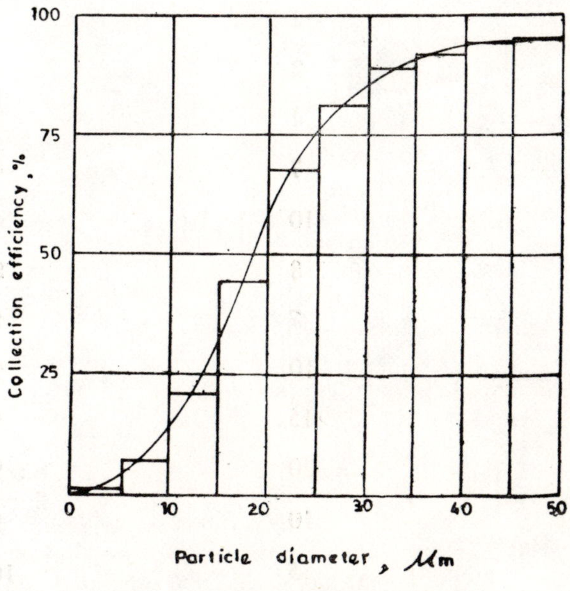

Fig. 5.1 Fractional efficiency curve.

particle size spectrum. A typical fractional efficiency curve is shown in Fig. 5.1. From this information, the efficiency in each size range can be computed by the equation,

$$\eta_i = \frac{\text{fraction in range ``}i\text{'' collected}}{m_i} \times 100 \tag{5.2}$$

where m_i = amount in range "*i*" entering the collector.

The overall efficiency η_T can then be calculated over n number of size fractions as:

$$\eta_T = \frac{\sum_{i=1}^{n} m_i \eta_i}{M} \text{ per cent} \tag{5.3}$$

where M = total amount entering the collector.

Example 1

The following table shows the size distribution of a dust sample and the fractional efficiency of removal in a gas cleaning equipment. Calculate the overall collector efficiency.

Dust size	Weight per 100 g of dust (g)	Fractional efficiency η_i (%)
< 5	2	1
5 – 10	2	7
10 – 15	4	16
15 – 20	7	44
20 – 25	10	67
25 – 30	8	81
30 – 35	7	88
35 – 40	10	92
40 – 50	15	93
50 – 60	20	95
60 – 70	10	98
> 70	5	100

Solution

Using Eq. (5.3), we can calculate the overall collection efficiency η_T as

$$\eta_T = \frac{\sum_{i=1}^{n} m_i \eta_i}{M} = \sum_{i=1}^{n} \omega_i \eta_i$$

where ω_i is the weight fraction in each size range

$$\eta_T = \sum_{i=1}^{n} \omega_i \eta_i = (0.02 \times 1) + (0.02 \times 7) + (0.04 \times 16)$$
$$+ (0.27 \times 44) + (0.1 \times 67) + (0.08 \times 81)$$
$$+ (0.07 \times 88) + (0.1 \times 92) + (0.15 \times 93)$$
$$+ (0.2 \times 95) + (0.1 \times 98) + (0.05 \times 100)$$
$$\eta_T = 80.17\%$$

Particulate Control Equipment

The basic mechanisms of removing particulate matter from gas streams may be classified as: (1) gravitational settling, (2) centrifugal impaction, (3) inertial impaction, (4) direct interception, (5) diffusion, and (6) electrostatic precipitation.

Equipment presently available, which make use of one or more of the above mechanisms, fall into the following five broad categories: (1) gravitational settling chambers, (2) cyclone separators, (3) fabric filters, (4) electrostatic precipitators, and (5) wet collectors (scrubbers).

The gravitational settling ehambers and cyclone separators will not generally achieve high efficiencies for removing small-size particles. For most practical applications, only fabric filters, electrostatic precipitators and high energy scrubbers are capable of meeting the rigorous air pollution control regulations.

Gravitational Settling Chambers

Gravitational settling chambers are generally used to remove large, abrasive particles (usually > 50 μm) from gas streams. They offer low pressure

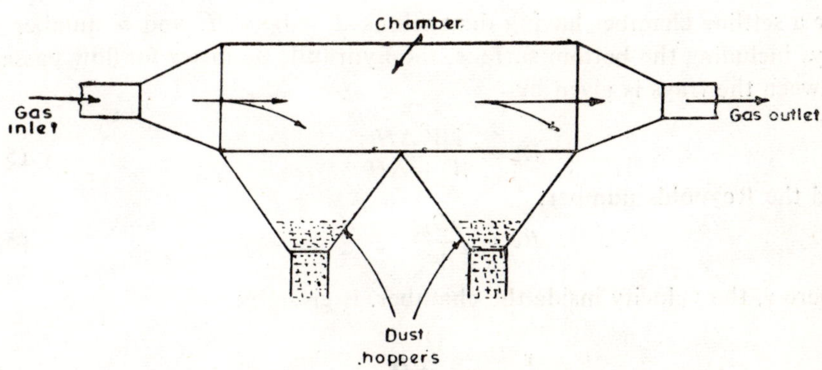

Fig. 5.2 Horizontal flow settling chamber

drop and require simple maintenance, but their efficiencies are quite low for particles smaller than 50 μm. Since most of the troublesome particles have much smaller sizes than 50 μm, these devices are normally used as precleaners prior to passing the gas stream through high efficiency collection devices. The simplest form of horizontal type settling chamber is shown in Fig. 5.2.

In the settling chamber the gas stream, with its entrained particles, is allowed to flow at a low velocity, allowing sufficient time for the particles to settle down. A more elaborate settling chamber is the Howard type whose simplified diagram is shown in Fig. 5.3. By inserting several trays, the collection efficiency of the device is improved since the gas flow velocity in the chamber remains substantially the same and yet each particle has a much shorter distance to fall before reaching the bottom of the passage between trays.

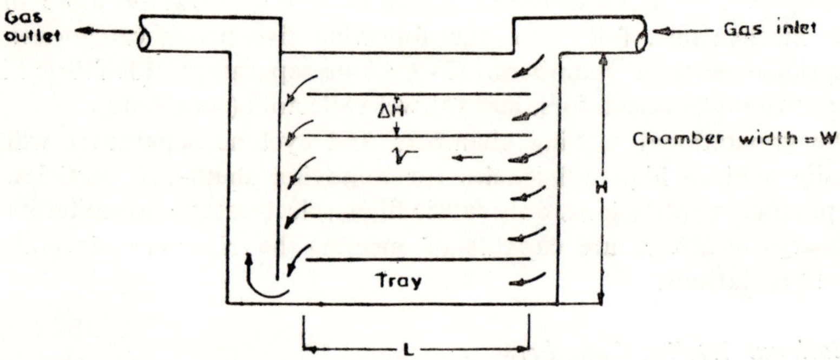

Fig. 5.3 Howard settling chamber

Design Concepts

For a settling chamber having dimensions $L \times W \times H$, and n number of trays including the bottom surface, the hydraulic diameter for flow passage between the trays is given by

$$D_h = \frac{2W\,\Delta H}{W + \Delta H} \tag{5.4}$$

and the Reynolds number,

$$Re = \frac{vD_h\rho}{\mu} \tag{5.5}$$

where v, the velocity inside the chamber, is given by

$$v = \frac{Q}{nW\,\Delta H}$$

where Q = volumetric flow rate of the gas stream.

Substituting for v and D_h in the Eq. (5.5), we get

$$Re = \frac{2Q\rho}{n\,\mu(W + \Delta H)} \tag{5.6}$$

The spacing between the trays, ΔH, is given by

$$\Delta H = \frac{H}{n}$$

provided there is no dust layer initially present on the tray surface. Substituting for ΔH in Eq. (5.6) we have

$$Re = \frac{2Q\rho}{\mu(nW + H)} \tag{5.7}$$

If a layer of dust H_d is initially present, then

$$\Delta H = \frac{H}{n} - H_d.$$

With this modification,

$$Re = \frac{2Q\rho}{\mu(nW + H - nH_d)} \tag{5.8}$$

For laminar flow conditions ($Re < 2300$) within the trays, particles of size d_p of a particular material will settle a distance y with a terminal velocity v_t in time t. During this time the particles are transported a distance L with the velocity of the gas stream. Equating these two, we have:

$$\frac{y}{v_t} = \frac{L}{v}, \tag{5.9}$$

the value of y can be found from a knowledge of the particle settling velocity.

If the particles are uniformly distributed over the incoming stream, the efficiency of collection $\eta = y/\Delta H$ or

$$\eta = \frac{Lv_t}{v\Delta H} = \frac{nWLv_t}{Q} \tag{5.10}$$

When $y \geqslant \Delta H$, all particles of that size (or larger) will be collected in the settling chamber.

The terminal settling velocity of the particles is found by equating the drag force on the particle with the weight minus the buoyancy of the particle :

$$F = C_D\left(\frac{\pi}{4}d_p^2\right)\left(\frac{1}{2}\rho_g v_t^2\right) = \frac{\pi}{6}d_p^3(\rho_p - \rho_g)\,g \tag{5.11}$$

where ρ_p = density of the particle, ρ_g = density of gas, and d_p = particle diameter. Rearranging, we get

$$v_t = \sqrt{\frac{4gd_p(\rho_p - \rho_g)}{3C_D\rho_g}} \tag{5.12}$$

where C_D is the drag coefficient which is related to the particle Reynolds number, $Re_p = d_p\rho_g v_t/\mu_g$. The general drag coefficient curve for spherical particles may be represented by three relationships. In the Stokes' law region,

$$C_D = \frac{24}{Re_p} \quad \text{for} \quad Re_p < 1 \tag{5.13}$$

In the turbulent region ($Re_p > 1000$), the drag coefficient C_D becomes almost constant with a value of 0.45.

$$C_D = 0.45 \quad \text{for} \quad Re_p > 1000 \tag{5.14}$$

Between this region and the Stokes' law region of laminar flow around the particle, there is the transition region. Several empirical equations have been suggested for this region, such as that given by Schiller and Naumann[2]:

$$C_D = \frac{24}{Re_p}(1 + 0.15\, Re_p^{0.687}) \tag{5.15}$$

Substituting Eq. (5.13) into Eq. (5.12), we can calculate the terminal settling velocity in the Stokes' region.

$$v_t = \frac{gd_p^2(\rho_p - \rho_g)}{18\,\mu_g} \tag{5.16}$$

In the transition region, where C_D is a function of Re_p, Eq. (5.12) is difficult to solve except by a trial-and-error method. This problem has been overcome by rearranging Eq. (5.12) as

$$C_D\, Re_p^2 = \frac{4gd_p^3(\rho_p - \rho_g)\rho_g}{3\mu_g^2} \tag{5.17}$$

Values of $C_D\, Re_p^2$, C_D, and Re_p are available in the *Chemical Engineering Handbook*[3] from which v_t can be calculated.

For particulates most often encounted in air pollution control, Stokes' law is a reasonable approximation for calculating the settling velocity. Table 5.1 shows typical values of terminal settling velocity calculated using Stokes' law, together with actual measured values, from which it is apparent that Stokes' law is accurate for particles smaller than 100 μm. The minimum particle size that can be removed with 100% efficiency can be found from the equation,

$$d_{p,\,\min} = \sqrt{\frac{8\,Q}{nWLg(\rho_p - \rho_g)}} \tag{5.18}$$

Table 5.1 Settling velocities of spherical particles in air [Particle density: 1000 kg/m^3, air temperature: 20°C. pressure: 760 mm Hg]

Particle size (μm)	Exp. (m/s)	Calc. from Eq. (5.12) (m/s)
0.1	8.7×10^{-7}	8.71×10^{-7}
0.2	2.3×10^{-6}	2.27×10^{-6}
0.4	6.8×10^{-6}	6.85×10^{-6}
1.0	3.5×10^{-5}	3.49×10^{-5}
2	1.19×10^{-4}	1.19×10^{-4}
4	5.0×10^{-4}	5.0×10^{-4}
10	3.06×10^{-3}	3.06×10^{-3}
20	1.2×10^{-2}	1.2×10^{-2}
40	4.8×10^{-2}	5.0×10^{-2}
100	0.246	0.25
400	1.57	4.83
1000	3.82	30.50

These equations should be used only as guides to the collection efficiency of the settling chamber, as several factors cause deviations. These include hindered settling at high particle concentrations, non-uniform gas velocity over the settling height and width, particle re-entrainment, and turbulence. As a general rule, chamber velocities below 3 m/s are satisfactory for avoiding re-entrainment of most materials.

Although the efficiency relationship is based on laminar flow conditions within the unit, it is practically impossible to achieve laminar conditions without a very large particle size or an inordinately large number of trays combined with an awkward shape of the chamber. Hence, the flow in the settling chamber will probably be turbulent rather than laminar. For turbulent flow, the following theoretical equation can be used for calculating the collection efficiency[4]

$$\eta = 1 - \exp\left(-\frac{nWLv_t}{Q}\right) \tag{5.19}$$

Example 2

A multi-tray settling chamber having 8 trays, including the bottom surface, handles 6 m^3/s of air at 20°C. The trays are spaced 0.25 m apart and the chamber is to be 1 m wide and 4 m long. What is the minimum particle size of density 2000 kg/m³ that can be collected with 100% efficiency? What will be the efficiency of the settling chamber if 50 μm particles are to be removed? Laminar flow condition withhin the chamber and presence of no dust initially on trays may be assumed.

Solution

From Eq. 5.18

$$d_{p,\min} = \sqrt{\frac{18\,\mu_g Q}{nWLg\,(\rho_p - \rho_g)}}$$

μ_g at 20°C $= 1.81\times10^{-5}$ kg/m-s and $\rho_p = 2000$ kg/m³. Since $\rho_p \gg \rho_g$, ρ_g may be neglected in the above equation. Substituting for $n = 8$, $W = 1$ m, $L = 4$ m, $g = 9.81$ m/s², and $Q = 6$ m³/s, we have

$$d_{p,\min} = \sqrt{\frac{18\,(1.81\times10^{-5})\,6}{8\,(1)\,(4)\,(9.81)\,(2000)}}$$

$$d_{p,\min} = 56\ \mu\text{m}$$

$$\eta_{dp} = 50 = \left(\frac{dp}{d_{p,\min}}\right)^2$$

$$= (50/56)^2 = 80\%$$

Example 3

In the above example, is the laminar flow assumption justified ? If not, what is the collection efficieney for 56 and 50 μm particles?

Solution

The Reynolds number may be calculated as,

$$Re = \frac{2Q\rho}{\mu(nW + H)}$$

The kinematic viscosity of air at 20°C, is 1.51×10^{-5} m²/s and $H = 2$ m.

$$Re = \frac{2\times6}{1.51\times10^{-5}\,(8\times1+2)} = 79{,}470$$

We see that the flow is turbulent and the laminar flow assumption in example 2 was not justified. Hence, using Eq. (5.19), we get

$$\eta = 1 - \exp\left(-\frac{nWL\,v_t}{Q}\right)$$

$$\eta_{dp=56} = 1 - \exp\left(-\frac{8\times1\times4\times0.188}{6}\right)$$

$$= 1 - 0.37 = 0.63 \text{ or } 63\%$$

$$\eta_{dp-50} = 1 - \exp\left(-\frac{8 \times 1 \times 4 \times 0.15}{6}\right)$$

$$= 1 - 0.45 = 0.55 \text{ or } 55\%$$

The terminal settling velocity v_t was calculated using the simplified relation for particles settling in air at 20°C;

$$v_t = 30{,}000\ \rho_p\ d_p^2$$

where d_p is in metres.

$$v_t\ (56\ \mu\text{m}) = 0.118\ \text{m/s and}$$

$$v_t\ (50\ \mu\text{m}) = 0.15\ \text{m/s}$$

Cyclone Separators

Cyclone separators utilize a centrifugal force generated by a spinning gas stream to separate the particulate matter from the carrier gas. The centrifugal force on particles in a spinning gas stream is much greater than gravity; therefore, cyclones are effective in the removal of much smaller particles than gravitational settling chambers, and require much less space to handle the same gas volumes.

There is a variety of cyclone designs, differing in the manner in which the rotating motion is imparted to the gas stream. The simple reverse-flow type depicted in Fig. 5.4 consists of a vertical cylinder having a conical bottom and is fitted with a tangential inlet located near the top, and an outlet at the bottom of the cone for discharging separated particles. The gas outlet pipe is extended into the cylinder to prevent short-circuiting of gas from inlet to outlet.

In operation, the particle-laden gas upon entering the cyclone cylinder receives a rotating motion. The vortex so formed develops a centrifugal force, which acts to throw the particles radially towards the wall. The gas spirals downward to the bottom of the cone, and at the bottom the gas flow reverses to form an inner vortex which leaves through the outlet pipe.

THEORY

In a cyclone, the inertial separating force is the radial component of the simple centrifugal force and is a function of the tangential velocity. The centrifugal force can be expressed by F_c where

$$F_c = \frac{mv_\theta^2}{r} \tag{5.20}$$

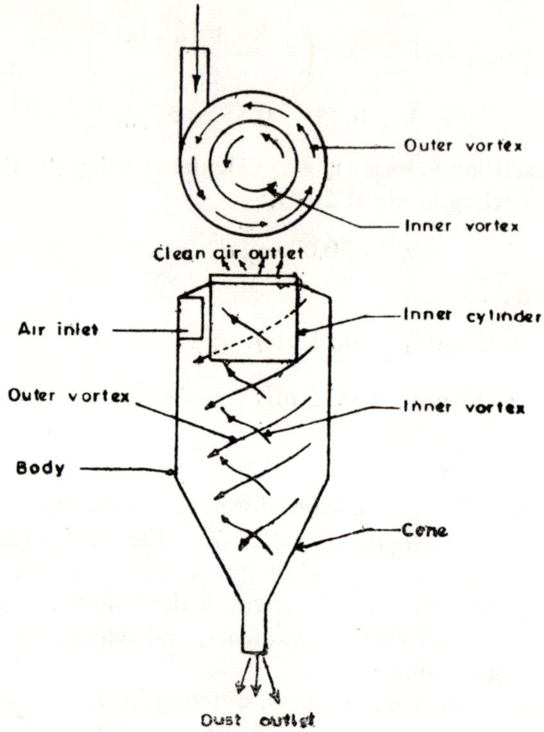

Fig. 5.4 Reverse-flow cyclone separator

where m = mass of the particle, v_θ = tangential velocity of the particle (and gas) at radius r, and r = radius of rotation.

The magnitude of the centrifugal force is frequently described in terms of the number of times this force exceeds the force of a gravity. The separation factor S is given by

$$S = \frac{v_\theta^2}{gr} \tag{5.21}$$

The separation factor varies from 5 in large, low velocity units to 2500 in small, high pressure units. Obviously, the higher the separation factor, the better is the performance of the cyclone.

In a cyclone, the gas, in addition to moving in a circular path, also moves radially inwards between the inlet on the periphery and the exit on the axis. Since the tangential velocities of the particle and the gas are the same, the relative velocity between the gas and particle is simply equal to the radial velocity of the gas. This results in a drag force on the particle towards the centre, and the equilibrium radius of rotation of the particle can be obtained by balancing the radial drag force and the centrifugal force:

$$3\pi\mu_g d_p v_r = \frac{\pi}{6} d_p^3 (\rho_p - \rho_g) \frac{v_\theta^2}{r} \tag{5.22}$$

where d_p=particle diameter, and v_r=radial velocity of the gas at radius r. Arranging the above equation for v_r, we have

$$v_r = \frac{d_p^2 (\rho_p - \rho_g)}{18\mu_g} \frac{v_\theta^2}{r} \tag{5.23}$$

The tangential velocity of the particle in the vortex has been found experimentally to be inversely proportional to the radius of rotation according to equation,

$$v_\theta r^n = \text{constant} \tag{5.24}$$

where n the exponent is dimensionless.
For an ideal gas $n = 1$. The real values observed are between 0.5 and 1, depending upon the radius of the cyclone body and gas temperature. v_θ can be related to the tangential velocity at the inlet to the cyclone, $v_{\theta i}$ as

$$v_\theta = v_{\theta i} = \left(\frac{D}{2r}\right)^n \tag{5.25}$$

where D = diameter of the cyclone.
$v_{\theta i}$ may be taken as the velocity of the gas through the inlet pipe, i.e.,

$$v_{\theta i} = \frac{Q}{A_i}$$

where Q = gas volumetric flow rate and A_i = cross-sectional area of the inlet.
Substituting in Eq. (5.25), we have

$$v_\theta = \frac{Q}{A_i}\left(\frac{D}{2r}\right)^n \tag{5.26}$$

Now from Eq. (5.23),

$$v_r = \frac{d_p^2 (\rho_p - \rho_g)}{18\mu_g} \left(\frac{Q}{A_i}\right)^2 \left(\frac{D}{2}\right)^{2n} \frac{1}{r^{(2n+1)}} \tag{5.27}$$

Generally, the collection efficiency is an increasing function of v_r. The relationship cannot be defined explicitly, principally because of factors such as re-entrainment, bounce, and particle interactions. However, it is clear from Eq. (5.27) that as the diameter of the particle increases or as the cross-sectional area of the inlet decreases, v_r will increase appreciably, resulting in an increase in efficiency but with a higher pressure drop.

The most satisfactory expression for cyclone performance is still the empirical one. Lapple[5] correlated collection efficiency in terms of the cut size d_{pc}, which is the size of those particles that are collected with 50% efficiency. Particles larger than d_{pc} will have a collection efficiency greater

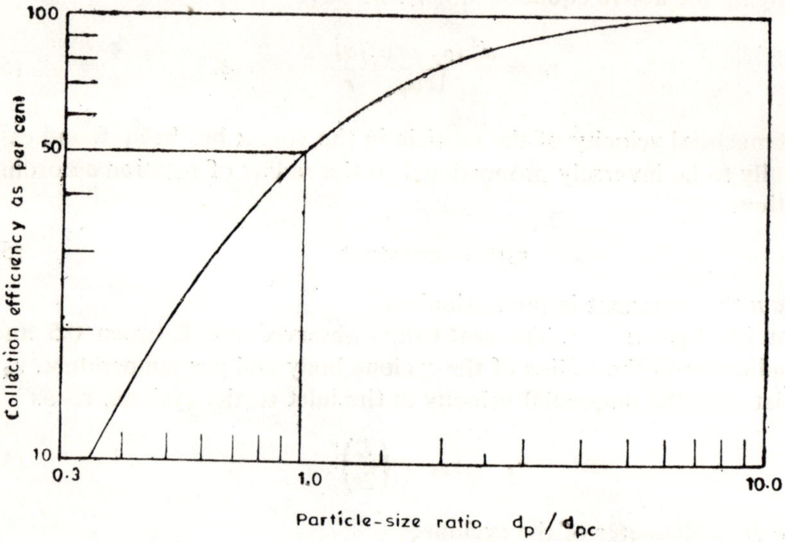

Fig. 5.5 Lapple's correlation for cyclone efficiency

than 50% while the smaller particles will be collected with lesser efficiency. The cut size is given by

$$d_{pc} = \sqrt{\frac{9\mu_g b}{2\pi N_e v_i (\rho_p - \rho_g)}} \tag{5.28}$$

where b = inlet width, v_i = gas inlet velocity and N_e = effective number of turns a gas makes in traversing the cyclone (5 to 10 in most cases). By plotting the ratio of the actual particle size (d_p) to the cut size (d_{pc}), a generalised curve of collection efficiency may be obtained. Such a curve is given in Fig. 5.5.

Fig. 5.6 shows a typical cyclone with all the necessary dimensions and gives values of the relative dimensions for conventional and high-efficiency designs. Representative overall efficiencies are presented for conventional as well as for high efficiency type cyclones in Table 5.2

Table 5.2 Efficiencies of Cyclones

Particle size (μm)	Efficiency of Cyclone	
	Conventional	High efficiency
< 5	< 50	50—80
5—20	50—80	80—95
15—40	80—95	95—99
> 40	95—99	95—99

Pressure Drop

A correct estimate of the pressure drop across a cyclone is necessary, in addition to collection efficiency, so that cost effectiveness may be calculated. A magnitude of 25 cm H_2O, or less, is the generally acceptable operating range. The pressure drop may be calculated from several formulas. The approach of Shepherd and Lapple[8], however, is the simplest and is of comparable accuracy to the more complex ones. The pressure drop ΔP may be estimated according to the following equation,

$$\Delta P = \frac{K \rho_g v_i^2 (ab)}{2D_e^2} \tag{5.29}$$

where K = a constant, which averages 13 and ranges from 7.5 to 18.4, ΔP = pressure drop, a, b, and D_e = cyclone dimensions, v_i = inlet gas velocity, and ρ_g = gas density.

Since the pressure drop is proportional to the square of the inlet velocity, it is obvious that high velocities cause not only re-entrainment but also excessive pressure drop.

Other Cyclone Types

Many types of commercial cyclones have been developed and information on their performance for specific applications is generally available. In straight through flow cyclone, the inner vortex of the air leaves at the bottom (rather than reversing direction), with initial centrifugal motion being imparted by vanes at the top. The chief advantages of this unit are low pressure drop and high volumetric flow rates. In the impeller collector, gases enter normal to an impeller and are swept out by the impeller around its circumference while the particles are thrown into an annular slot around

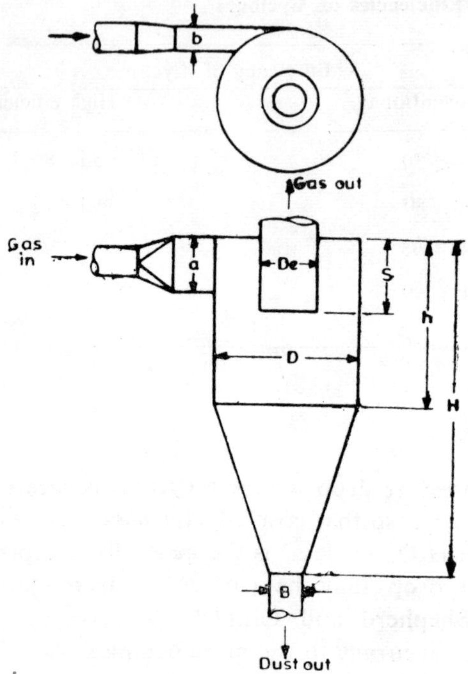

Nomenclature		Conventional (ref. 6)	High efficiency (ref. 7)
D	cyclone diameter	1·0	1·0
a	entrance height	0·5	0·5
b	entrance width	0·25	0·2
S	exit length	0·625	0·5
De	exit diameter	0·5	0·5
h	cylinder height	2·0	1·5
H	overall height	4·0	4·0
B	dust exit diameter	0·25	0·375

Fig. 5.6 Generalised cyclone design configurations

the periphery of the device. The main advantage of this unit is its compactness; its principal drawback is a tendency toward plugging from solid buildup in the unit.

Both these devices are used frequently as pre-cleaners to remove fly ash and large particles. For high efficiency at reasonable capacity, a battery of smaller cyclones operating in parallel is used in preference to a large unit. These multicellular units usually require less space.

Example 4.

A conventional cyclone with diameter 1.0 m handles 3.0 m³/s of standard

air carrying particles with a density of 2000 kg/m³. For Ne = 6, determine the cut size and the efficiency as a function of particle diameter.

Solution

From Fig. 5.6, $b = 0.25$ m, and $a = 0.5$ m.

$$\rho_p = 2000 \text{ kg/m}^3$$

$$v_i = \frac{Q}{ab} = \frac{3.0}{(0.5)(0.25)} = 24 \text{ m/s}$$

$$\mu_g = 1.81 \times 10^{-5} \text{ kg/m-s}$$

For the cut size d_{pc}, Eq. (5.28) gives

$$d_{pc} = \sqrt{\frac{9(1.81 \times 10^{-5})(0.25)}{2\pi(6)(24)(2000)}} = 4.7 \ \mu\text{m}$$

The empirical efficiency is then obtained from Fig. 5.5 and the values are plotted for various particle sizes in Fig. 5.7.

Fabric Filters

Filtration is the oldest and generally one of the most versatile and efficient methods for removing particulate matter from industrial gases. In principle, the dust-laden gas passes through a fabric in which the particles are

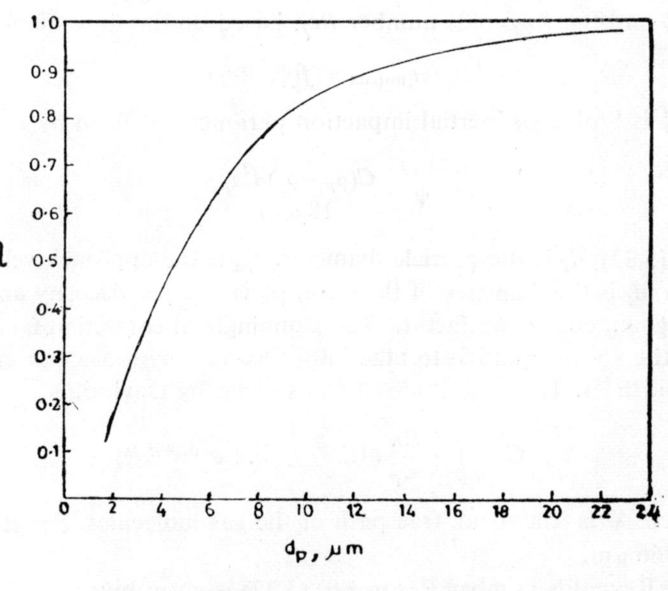

Fig. 5.7 Efficiency curve for example 4

trapped due to different physical mechanisms. The important mechanisms by which the particles are separated from the gas stream are usually inertial impaction, direct interception, and diffusion. These three mechanisms are illustrated in Fig. 5.8.

Inertial Impaction

Inertial impaction occurs when the particle with high inertia follows a fluid streamline, but as it approaches the fibre its trajectory departs from the fluid streamline around the fibre. The departure of the particle from the streamline may be sufficient to allow the particle to deposit on the curved surface. Fig. 5.8(a) shows that the particles approaching a fibre within a distance of y_0 from the plane of symmetry collide with the fibre. The ratio $2y_0/d_f$ is called the collision factor ϕ_c:

$$\phi_c = \frac{2y_0}{d_f} \tag{5.30}$$

where d_f is the diameter of the fibre. Since all the particles that collide will not adhere to the fibre, an adherence factor ϕ_a is defined such that the collection efficiency of a single fibre due to inertial forces, η_{impact}, is the product of ϕ_c and ϕ_a:

$$\eta_{\text{impact}} = \phi_c \phi_a \tag{5.31}$$

Theoretical investigations based on potential flow theory and experimental results reveal that the collection efficiency η_{impact} is a function of the factor ψ and the Reynolds number Re_f, based on the fibre diameter:

$$\eta_{\text{impact}} = f(\psi, Re_f) \tag{5.32}$$

where ψ is Stokes or inertial impaction parameter defined by

$$\psi = \frac{C(\rho_p - \rho_g)d_p^2 v_{p0}}{18\mu_g d_f} \tag{5.33}$$

In Eq. (5.33), d_p is the particle diameter. v_{p0} is the approach velocity of the particle, d_f is the diameter of the fibre, μ_g is the gas viscosity and C is the Cunningham correction factor. The Cunningham correction factor is included in the above equation to take into account any molecular slip for very small particles. The magnitude of C, as given by Davies[9], is

$$C = 1 + \frac{2\lambda}{d_p}(1.257 + 0.4\, e^{-0.55\, d_p/\lambda}) \tag{5.34}$$

in which λ is the mean free path of the gas molecules. For standard air, $\lambda = 0.066\ \mu\text{m}$.

The Reynolds number Re_f in Eq. (5.32) is given by:

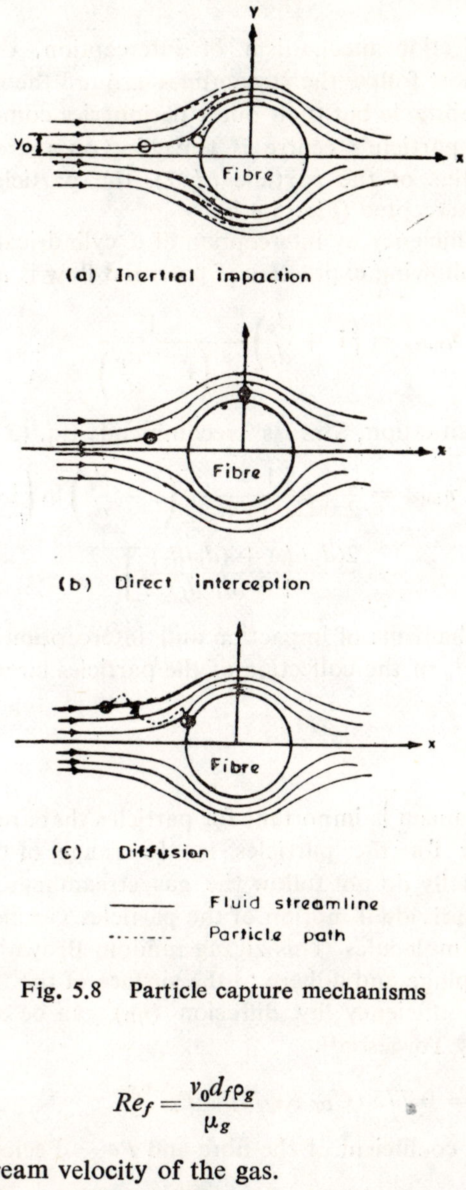

Fig. 5.8 Particle capture mechanisms

$$Re_f = \frac{v_0 d_f \rho_g}{\mu_g} \tag{5.35}$$

where v_0 = free stream velocity of the gas.

In Eq. (5.33), the particle approach velocity is generally assumed to be equal to the free stream gas velocity.

Extensive measurements of collection efficiencies have shown that the particle diameter d_p should be larger than about 1 μm, for inertial effects to play an important role in particle collection. Also, to achieve the greatest possible collection efficiency, the filter medium should be composed of smallest possible fibres consistent with the allowable pressure.

Direct Interception

In the particle collection mechanism of interception, the particles have less inertia and almost follow the streamlines around the obstruction. The particles clear the obstacle but their outer peripheries come in contact with the fibre. Thus, if particle's centre is travelling on a streamline which is closer than the radius of the particle ($d_p/2$), the particle will touch the fibre and will be intercepted (Fig. 5.8 b).

The collection efficiency by interception of a cylindrical target may be calculated by the following expression if potential flow is assumed:

$$\eta_{\text{inter}} = \left(1 + \frac{d_P}{d_f}\right) - \frac{1}{\left(1 + \frac{d_P}{d_f}\right)} \tag{5.36}$$

For a viscous flow situation, Strauss[10] recommends Eq. (5.37).

$$\eta_{\text{inter}} = \frac{1}{2.002 - \ln Re_f}\left[\left(1 + \frac{d_P}{d_f}\right)\ln\left(1 + \frac{d_P}{d_f}\right) - \frac{2(d_P/d_f) + (d_P/d_f)^2}{2 + 2\,(d_P/d_f)}\right] \tag{5.37}$$

The combined mechanisms of impaction and interception usually account for more than 99.9% of the collection of the particles larger than 1 μm.

Diffusion

Diffusion mechanism is important for particles that are in the submicron range—particularly for the particles in the range of 0.001 to 0.05 μm. These particles usually do not follow the gas streamlines surrounding the fibre because the individual motion of the particles can be affected by their collisions with gas molecules. This zigzag random Brownian motion causes the particles to impinge and adhere to the surface of the fibre (Fig. 5.8c).

The collection efficiency by diffusion (η_D) can be obtained from the expression given by Torgeson[11]:

$$\eta_D = 0.775\,(C_{Df}\,Re_f/2)^{0.4}\,Pe^{-0.6} \tag{5.38}$$

where C_{Df} = drag coefficient of the fibre and Pe = Peclet number, defined by Eq. (5.39)

$$Pe = Re_f \cdot Sc = \frac{v_0 \rho_g d_f}{\mu_g} \cdot \frac{\mu_g}{\rho_g \mathcal{D}} = \frac{v_0 d_f}{\mathcal{D}} \tag{5.39}$$

In Eq. (5.39). Sc is the Schmidt number and $\mathcal{D}$ is particle diffusivity given by:

$$\mathcal{D} = \frac{kTC}{3\pi\mu_g d_P} \tag{5.40}$$

where k = Boltzmann constant ($= 1.38 \times 10^{-23}$ J/K) and T = absolute temperature.

In practice the collection mechanisms of impaction, interception and diffusion are not independent and, hence, all three mechanisms must be combined for calculating the fibre efficiency. A reasonable estimate for the combined efficiency of collection η_f is

$$\eta_f = 1 - (1 - \eta_{\text{impact}})\,(1 - \eta_{\text{inter}})\,(1 - \eta_D) \tag{5.41}$$

In addition to the above mentioned three collection mechanisms, electrostatic forces between particles and fibres increase the collection efficiency. The generation of electrostatic charges in filter fabrics may be due to friction between gas and fabrics as well as that between particles and fabric at high gas velocities (1.5 to 2.0 m/s). The increase in collection efficiency on account of electrostatic forces becomes greater with increase in the strength of the electric charge.

In fabric filters, the open spaces in woven fabrics are usually many times the size of the individual particle. Because of this, the collection efficiency in the initial stages is rather poor. After a short period of operation, the captured particles form a bridge across the cloth openings and a particle layer forms which improves the collection efficiency greatly. Fabric filters usually provide collection efficiencies exceeding 99% and even efficiencies of 99.99% are not uncommon. Careful maintenance is necessary to prevent a deterioration of efficiency. Periodic mechanical cleaning is necessary and this usually does not remove all of the interstitial layer so that collection efficiencies remain high throughout the useful life of the filter.

Fabric and Fibre Characteristics

Fabric filters may be classified according to the filtering media: woven fabric or felt cloth. Woven fabrics have a definite long range repeating pattern and have considerable porosity in the direction of gas flow. These open spaces must be bridged by impaction of interception to form a ture filtering surface. Felted cloth consists of randomly oriented fibres, compressed into a mat and needled to some loosely woven backing material to improve mechanical strength. Each one of the randomly oriented fibres acts as a target for impaction and interception. This type of filter usually requires substantially lesser cloth area for a given loading than a woven type fabric filter. Felted filters are more expensive than the woven type filters and they cannot be used in high humidity service, especially if the particles are hygroscopic because of the high tendency for clogging.

The choice of fabric fibre is based primarily on operating temperature and the corrosiveness or abrasiveness of the particles. Some properties of common fibres are shown in Table 5.3. Cotton is the least expensive fibre, and is preferably used in low temperature dust collection service where no special problems are anticipated. Silicon coated glass fibre cloth is commonly employed in high-temperature applications. The glass fibres must be lubricated to prevent abrasion. All fibres may be applied to the manufacture of woven and felt type fabrics.

Fabric Filter Systems

Fabric filter systems typically consist of a tubular bag or an envelope, suspended or mounted in such a manner that the collected particles fall into a hopper when dislodged from the fabric. The structure in which the bags hang is known as a baghouse. Generally, particle-laden gas enters the bag at the bottom and passes through the fabric while the particles are deposited on the inside of the bag. Fig. 5.9 shows a typical baghouse design in which the cleaning is accomplished by shaking at fixed intervals of time.

Large baghouses are constructed with several compartments, so that one compartment may be isolated for cleaning as needed while the other compartments are operating. The bags should be arranged in each compartment in such a manner that the available space is utilised effectively and proper access to each bag is provided for its replacement. Hoppers are

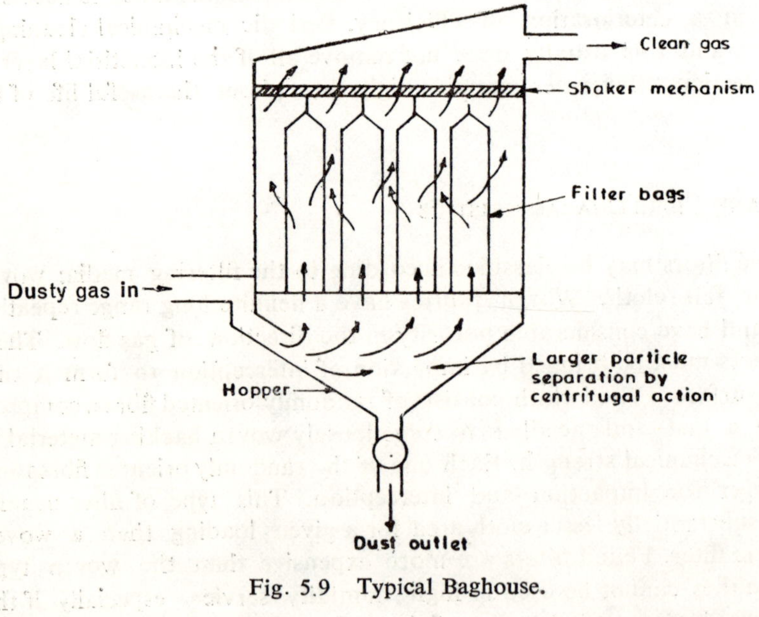

Fig. 5.9 Typical Baghouse.

Table 5.3 Properties of some fibre materials (ref. 12)

Fibre	Relative strength	Maximum usable temp. (°C)	Relative resistance to attack by			Other properties
			Acid	Base	Organic solvent	
Cotton	Strong	80	Poor	Medium	Good	Low cost
Wool	Medium	95	Medium	Poor	Good	Fair resistance to abrasion
Polyamide (nylon)	Strong	100	Medium	Good	Good	Easy to clean, excellent resistance to abrasion
Polyester (dacron)	Strong	135	Good	Medium	Good	Easy to clean
Tetrafluoroethylene (Teflon)	Medium	260	Good	Good	Good	Expensive
Glass	Strong	280	Medium	Medium	Good	Poor resistance to abrasion
"Nomex" nylon	Strong	230	Good	Medium	Good	Poor resistance to moisture

provided for dust collection and the dust is removed usually by rotary or screw valves.

The method of cleaning fabric filters varies according to manufacturer and type of filter. The cleaning may be accomplished by shaking the bags or by increasing the air pressure on the bag in a manner that causes the bag to collapse or otherwise deform sufficiently to dislodge the accumulated dust. Cleaning generally takes less than one minute with the bulk of the dust being removed in a few seconds. A good cleaning schedule ensures longer bag life and efficient filtration.

The following advantages make fabric filters the best choice in many cases: (i) very high efficiencies, (ii) retention of finest particles, (iii) collection of particulates in dry form, and (iv) relatively low pressure drop. The main disadvantages are: (i) their large size, (ii) high construction costs, and (iii) their application only to process temperatures generally below 285°C.

Electrostatic Precipitators

The electrostatic precipitator is one of the most widely used devices for controlling particulate emissions at industrial installations ranging from power plants, cement and paper mills to oil refineries. In most cases, the particulates to be collected are by-products of combustion. In others, they are dust fibres or other small particles such as acid mists from process industries.

Electrostatic precipitation is a physical process by which particles suspended in gas stream are charged electrically and, under the influence of the electrical field, separated from the gas stream. A typical wire and pipe precipitator is illustrated in Fig. 5.10.

The precipitation system consists of a positively charged (grounded) collecting surface and a high-voltage discharge electrode wire suspended from an insulator at the top and held in position by a weight at the bottom. At a very high DC voltage, of the order of 50 kV, a corona discharge occurs close to the negative electrode, setting up an electric field between the emitter and the grounded surface. The particle-laden gas enters near the bottom and flows upward. The gas close to the negative electrode is, thus, ionized upon passing through the corona. As the negative ions and electrons migrate toward the grounded surface, they in turn charge the passing particles. The electrostatic field then draws the particles to the collector surface where they are deposited.

Periodically, the collected particles must be removed from the collecting surface. This is done by rapping or vibrating the collector to dislodge the particles. The dislodged particles drop below the electrical treatment zone and are collected for ultimate disposal. Usually, a large number of these collectors are placed in parallel in a single housing forming a multiple cylindrical unit. Although the multiple unit is not as widely used as the

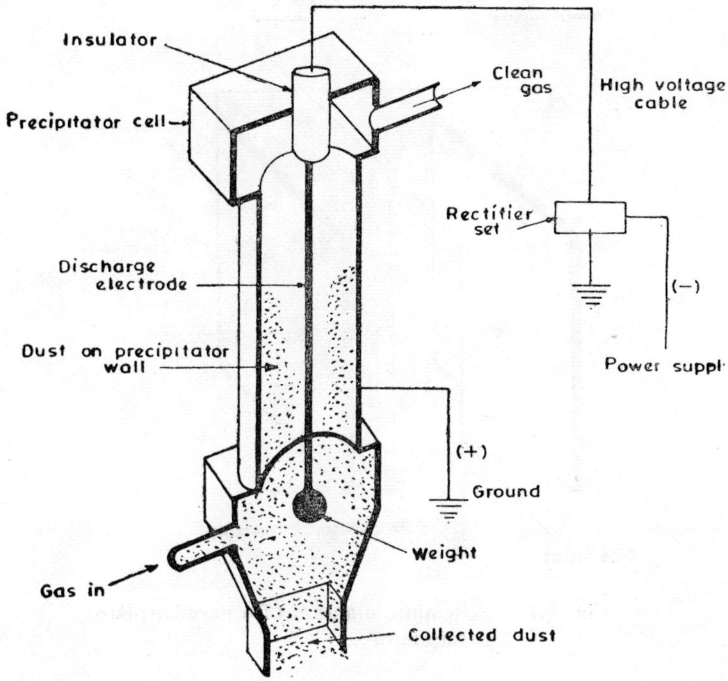

Fig. 5.10 Schematic diagram of a wire and pipe precipitator

parallel-plate precipitator described in the next paragraph, it finds frequent use in collecting liquid particles.

For industrial applications, vertical plates exposed to horizontal gas flow are normally used. In this type of collector, the gas flows between two vertical parallel plates between which are suspended a number of vertical wires held in place by weights attached at the bottom. These wires form the discharge electrode, while the vertical plates form the collection electrode. Fig. 5.11 shows the schematic diagram of a typical parallel-plate arrangement.

Collection Efficiency

The collection efficiency of an electrostatic precipitator as a function of gas flow rate and precipitator size is given by the Deutsch equation (Eq. 5.46) which is applicable to both cylindrical and parallel plate type precipitators. Consider a precipitator of an arbitrary geometry with a cross section A and length L, as shown in Fig. 5.12. It is assumed that the particle distribution over the cross-sectional area is constant. A particle that has entered

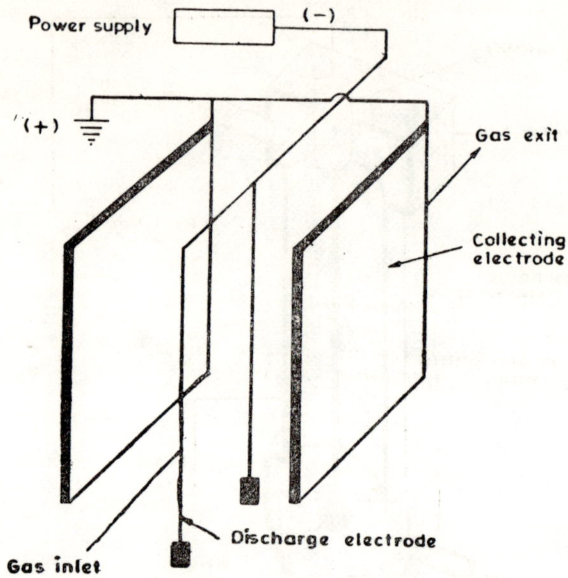

Fig. 5.11 Schematic diagram of a parallel-plate precipitator

the precipitator and received an electric charge moves along a trajectory towards the collection electrode as shown in Fig. 5.12. This trajectory can be resolved into the axial z-component with a velocity v_{pz}, and a y-component with a velocity v_{pm}. The particle velocity v_{pz} is equal to the gas velocity v in the axial direction.

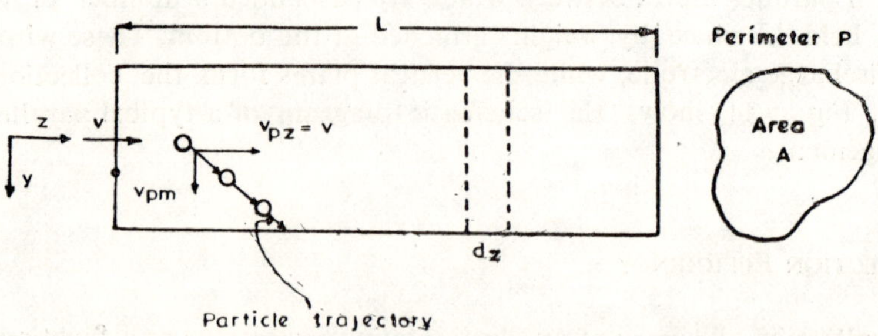

Fig. 5.12 Collection efficiency derivation

Writing a mass balance across the differential element dz for the steady operation with no net accumulation of particles in the gas space, we get

$$-\frac{dC}{C} = \frac{v_{pm}}{v}\frac{p}{A}dz \tag{5.42}$$

where C = particle concentration, mass per unit volume of gas stream; p = perimeter of the geometry; v_{pm} = particle migration velocity; and v = gas velocity.
Integrating Eq. (5.42) over the length L for concentrations from C_0 to C_L, we get

$$\frac{C_L}{C_0} = \exp\left(-\frac{v_{pm}}{v}\frac{p}{A}L\right) \tag{5.43}$$

The collection efficiency η is defined as

$$\eta = 1 - \frac{C_L}{C_0} \tag{5.44}$$

Substitution of Eq. (5.43) into Eq. (5.44) gives:

$$\eta = 1 - \exp\left(-\frac{v_{pm}}{v}\frac{p}{A}L\right) \tag{5.45}$$

Noting that $Av = Q$, the volumetric flow rate of the gas, and that the product pL is equal to the collector surface area A_c we may write Eq. (5.45) as

$$\eta = 1 - \exp\left(-\frac{v_{pm}A_c}{Q}\right) \tag{5.46}$$

This is the well known equation for the collection efficiency of a precipitator obtained by Deutsch and it can be used for both tubular and plate type precipitators. Equation (5.46) may be written in the following form:

$$\eta = 1 - \exp\left(\frac{-v_{pm}}{v}\frac{A_cL}{V}\right) \tag{5.47}$$

where V = volume of the precipitator.

For a cylindrical type collector $A_c/V = 4/D_c$, and for a parallel-plate type collector $A_c/V = 2/S$, S being the distance between the two parallel plates.

The particle migration velocity is probably the most important parameter in Eq. (5.47). It is a function of a large number of operational quantities such as the electric field strength, particle size, gas viscosity, dielectric and resistivity properties of the dust. Typical values of the migration velocity encountered in practice for use in Eq. (5.47) are given in Table 5.4.

Example 5

A plate-type electrostatic precipitator for use in a cement plant for removing dust particles consists of 10 equal channels. The spacing between the

Table 5.4 Effective migration velocities (ref. 13)

Application	v_{pm} (m/s)
Pulverized coal fly ash	0.10—0.13
Paper mills	0.08
Open-hearth furnaces	0.06
Secondary blast furnace	0.12
Gypsum	0.16–0.20
Hot phosphorus	0.03
Sulphuric acid mist	0.06–0.08
Cement	
Wet process	0.10—0.11
Dry process	0.06–0.07
Catalyst dust	0.08

plates is 0.15 m, and the plates are 2 m high and 2 m long. The unit handles 10,000 m³/hr of gas. What is the efficiency of collection? What should be the length of the plates for achieving 99% collection efficiency if other condition are the same?

Solution

From Eq. (5.46),

$$\eta = 1 - \exp\left(-\frac{v_{pm}\,A_c}{Q}\right)$$

$A_c = 2(2\times 2) = 8\ m^2$, $v_{pm} = 0.10$ from Table 5.4, and

$Q = (10{,}000)/(10)\,(3600) = 0.278\ m^3/s$ per channel.

Therefore, $\eta = 1 - \exp\left(-\dfrac{(0.10)\,(8)}{0.278}\right) = 94.4\%$

$A_c/V = 2/S$. The spacing between the plates, S, is equal to 0.15 m.

Therefore, $A_c/V = \dfrac{2}{0.15} = 13.33\ m^{-1}$

The velocity of the gas in each channel,

$$v = \frac{Q}{A} = \frac{0.278}{0.15)\,(2)} = 0.927\ m/s$$

From Eq. (5.47), for $\eta = 0.99$,

$$\frac{v_{pm}}{v}\left(\frac{A_c}{V}\right)L = 4.6$$

$$L = \frac{4.6\,(0.927)}{13.33\,(0.1)} = 3.2 \text{ m}$$

A unique feature of electrostatic precipitators is that the separating force is applied directly to the particles without the necessity of accelerating the gas as is done for all other particulate collection devices. This unique feature leads to an extremely low pressure loss for the collection of fine particles compared with that in any other type of collector.

The most economical design and operation of dry electrostatic precipitators is obtained when the resistivity of particles falls within certain desired limits. If the resistivity is low, particles give up their charge too easily and become re-entrained in the gases flowing past the collecting electrode. On the other hand, high-resistivity particles cannot lose their charge upon reaching the collector electrode because of the low conductivity of the material deposited earlier. As a result, the effective field strength is reduced and the efficiency falls. Particle resistivity can be controlled in many cases by selecting the proper operating temperature or by injecting water to reduce the resistivity. Such irrigated precipitators are used for the collection of fumes from electric arc furnaces.

Table 5.5 summarizes the advantages and disadvantages of the electrostatic precipitator as a particulate control device. These devices are often used in series with cyclones; the coares particles are first removed in cyclones before the gas enters the precipitator.

Table 5.5 Advantages and disadvantages of electrostatic precipitator.

Advantages	Disadvantages
1. Pressure drop and hence power requirement is small compared to that in other devices; economical and simple to operate	1. Relatively high initial cost and large space requirement
2. 99+ per cent efficiency obtainable very small particles can be collected wet or dry	2. Sensitive to variable particulate loadings or flow rates
3. Can handle both gases and mists for high volume flow	3. Safeguard of operating personnel from high voltage is necessary
4. Few moving parts; can be operated at high temperatures and pressures	4. Collection efficiency can deteriorate gradually

Wet Scrubbers

In Chapter 2, it was pointed out that wet precipitation is the principal mechanism by which atmospheric particles are removed by nature. This idea has been exploited by industry to develop a variety of liquid scrubbing equipment. Wet collectors have a number of advantages over dry collectors, such as simultaneous removal of particles and gaseous pollutants but suffer from the problems of corrosion and liquid waste disposal. These are summarized in Table 5.6.

The basic function of wet scrubbers is to provide contact between the scrubbing liquid, usually water, and the particulate to be collected. This contact can be achieved in a variety of ways as the particles are confronted with so-called impaction targets—which can be wetted surfaces or individual droplets. Whether the particles encounter wetted surfaces as in packed scrubbers or individual droplets as in spray scrubbers, the basic collection mechanisms are the same as in filters: inertial impaction, interception and diffusion, which were discussed in an earlier section. Generally, impaction and interception are the predominant mechanisms for particles of diameter above 0.3 μm, and for particles of diameter below 0.3 μm, diffusion begins to prevail. In addition, diffusiophoretic deposition and electrostatic precipitation also effect particle collection.

Diffusiophoretic deposition can be significant when mass transfer within the scrubber, caused by condensation of water vapour from the gas onto a cold liquid surface, exerts a force upon the particles and causes them to get deposited on the surface. The electrostatic charge produced on droplets during atomization is too small to have any influence on particle collection except when a charge is deliberately induced on the droplets[14]. In addition, agglomeration, another phenomenon that occurs in all particle collectors, also enchances particle removal.

There are many scrubber designs presently available where the contact between the scrubbing liquid and the particles is achieved in a variety of ways. The major types are: (i) spray towers, (ii) centrifugal scrubbers, (iii) packed beds and plate columns, and (iv) venturi scrubbers.

Each type of design has a certain range of applicability. Thus, low-energy scrubbers such as spray towers are most often used to handle particles largely above 5–10 μm in diameter, while the high-energy units such as the venturi scrubbers are very effective with fine particles (smaller than 3 μm).

Spray Towers

The simplest type of wet scrubber is a spray tower into which water is introduced by means of spray nozzles (Fig. 5.13). The polluted gas flows upward and the particle collection results because of inertial impaction and interception on the droplets. The maximum collection efficiency by inertial

Table 5.6 Advantages and disadvantages of wet collectors

Advantages	Disadvantages
1. Simultaneous removal of gases and particulates	1. Relatively high energy costs
2. Effective performance over a wide loading range	2. Problem of wet sludge disposal
3. Equipment occupies only a moderate amount of space compared to dry collectors such as bag houses	3. Corrosion problems
4. Hazards of explosive dust-air mixtures are reduced	4. Visible wet plume, reduction in buoyancy
5. Indifference to the temperature and moisture content of gas	5. Very small particles (sub-micron sizes) may not be captured
6. Corrosive gases may be neutralised by proper choice of scrubbing liquid	

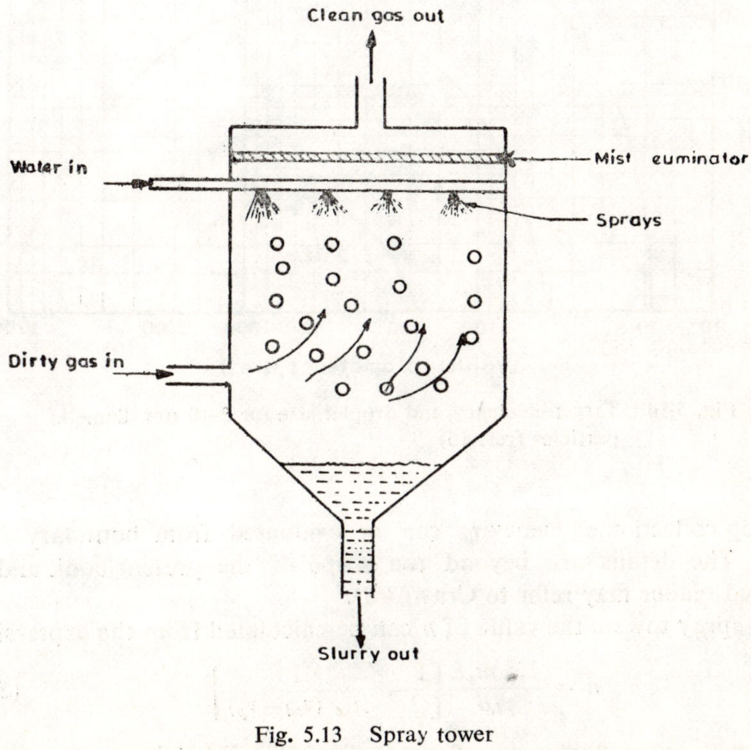

Fig. 5.13 Spray tower

impaction has been found by Stairmand[15] to be independent of particle size, when droplets fall under gravity through still air, and this maximum occurs for droplets of the diameter of about 800 μm (0.8 mm) as shown in Fig. 5.14. Thus, a nozzle producing a coarse spray with droplet size just below 1 mm is satisfactory.

The efficiency of a spray tower is a complex function of droplet size, flow velocity of the gas, liquid : gas ratio, and droplet trajectories. For identical spherical droplets falling through a slow upward moving gas stream, the overall collection efficiency is given by

$$\eta_{sc} = 1 - (1 - \eta_d)^n \tag{5.48}$$

where n = number of collecting droplets encountered by a group of particles and η_d = individual drop collection efficiency.

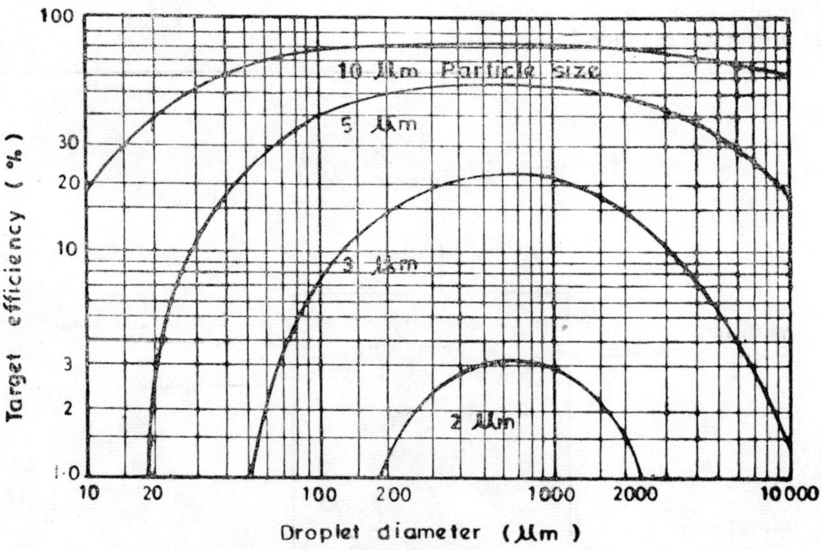

Fig. 5.14 Target efficiency and droplet size for 2–10 μm diameter particles (ref. 15)

The drop collection efficiency η_d can be evaluated from boundary layer theory. The details are beyond the scope of the present book and the interested reader may refer to Crawford[4].

For spray towers the value of n can be calculated from the expression[4],

$$n = \frac{1.5\, m_s L}{\rho_L d}\left[\frac{1}{Q} + \frac{1}{A_{sc}\,(v_{td} - v_p)}\right] \tag{5.49}$$

where m_s = mass flow rate of spray liquid, L = height of spray tower,

d = average droplet size, Q = volumetric flow rate of gas, ρ_L = spray liquid density, A_{sc} = tower cross-sectional area, v_{td} = terminal free falling velocity of the droplet, and v_p = particle velocity (taken to be the same as that of gas velocity = Q/A_{sc}).

The effectiveness of the conventional spray tower ranges from 94 per cent for 5 μm particles to 99 per cent for 25 μm particles.

Centrifugal Scrubbers

The collection efficiency for particles smaller than those recovered in spray towers can be increased through the use of centrifugal scrubbers. The simplest type of centrifugal scrubber can be constructed by inserting banks of nozzles inside a conventional dry cyclone. The spray acts on the particles in the outer vortex, and the droplets loaded with particles are thrown outward against the wet inner wall of the cyclone. In the absence of spray, the efficiency will be the same as that for a dry collector.

Another version of a centrifugal scrubber, shown in Fig. 5.15, has the polluted gas introduced tangentially into the lower portion of the vertical

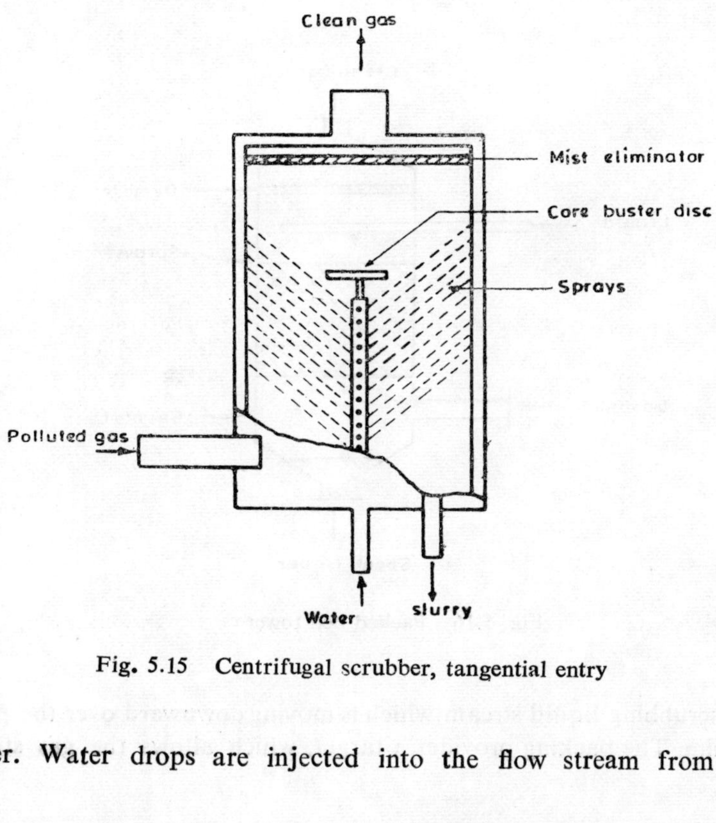

Fig. 5.15 Centrifugal scrubber, tangential entry

cylinder. Water drops are injected into the flow stream from sprays directed outward from a central manifold. These droplets are caught in the spinning gas stream and are thrown upward toward the wall by centrifugal force. During their motion, the droplets collide with the particles and capture them. The scrubbing liquid along with the particles flows down the wall to the bottom of the scrubber. The cleaned gas exits through a demister and is processed for the removal of any entrained water droplets.

The particle cut diameter (d_{pc}) for centrifugal spray scrubbers is between 2 and 3 μm. Commercial scrubbers have operating efficiencies of 97 per cent or better for particles larger than 1 μm. The centrifugal acceleration involved may range from 25 to 300 g. In a centrifugal field of 100 g, for example, the most effective droplet size is 100 μm (ref. 16). Smaller droplets can be entrained in the gas stream whereas the bigger droplets are found to have smaller inertial impaction parameters.

Packed Beds and Plate Columns

Packed beds and plate columns, well known to chemical engineers as absorbers or fractionators, can also be used to scrub particulate matter from polluted gases. In a typical countercurrent-flow packed scrubber (Fig. 5.16) the polluted gas stream moves upward and comes in contact

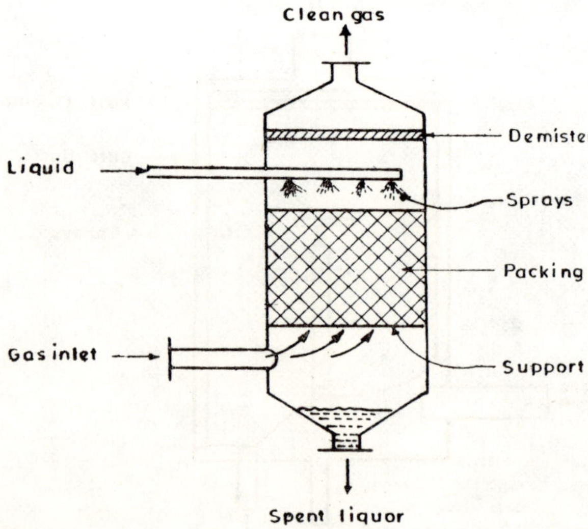

Fig. 5.16 Packed bed tower

with the scrubbing liquid stream which is moving downward over the packing in a film. The packing provides a target, which allows the gas stream to take a curved path through the pore spaces while the particles carried by the gas stream are captured by inertial impaction. Because of the good mass transfer characteristics of the packing, efficient collection of fine particles by diffusion is also possible. The scrubber can be packed with a variety of materials such as Raschig rings, saddles, coke or broken stone. Smaller packing increases the efficiency of collection but its shape does not appear to affect the collection efficiency.

Packed towers can be chosen for scrubbing particles that are soluble in the scrubbing liquid; otherwise, the packings will encounter plugging problems. The plugging problems can be reduced by employing sprays to wash the packing or by using low density spheres, agitated by the gas flow. More recently, a moist chemical foam packing has been employed, which drains slowly from the scrubber with captured particles and is replaced with fresh material.

In the impingement baffle plate scrubber, the gas stream is passed through a flooded, perforated plate for scrubbing (Fig. 5.17). The impingement baffles are situated directly above each perforation on to which the

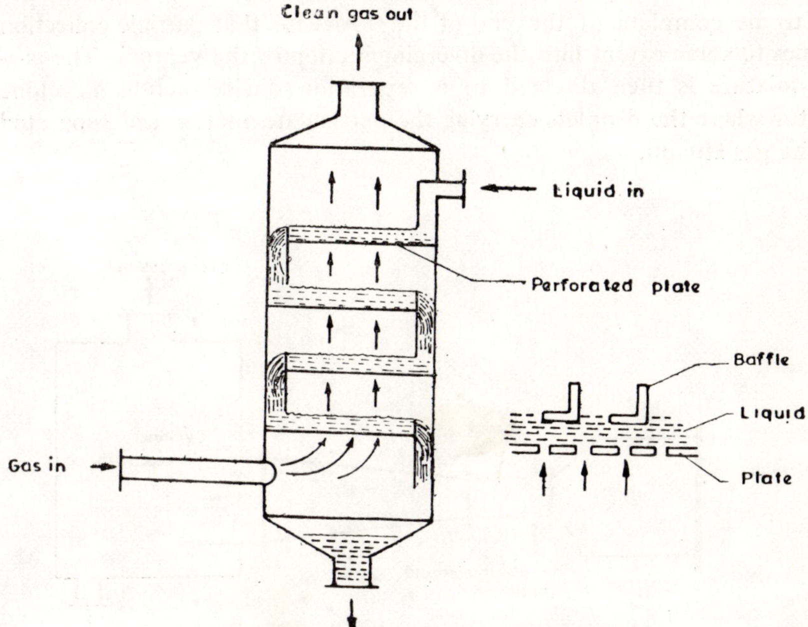

Fig. 5.17 Impingement baffle-plate scrubber

gas jet is impinged. The particles are collected by inertial impaction. Particle collection may be aided by atomization of liquid flowing past the perforations in the plate.

In impingement type scrubbers the superficial gas velocities are 5 times the velocities generally used for mass transfer operations such as distillation. The main disadvantage of impingement plate systems is their tendency towards scaling, resulting in plugging of the perforations.

VENTURI SCRUBBERS

Venturi scrubbers offer high performance collection of fine particles, usually smaller than 2 to 3 μm in diameter. They are particularly suitable when the particulate matter is sticky, flammable, or highly corrosive. The high performance of the venturi scrubbers is achieved by accelerating the gas stream to very high velocities, of the order of 60–120 m/s The high-speed action atomizes the feed liquid, generally introduced in a uniform fashion across the throat through several low-pressure spray nozzles directed radially inward as shown in Fig. 5.18.

The droplets accelerate in the throat section and due to the velocity difference between the particles and the droplets the particles are impacted against the slow-moving droplets. This acceleration of the droplets is not likely to be complete at the end of the throat, so that particle collection continues to some extent into the diverging section of the venturi. The gas-liquid mixture is then directed to a separation device such as a cyclone separator where the droplets carrying the particulate matter are separated from the gas stream.

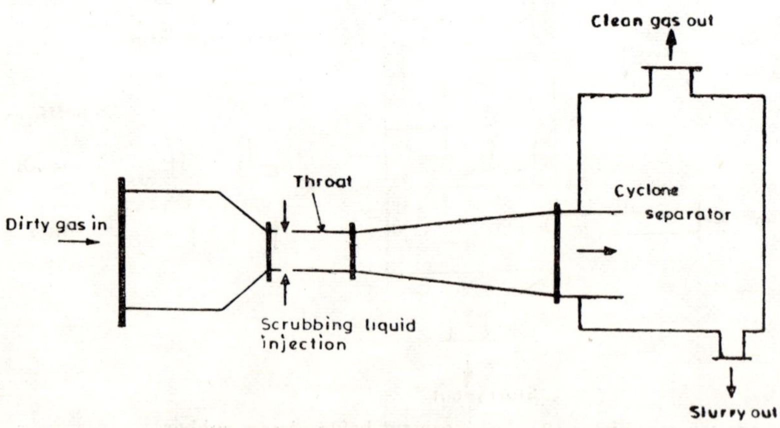

Fig 5.18 Venturi scrubber followed to cyclone separator

The mechanisms affecting the collection of particulates in the venturi scrubber are inertial impaction, diffusion, electrostatic-phenomenon, and condensation and agglomeration. Although condensation plays a part in the effectiveness of the scrubber by helping the particles to agglomerate, the predominant mechanism is inertial impaction. Johnstone *et al.*[17] correlated the collection efficiency of the venturi scrubber with the impaction parameter (ψ) as :

$$\eta_{sc} = 1 - \exp(-KL\sqrt{\psi}) \tag{5.50}$$

where K is an empirical factor determined by throat geometry and other parameters. The value of K may be of the order of 0.1 to 0.2. The inertial impaction parameter can be evaluated from Eq. 5.33 by replacing d_f, the fibre diameter with the droplet diameter d_0, and v_{p0}, the particle velocity with the relative velocity of gas to liquid, v_r.

The diameter of the droplet obtained from gas atomization in the venturi throat can be evaluated from the following equation :

$$d_0 = 16{,}400/v_r + 1.45L^{1.5} \tag{5.51}$$

where d_0 is in μm, L is the liquid flow rate (gal/1000 ft³ of gas), and v_r is the relative velocity of gas to liquid at venturi throat in ft/sec.

Example 6.

A venturi scrubber is to be used to collect particulate matter from an industrial operation. The liquid flow rate through the scrubber is 10 gpm per 1000 ft³ per min of the gas and the relative velocity of the gas to liquid is 300 ft/sec. The gas is air at standard temperature of 298 K and pressure of 1 atm carrying particles of density 1000 kg/m³. Determine the efficiency of the scrubber as a function of particle diameter.

Solution:

The droplet diameter from gas atomization (d_0) is obtained from Eq. (5.51).

$$d_0 = \frac{16{,}400}{v_r} + 1.45L^{1.5}$$

where L = 10 gpm/1,000 ft³ per min, and v_r = 300 ft/sec = 91.5 m/s.

Substituting the values, we get

$$d_0 = \frac{16{,}400}{300} + 1.45(10)^{1.5} = 100\ \mu\text{m} = 100 \times 10^{-6}\ \text{m}$$

The inertial impaction parameter ψ is given by the equation,

$$\psi = \frac{C\rho_p d_p^2 v_r}{18\mu_g d_0}$$

where $C = 1 + 2\dfrac{\lambda}{d_p}(1.257 + 0.4e^{-0.55\, d_p/\lambda})$

$\lambda = 0.066\ \mu m$ for standard air at 298 K and 1 atm.

$\rho_p = 1000\ kg/m^3$, and $\mu_g = 1.8 \times 10^{-5}$ kg/m-s,

Substituting, we get

$$\psi = \frac{C(1000)d_p^2(91.5)}{18(1.8 \times 10^{-5})(100 \times 10^{-6})} = 2.824Cd_p^2$$

The collection efficiency η_{sc} of the scrubber can be obtained from Eq. (5.50):

$$\eta_{sc} = 1 - \exp(-KL\sqrt{\psi})$$

Assuming $K = 0.2$, we get

$$\eta_{sc} = 1 - \exp(-0.2\,(10)\sqrt{\psi})$$
$$= 1 - \exp(-2\sqrt{2.824\,Cd_p^2})$$
$$= 1 - \exp(-3.361\sqrt{C}d_p)$$

The values of η_{sc} calculated from the above equation for various particle sizes are tabulated below:

d_p (μm)	C	η_{se} (%)
0.1	2.911	43.64
0.2	1.890	60.32
0.3	1.574	71.78
0.5	1.337	85.68
0.7	1.240	92.72
1.0	1.168	97.36
2.0	1.084	99.91
5.0	1.034	100

The results are plotted in Fig. 5.19.

The pressure drop through the venturi scrubber can be predicted on the basis of the equation suggested by Calvert[18]

$$\Delta P = 5 \times 10^{-5}\, v_r^2\, L \tag{5.52}$$

where Δp is the pressure drop in inches of water. For the problem of example 6, the pressure drop, ΔP, becomes equal to 45″ H_2O. Fig. 5.19 shows typical collection efficiency curves at two other pressure drops for

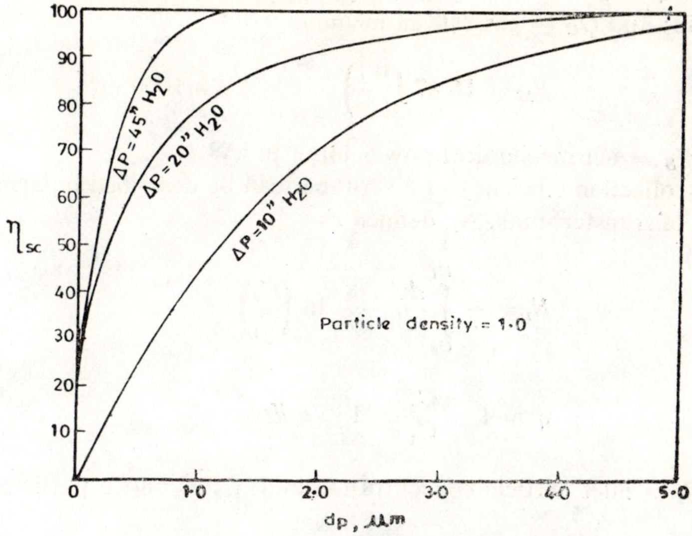

Fig. 5.19 Venturi scrubber efficiency curves

particulate matter having a specific gravity of 1.0 in a well-designed scrubber[19]. As is shown in Fig. 5.19, for particles of a particular size higher efficiencies require greater pressure drop and consequently greater power consumption. Similarly, for smaller particles, the same efficiency requires a higher pressure drop or greater power consumption.

Scrubber Performance and Efficiency Correlation

The above ideas have led Semrau[20] to correlate the scrubber efficiency in terms of "contacting power." It is a concept used to describe the power expended in a wet scrubber to force contact between the liquid and solid phases and is usually defined in terms of the power per unit of volumetric gas flow rate that is dissipated in contacting and is ultimately converted into heat. The total contacting power P_T, is the sum of the gas-phase contacting power, P_G, the liquid-phase contacting power, P_L, and mechanical contacting power, P_M (for those devices which are mechanically aided).

$$P_T = P_G + P_L + P_M \tag{5.53}$$

These terms may be determined from Eqs. (5.54–5.56)

$$P_G = 0.02724\, \Delta P_g \qquad \text{kWh/1000 m}^3 \tag{5.54}$$

where ΔP_g = gas pressure loss in cm H_2O.

$$P_L = 0.02815\, \Delta P_l \left(\frac{Q_L}{Q_G}\right) \qquad \text{kWh/1000 m}^3 \tag{5.55}$$

where ΔP_l = pressure drop for liquid input in atm; Q_L = liquid rate in litres/min; and Q_G = gas rate in m^3/min.

$$P_M = 16.67 \left(\frac{W_S}{Q_G}\right) \qquad \text{kWh/1000 m}^3 \qquad (5.56)$$

where W_S = net mechanical power input in kW.

The collection efficiency of a scrubber can be described in terms of the number of transfer units, N_t, defined as

$$N_t = -\int_{C_2}^{C_1} \frac{dC}{C} = -\ln\left(\frac{C_2}{C_1}\right) \qquad (5.57)$$

$$\eta = 1 - \frac{C_2}{C_1} = 1 - e^{-N_t} \qquad (5.58)$$

where C_1 = inlet particle concentration and C_2 = outlet particle concentration.

The number of transfer units was plotted against total power consumption (P_T) for a series of scrubbers and particulates when the following expression was obtained:

$$N_t = \alpha P_T^{\beta} \qquad (5.59)$$

where α and β are empirical constants that depend on the particulate size and characteristics, and are little affected by the scrubber size or geometry, or the manner of applying the contacting power. Some representative values of α and β are given in Table 5.7.

In Table 5.7, α, the coefficient of P_T in Eq. (5.59), has units of $hp/(1000$ ft^3/min)$^{-\beta}$ and the exponent β is dimensionless. Since the contacting power P_T is given in the units of kWh/1000 m^3, it has to be converted into units of hp/(1000 ft^3/min) when using in Eq. (5.59) along with α and β.
(Note; 1.0 kWh/1000 m^3 = 2.278 hp/1000 ft^3/min).

5.5 SELECTION OF A PARTICULATE COLLECTOR

The selection of a suitable control device for a specific application requires consideration of particle size, its concentration, desired efficiency of collection, costs, space available, and maintenance factors. Fig. 5.20 gives the efficiencies of various gas cleaning devices for three broad particulate size categories: coarse, fine and superfine dusts[22]. This can be used conveniently for a preliminary selection of devices. For example, for 99% removal of fine dust (50% < 25 μm), preliminary selection of an irrigated target type scrubber, an electrostatic precipitator, and an irrigated electro-

Table 5.7 Scrubber efficiency parameters (ref. 21)

Scrubber	Aerosol	Correlation parameter	
		α	β
Venturi	Talc dust	2.97	0.362
	Phosphoric acid mist	1.33	0.647
	Foundry cupola dust	1.35	0.621
	Open hearth steel furnace fume	1.26	0.569
Venturi and cyclonic spray	Raw gas (lime dust and soda fume)	1.47	1.05
	Black liquor recovery furnace fume	1.75	0.620
	Ferrosilicon furnace fume	0.870	0.459
Venturi pipeline and cyclonic spray	Pre-washed gas (soda fume)	0.915	1.05
	Hot fume solution for scrubbing	0.740	0.861
Venturi condensation scrubber with			
(i) mechanical spray generator	Copper sulphate	0.390	1.14
(ii) hydraulic nozzles	Copper sulphate	0.562	1.06
Cyclone	Talc dust	1.16	0.655

static precipitator may be made. After the preliminary selection, a comparative cost analysis must be conducted including the operating and the maintenance costs After a general selection of equipment is made to meet specific emission limitations, the exact costs are determined, and the actual operating efficiency and pressure drop are obtained on the installed system.

It is important that the inter-related process and environmental factors must be weighed carefully before taking the final decision. Often the collection equipment eventually selected represents the most desirable compromise. Alternatively, two collectors may be combined to achieve the desired results. Table 5.8 provides a summary of some successful applications of particulate control devices in various industries.

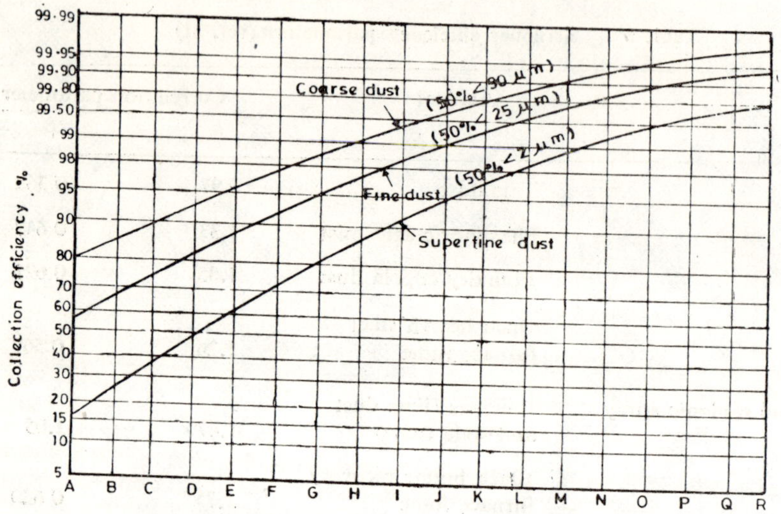

A - Inertial collector
B - Medium efficiency cyclone
C - Low resistance cellular cyclones
D - High efficiency cyclone
E - Impingement scrubber
F - Self-induced spray scrubber
G - Void spray tower
H - Fluidized bed scrubber
I - Irrigated target scrubber (Peabody type)
J - Electrostatic precipitator
K - Irrigated electrostatic precipitator
L - Flooded-disc scrubber (low energy)
M - Flooded-disc scrubber (medium energy)
N - Venturi scrubber (medium energy)
O - High-efficiency electrostatic precipitator
P - Venturi scrubber (high energy)
Q - Shaker-type fabric filter
R - Reverse-jet fabric filter

Fig. 5.20 Efficiency of various types of gas cleaning devices (ref. 22, copyright (c) 1974 by Litton Educational Publications Inc., copyright (c) 1979 renewed by Van Nostrand Reinhold, all rights reserved, reproduced with permission.)

5.6 CONTROL OF GASEOUS EMISSIONS

As noted earlier, there are essentially two classes of techniques by which gaseous pollutants may be removed from an effluent stream: (1) sorption of pollutant, for example, through absorption in a liquid or adsorption on a solid surface, and (2) chemical alteration of the pollutant, usually, through combustion or catalytic treatment.

Absorption by Liquids

Absorption is one of the most frequently used techniques for controlling the concentrations of gaseous pollutants before they are discharged into the

Table 5.8 Use of particulate collectors in industry

Industry or process	Particulate matter	Collection devices
Steel	Iron oxide, dust, smoke	Cyclones fabric filters, electrostatic precipitators, wet scrubbers
Metallurgical (non-ferrous)	Smoke, metal fumes, oil, grease	Electrostatic precipitators, fabric filters, wet scrubbers
Cement, phosphate, gypsum, alumina processing	Process dusts	Electrostatic precipitators, mechanical collectors, fabric filters, wet scrubbers
Kraft paper mills	Chemical dusts	Electrostatic precipitators, venturi scrubbers
Sulphuric, phosphoric, nitric acid manufacture	Acid mist, dust	Electrostatic precipitators, mesh mist eliminators
Petroleum refineries	Catalyst dust, ash from sludge	High-efficiency cyclones, electrostatic precipitators, scrubbing towers
Power generation	Coal dust, fly ash	Mechanical collectors, electrostatic precipitators

atmosphere. It involves the transfer of the pollutant from the gas phase to the liquid phase across the interface in response to a concentration gradient, with the concentration decreasing in the direction of mass transfer.

A two-resistance theory, initially suggested by Whitman[23], is often used to explain this process. According to the theory, the interface offers no resistance to mass transfer, and the mass transfer rate between the two phases is controlled by the rates of diffusion through the phases on each side of the interface. The transfer of component A from the gas phase to the liquid phase is illustrated in Fig. 5.21, with a concentration gradient in the gas phase from the bulk gas concentration of component A, p_{AG}, to the interfacial concentration, p_{Ai}, and a concentration gradient in the liquid phase from, C_{Ai} at the interface to the bulk liquid concentration, C_{AL}. The bulk concentrations p_{AG} and C_{AL} are not equilibrium values; because if they were, diffusion of component A will not take place. If no resistance to mass transfer exists at the interface, p_{Ai} and C_{Ai} are equilibrium concentrations; these are the values which would be obtained if the two phases had

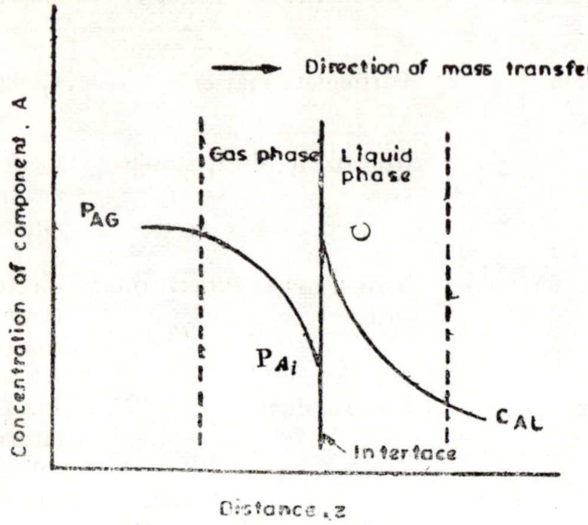

Fig. 5.21 Concentration profiles

been in contact for an infinite period of time. p_{Ai} and C_{Ai} are directly related through the equilibrium data.

For dilute solutions the equilibrium relation between p_A and C_A can be expressed in terms of Henry's law:

$$p_A = HC_A \tag{5.60}$$

where H is the Henry's constant, which depends on the definition of concentration units. For example, if the Henry's law is expressed in terms of the mole fraction units, Eq. (5.60) becomes

$$p_A = H' x_A \tag{5.61}$$

where x_A is the mole fraction of component A in the liquid. Table 5.9 shows values of Henry's constant for several gases in water. Smaller values of H' represent higher solubilities of the gas. That is, for a given partial pressure of the gas, a higher liquid concentration will be reached at equilibrium.

Selection of the Scrubbing Medium

The effectiveness of an absorption process for air pollution control mostly depends on the selection of a suitable scrubbing medium. Some of the properties of solvents, which are important in industrial applications, are the following: high gas solubility and high selectivity between solute species, low vapour pressure, non-corrosiveness, easy regeneration and recovery, low costs, low viscosity at absorption temperature, nontoxicity, nonflammability, and chemical stability.

In addition, the chosen scrubbing medium should not cause water pollution or waste disposal problems. Table 5.10 summarizes the more com-

Table 5.9 Values of Henry's constant (H') for several gases in water (ref. 3)

Gas	$H' \times 10^{-3}$ (atm/mole fraction) at		
	10°C	20°C	30°C
CO	44.2	53.6	62.0
H_2S	0.367	0.483	0.609
NO	21.8	26.4	31.0
N_2	66.8	80.4	92.4
O_2	32.7	40.1	47.5
SO_2	0.007	0.014	0.016
CO_2	1.04	1.42	1.86

monly used solvents in industry for the removal of various gaseous pollutants.

Table 5.10 Suitable solvents for various gaseous pollutants

Gaseous pollutants	Suitable solvents
Sulphur dioxide	Sodium hydroxide, sodium sulphite, magnesium oxide, calcium carbonate, calcium oxide and calcium hydroxide solutions
Nitrogen oxides	Ammonium bicarbonate, ammonium bisulphite, calcium hydroxide, magnesium hydroxide and sodium hydroxide solutions
Hydrogen sulphide	Sodium hydroxide, potassium hydroxide solutions
Hydrogen chloride	Water, ammonia, calcium and magnesium hydroxide solutions
Chlorine	Solutions of sodium hydroxide, sodium sulphite, sodium thiosulphite and water
Phosgene	Sodium hydroxide and water
Ammonia	Sulphuric acid, nitric acid
Mercaptans	Sodium hypochlorite solution

Gas Absorption Equipment

Gas absorption is commonly conducted in equipment which are designed to provide intimate contact between the two phases. The contact between gas and liquid can be accomplished by dispersing the liquid in the gas or vice versa. Some of the commonly used absorbers in air pollution control are packed towers, plate and spray towers and venturi scrubbers.

Packed towers are very efficient absorption devices involving a continuous contact of two phases. These use a variety of packing materials ranging from specially designed ceramic packing to crushed rock. The liquid is distributed over the packing, which provides high interfacial surface area, and flows down the packing surface in the form of thin films or subdivided streams. Normally, the liquid and gas flow countercurrent to each other, the gas flowing upward and the liquid flowing downward. The use of packed towers is limited to clean gases, as any precipitate or slurry will cause plugging of packing.

In *tray or plate towers,* stepwise contact between the liquid and gas occurs. Liquid introduced at the top cascades from one stage to another and eventually is withdrawn from the bottom of the tower. The gas flows in a countercurrent direction to the dropping liquid. At each tray, gas passes through small holes drilled in the tray or under slotted caps immersed in the liquid, forming bubbles. Absorption takes place during the bubble formation and their ascent through the liquid. In the *spray towers,* the flow arrangements are normally countercurrent and cross-flow. The liquid is introduced by spray nozzles or other atomizing devices which form droplets. These droplets provide the mass transfer surface. Spray towers are used for absorption of highly soluble gases and solid particles when these are present in the gas.

In *venturi scrubbers* the gas and liquid flow in the same direction. The absorbing liquid is introduced at the throat of the venturi and the dispersion of the liquid is accomplished by the high velocity gas stream. The resulting large interfacial area makes the venturi scrubber an extremely efficient absorber. However, the energy required for achieving high efficiencies is considerable.

In air pollution control, the most popular means of gas cleaning is the packed tower; however, tray columns, spray and venturi scrubbers are used in nearly 10% of the cases. Absorption is treated extensively in most chemical engineering texts (ref. 24-27), to which reference can be made for detailed information regarding the equipment design and construction. Here the discussion will be limited to the basic concepts related to the design of packed columns.

BASIC DESIGN CONSIDERATIONS

Consider a steady-state operation involving a countercurrent absorber as shown in Fig. 5.22.

A solute material balance gives,

$$G_1 y_1 + L_2 x_2 = G_2 y_2 + L_1 x_1 \tag{5.62}$$

where G = total molal flow rate of the gas phase per cross-sectional area of tower, L = total molal flow rate of the liquid phase per cross-sectional area of tower, y = mole fraction of solute in the gas phase and x = mole fraction of solute in the liquid phase.

Since G and L vary along the height of the tower, it is much simpler to express material balance equations in terms of solute-free concentration units. The material balance then becomes:

$$G_S(Y_1 - Y_2) = L_S(X_1 - X_2) \tag{5.63}$$

where G_S = molal flow rate of non-diffusing material in gas phase/cross-sectional area of tower, L_S = molal flow rate of non-diffusing material in liquid phase/cross-sectional area of tower, Y = moles of solute per mole of non-solute in gas phase and X = moles of solute/mole of non-solute in liquid phase.

Eq. (5.63) is the equation for the operating line with slope,

$$\frac{L_S}{G_S} = \frac{Y_1 - Y_2}{X_1 - X_2} \tag{5.64}$$

The gas concentration varies from Y_1 at the bottom to Y_2 at the top of the unit, while the liquid concentration varies from X_2 at the top to X_1 at the bottom of the unit during the period of contact. In general form, the operating line equation may be represented as

$$Y = Y_1 - \left(\frac{L_S}{G_S}\right)(X_1 - X) \tag{5.65}$$

The operating line together with the equilibrium curve is shown in Fig. 5.23.

A solute material balance over the differential length dz, is easily obtained by differentiating Eq. (5.65).

$$G_S\, dY = L_S\, dX \tag{5.66}$$

Referring to Fig. 5.21, we find that the rate of mass transfer of solute N_A per unit interfacial area from the gas phase and across the interface must equal to that in the liquid phase, i.e.,

$$N_A = k_G(p_G - p_i) = k_L(c_i - c_L) \tag{5.67}$$

where N_A = moles of solute transferred/(time) (interfacial area), k_G = mass transfer coefficient in the gas phase, moles of solute transferred/(time) (interfacial area) (Δp units of concentration), k_L = mass transfer coefficient

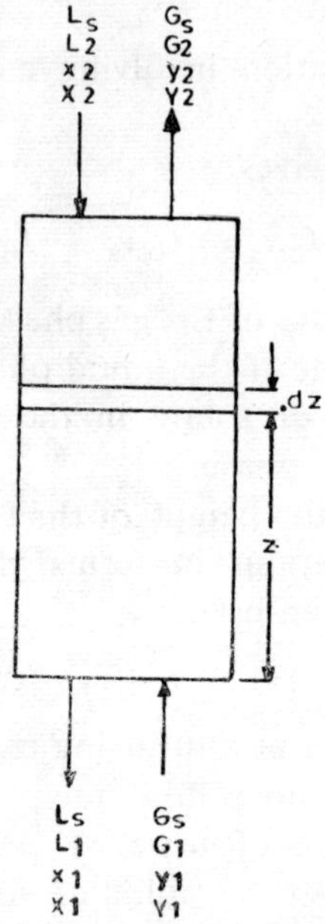

Fig. 5.22 Countercurrent absorber mass balance

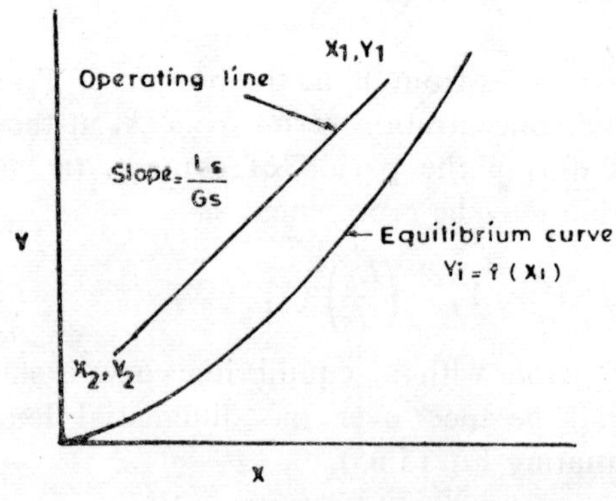

Fig. 5.23 Equilibrium and operating lines in a countercurrent tower

in the liquid phase, moles of solute transferred/(time) (interfacial area) (Δ *c* units of concentration) p_G = partial pressure of solute in the bulk of the gas stream, p_i = partial pressure at the interface, c = concentration in the bulk of the liquid and c_i = concentration at the interface.

Since it is impossible to know the interface partial pressures and concentrations, the rate equations are normally expressed in terms of overall coefficients based on an overall driving force between the bulk compositions, p_G and c_L. These are defined by,

$$N_A = K_G(p_G - p^*) = K_L(c^* - c_L) \tag{5.68}$$

where K_G = overall mass transfer coefficient based on partial pressure driving force, K_L = overall mass transfer coefficient based on liquid phase concentration driving force, p^* = partial pressure of solute in equilibrium with the bulk concentration in the liquid phace c_L and c^* = concentration of solute in equilibrium with p_G. These points are shown in the equilibrium diagram (Fig. 5.24).

The overall mass transfer coefficients K_G and K_L can be determined experimentally and used directly in the design equations. The point *B* on the equilibrium curve represents the interfacial concentrations, and the overall driving forces $p_G - p^*$ and $c^* - c_L$ are represented by the vertical line *AD* and the horizontal line *AC*, respectively.

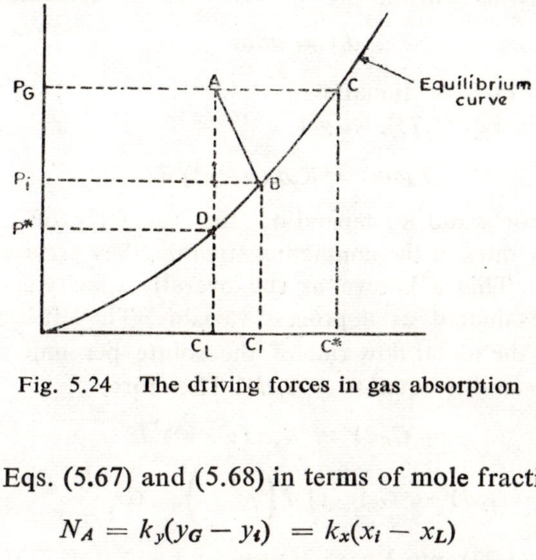

Fig. 5.24 The driving forces in gas absorption

Expressing Eqs. (5.67) and (5.68) in terms of mole fractions, we get

$$N_A = k_y(y_G - y_i) = k_x(x_i - x_L) \tag{5.69}$$

and

$$N_A = K_y(y_G - y^*) = K_x(x^* - x_L) \tag{5.70}$$

respectively. In Eqs. (5.69) and (5.70), k_y and k_x are the individual mass transfer coefficients in the gas and liquid phases respectively, and K_y and K_x are the overall mass transfer coefficients. Similarly, if the equilibrium relationship is of the form given by Eq. (5.61), the expression in terms of mole fractions is given by

$$y_i = m\, x_i \tag{5.71}$$

where $m = H'/P$, and P = total pressure.

Combining Eqs. (5.69), (5.70) and (5.71) we can show that the overall mass transfer coefficient K_y can be related to the individual coefficients k_y and k_x as,

$$\frac{1}{K_y} = \frac{1}{k_y} + \frac{m}{k_x} \tag{5.72}$$

Now for a differential section of the tower with height dz and interfacial area dA, the rate of absorption, from Eq. (5.70), is

$$N_A \, dA = K_y (y - y^*) \, dA \tag{5.73}$$

where y is the solute concentration in the gas phase. Note that the subscript G has been dropped for clarity.

In order to use Eq. (5.73) for the design of absorption units, we must know the interfacial area dA. However, in most systems this is either unknown or difficult to estimate. For example, in the case of packed towers only a fraction of the available surface may be absorbing. So, it is convenient to introduce a factor 'a' which represents the interfacial area per unit of tower volume. Then dA in Eq. (5.73) can be represented by,

$$dA = aSdz \tag{5.74}$$

where S = tower cross-sectional area

Substituting in Eq. (5.73), we get

$$N_A a dz = K_y a \, (y - y^*) \, dz \tag{5.75}$$

Since the factor a and K_y depend on the geometry of the absorption unit and the flow rates of the contacting streams, they are usually combined as a product $K_y a$. This is known as the overall capacity coefficient and is experimentally evaluated as a process variable. The left hand side of Eq. (5.75) represents the molal flow rate of the solute per unit cross-sectional area of the tower and is equal to $-G_S dY$. Therefore,

$$-G_S dY = K_y a \, (y - y^*) \, dz \tag{5.76}$$

Also,

$$G_S dY = G(1 - y) \, d\left(\frac{y}{1 - y}\right) = G \frac{dy}{1 - y} \tag{5.77}$$

Substituting Eq. (5.77) into Eq. (5.76) we get

$$-G \frac{dy}{1 - y} = K_y a \, (y - y^*) \, dz \tag{5.78}$$

Multiplying and dividing the right hand side of Eq. (5.78) by $(1 - y)_{lm}$ and assuming that the product of this and $K_y a$ is a constant, we can integrate Eq. (5.78) for the height of the tower Z.

$$Z = \frac{G}{K_y a \, (1 - y)_{lm}} \int_{y_2}^{y_1} \frac{(1 - y)_{lm} \, dy}{(1 - y)(y - y^*)} \tag{5.79}$$

The term outside the integral is the height (H_{OG}) of an overall transfer unit based on the gas side, while the integral term represents the number (N_{OG}) of overall transfer units required to achieve separation from y_1 to y_2. The tower height is then,

$$Z = H_{OG}N_{OG} \tag{5.80}$$

The evaluation of N_{OG} requires graphical integration. To simplify the graphical integration, the quantity $(1-y)_{lm}$, which is defined as

$$(1-y)_{lm} = \frac{(1-y^*)-(1-y)}{\ln \dfrac{1-y^*}{1-y}}$$

can be taken as the arithmetic average as follows:

$$(1-y)_{lm} = \frac{(1-y^*) + (1-y)}{2} \tag{5.81}$$

Substitution of this in the integral term of Eq. (5.79) results in

$$N_{OG} = \int_{y_2}^{y_1} \frac{dy}{y-y^*} + \frac{1}{2} \ln \frac{1-y_2}{1-y_1} \tag{5.82}$$

or, in terms of mole ratios,

$$N_{OG} = \int_{Y_2}^{Y_1} \frac{dY}{Y-Y^*} - \frac{1}{2} \ln \frac{1+Y_1}{1+Y_2} \tag{5.83}$$

The integrals in both Eqs. (5.82) and (5.83) must be evaluated by graphical method.

If the two contacting streams are relatively dilute, and both the operating and equilibrium lines are straight over the tower height, then it can be shown that

$$N_{OG} = \frac{y_1 - y_2}{(y-y^*)_{lm}} \tag{5.84}$$

where

$$(y-y^*)_{lm} = \frac{(y-y^*)_1 - (y-y^*)_2}{\ln \dfrac{(y-y^*)_1}{(y-y^*)_2}}$$

The determination of the height (H_{OG}) of an overall transfer unit is best done using data obtained from small pilot-plant scale absorbers filled with the packing to be used in the large-scale equipment. The pilot-plant equipment is operated under the same conditions of flow rates, concentrations, and temperatures which are expected to prevail in the large plant. In the absence of pilot plant data, H_{OG} may be estimated from individual phase transfer unit heights according to the relation,

$$H_{OG} = H_G + \left(\frac{mG}{L}\right) H_L \tag{5.85}$$

The heights of gas and liquid transfer units, in Eq. (5.85), have been correlated by

$$H_G = \frac{\alpha (G')^{\beta}}{(L')^{\gamma}} \sqrt{Sc_G} \tag{5.86}$$

$$H_L = \phi \left(\frac{L'}{\mu_L}\right)^{\eta} \sqrt{Sc_L} \tag{5.87}$$

where H_G and H_L are in meters, G' and L' are in kg/m²-sec, μ_L is in kg/m-sec and the gas and liquid Schmidt numbers are $Sc_G = (\mu_G/\rho_G D_G)$ and $Sc_L = (\mu_L/\rho_L D_L)$. D_G and D_L are the gas and liquid diffusivities respectively of the solute absorbed. Values of the constants α, β, and γ for several packing materials are presented in Table 5.11, and those of ϕ and η are given

Table 5.11 Values of α, β and γ for several packing materials (ref. 28)

Packing	α	β	γ	Range of G' (kg/m²–s)	Range of L' (kg/m²–s)
Raschig rings					
$\frac{3}{8}$ in	0.620	0.45	0.47	0.27–0.68	0.68–2.0
1 in	0.608	0.39	0.58	0.27–1.1	0.54–0.68
1 in	0.557	0.32	0.51	0.26–0.81	0.68–6.1
$1\frac{1}{2}$ in	0.830	0.38	0.66	0.27–0.95	0.68–2.0
$1\frac{1}{2}$ in	0.689	0.38	0.40	0.27–0.95	2.0–6.1
2 in	0.894	0.41	0.45	0.27–1.1	0.68–6.1
Berl saddles					
$\frac{1}{2}$ in	0.540	0.30	0.74	0.27–0.95	0.68–2.0
$\frac{1}{2}$ in	0.367	0.30	0.24	0.27–0.95	2.0–6.1
1 in	0.461	0.36	0.40	0.27–1.1	0.54–6.1
$1\frac{1}{2}$ in	0.652	0.32	0.45	0.27–1.36	0.54–6.1

Table 5.12 Values of ϕ and η for several packing materials (ref. 28)

Packing	ϕ	η	Range of L' (kg/m^2–s)
Raschig rings			
$\frac{3}{8}$ in	0.000321	0.46	0.54–20
$\frac{1}{2}$ in	0.000718	0.35	0.54–20
1 in	0.00235	0.22	0.54–20
$1\frac{1}{2}$ in	0.00261	0.22	0.54–20
2 in	0.00293	0.22	0.54–20
Berl saddles			
$\frac{1}{2}$ in	0.00146	0.28	0.54–20
1 in	0.00129	0.28	0.54–20
$1\frac{1}{2}$ in	0.00137	0.28	0.54–20

in Table 5.12. In Eqs. (5.85), (5.86) and (5.87), the values of G, L, G' and L' are the average values for the overall tower.

Tower Diameter

The packed-tower diameter is usually decided from a consideration of the flooding velocity. For a given liquid flow rate, as the gas flow rate is gradually increased, a situation is reached where the drag on the liquid is so high that it prevents the liquid from flowing freely down the column. The point at which some of the liquid is held in the packing is referred to as the loading point. At a higher gas rate the liquid accumulates and tends to block the entire free cross-section preventing gas flow. The pressure drop through the column increases very quickly and some liquid may even be carried back, towards the top of the column. This undesirable condition is known as flooding, and the superficial gas velocity at which it occurs is called the flooding velocity. The most economical gas flow rate is usually 60 to 70% of the flooding velocity.

The flooding velocity is a function of the liquid flow rate, the gas and liquid densities, the viscosity of the liquid, and the nature of the packing. These factors are correlated in Fig. 5.25. In the figure, L'=liquid flow rate

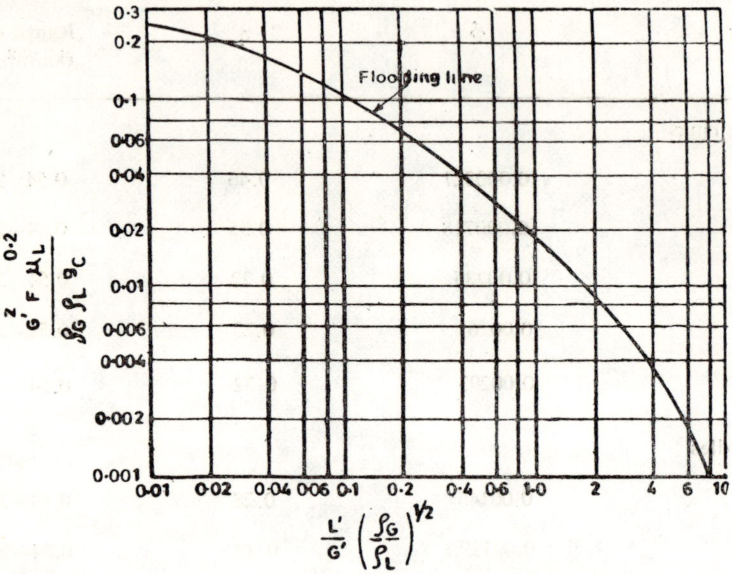

Fig. 5.25 Flooding in random-packed towers

(lb/sec. ft²), G' = gas flow rate (lb/sec. ft²) ρ_L = liquid density (lb/ft³), ρ_G = gas density (lb/ft³), F = packing factor, μ_L = viscosity of liquid (cps), and g_c = gravitational constant (32.2). The procedure for evaluating the tower diameter is as follows:

(1) Calculate $L'/G'\ (\rho_G/\rho_L)^{1/2}$; read $G'^2 F\mu_L^{0.2}/\rho_G\rho_L g_c$ at the flooding condition.
(2) Solve for G', the superficial gas mass flow velocity. This is G'_{flooding}.
(3) For proper design use $G' = 0.6$ to $0.7\ G'_{\text{flooding}}$.
(4) Calculate tower cross-sectional area,

$$S = \text{gas mass flow rate (lb/sec.)}/G'\ \text{(lb/m}^2\text{-sec.)}$$

(5) Calculate tower diameter, $D = (4S/\pi)^{1/2}$.

For a countercurrent absorber, the largest values of gas and liquid flow rate will be at the bottom of the tower, and these should be used for evaluating the diameter. A list of the values of packing factor, F, for various random packings used in industrial operations is presented in Table 5.13.

Table 5.13 Values of packing factor for various packings*

Packing	Nominal size, inches	Packing factor, *F*
Stoneware Raschig rings	1/4	768
	1/2	570
	1	155
	1 1/2	108
	2	46
Carbon Raschig rings	1/2	373
	1	170
	1 1/2	92
	2	56
Berl saddles [porcelain]	1/4	4,225
	1/2	574
	1	229
	1 1/2	79
Pall rings (polypropylene]	5/8	97
	1	52
	1 1/2	40
	2	25
Pall rings [metal]	5/8	70
	1	48
	1 1/2	28
	2	20
In talox saddles [ceramic]	1/2	200
	1	98
	1 1/2	52

*For use as a guide only. Manufacturers should be consulted for exact data.

Example 7

A gas mixture consisting of 5 per cent ammonia and 95 per cent air (by volume) is to be scrubbed with water to reduce the ammonia content to 0.5 per cent. The flow rate of the gas mixture entering the tower at the bottom at 25°C and 1 atm is 38.0 kg moles/hr while ammonia-free water enters at the top of tower at a rate of 47.4 kg moles/hr. The tower uses 1 in, stoneware Raschig rings packed at random and operates at 60 per cent of the flooding point. The equilibrium data for ammonia-water system at 20°C and 1 atm are as follows :

y_{NH_3}	0.000	0.0158	0.0239	0.0328	0.0416
x_{NH_3}	0.000	0.0208	0.0309	0.0405	0.0503

Calculate the scrubber diameter and the height of the packed section that are necessary to satisfy the given conditions.

Solution :

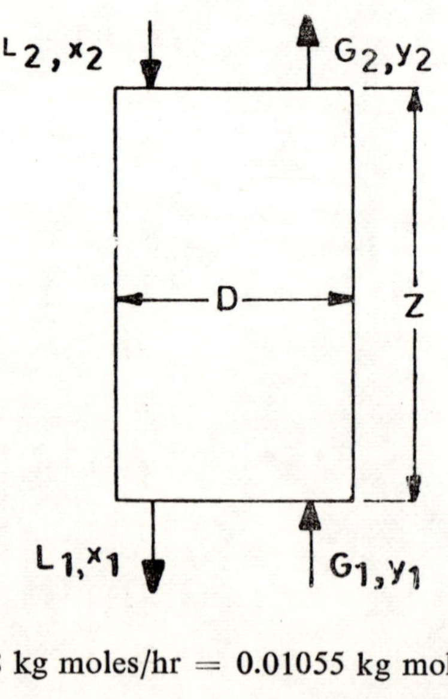

Given : $G_1 = 38$ kg moles/hr $= 0.01055$ kg moles/sec.

$L_2 = 47.4$ kg moles/hr $= 0.01316$ kg moles/sec.

$y_1 = 0.05,\ y_2 = 0.005$

$x_2 = 0.0$ (HN_3-free water)

The moles of NH_3-free gas entering the tower,

$$G_s = G_1(1-y_1) = 0.01055\ (1-0.05)$$

$$= 0.01 \text{ kg moles/sec.}$$

Similarly, the moles of NH_3-free water entering the tower,

$$L_s = L_2 = 0.01316 \text{ kg moles/sec.}$$

$$G_2 = \frac{G_s}{1-y_2} = \frac{0.01}{0.995} = 0.01005 \text{ kg moles/sec.}$$

The concentrations of the gas and liquid phases on a solute-free basis are obtained as,

$$Y_1 = \frac{y_1}{1-y_1} = \frac{0.05}{0.95} = 0.0526 \frac{\text{moles } NH_3}{\text{mole air}}$$

$$Y_2 = \frac{y_2}{1-y_2} = \frac{0.005}{0.995} = 0.005 \frac{\text{moles } NH_3}{\text{mole air}}$$

$$X_2 = 0.$$

From a solute material balance, X_1 can be evaluated.

$$G_s(Y_1 - Y_2) = L_s(X_1 - X_2)$$

$$X_1 = \frac{G_s}{L_s}(Y_1 - Y_2)\frac{0.01}{0.01316}\ (0.0526-0.005)$$

$$X_1 = 0.0362 \frac{\text{moles } NH_3}{\text{mole water}}$$

The molecular weight of the gas mixture, M, at the top :

$$M_2 = \Sigma Y_i M_i = 0.005(17) + 0.995(29)$$

$$= 28.94 \text{ kg/kg-mole}$$

At the bottom :

$$M_1 = 0.05(17) + 0.95(29) = 28.4 \text{ kg/kg mole}$$

Total gas mass flow rate at the bottom

$$G_{m_1} = 0{\cdot}01055(28.4) = 0.2996 \text{ kg/sec.}$$

Similarly,

$$G_{m_2} = 0.01005(28.94) = 0.2908 \text{ kg/sec.}$$

and $\quad L_{m_2} = L_2(18) = 0.01316(18) = 0.2369$ kg/sec.

From an overall material balance,

$$L_{m_1} = (G_{m_1} + L_{m_2}) - G_{m_2}$$

$$= (0.2996 + 0.2369) - 0.2908$$

$$L_{m_1} = 0.2457 \text{ kg/sec.}$$

Tower Diameter

The flooding correlation is given in English units and so these units are used wherever necessary. The correlation requires gas and liquid-phase densities at the top and bottom of the tower. Since the ammonia concentration in the liquid phase is quite small, the density of pure water, 62.3 lb/ft³, is taken as the solution density throughout the tower. For gas phase, on the assumption that the gas mixture acts as an ideal gas,

$$\rho_G = \frac{P}{RT} = \frac{MP}{R_u T}$$

where M = molecular weight of the gas mixture and R_u = universal gas constant.

$$\text{At the top, } \rho_G = \frac{28.94\ (14.7)}{10.73\ (528)} = 0.075 \text{ lb/ft}^3$$

Similarly, at the bottom $\rho_G = 0.0736$ lb/ft³

The value of G' at the top and bottom of the tower can now be ascertained from the flooding correlation.

$$\text{At the top, } \frac{L'}{G'}\sqrt{\frac{\rho_G}{\rho_L}} = \frac{L_{m_2}}{G_{m_2}}\sqrt{\frac{\rho_{G_2}}{\rho_L}} = \frac{0.2369}{0.2908}\sqrt{\frac{0.075}{62.3}} = 0.0281$$

From Fig. 5.25, the ordinate for flooding condition is found to be 0.19. Therefore,

$$\frac{(G')^2_{\text{flooding}}\ F\mu_L^{0.2}}{\rho_G \rho_L g_c} = 0.19$$

The packing factor for 1 in. stoneware Raschig rings,

$$F = 155, \text{ and } \mu_L = 0\ 982 \text{ centipoise}$$

Hence,

$$G'_{\text{flooding}} = \left[\frac{0.19\ (0.075)\ (62.3)\ (32.2)}{155\ (0.982)^{0.2}}\right]^{0.5}$$
$$= 0.43 \text{ lb/ft}^2\text{-sec.}$$

For 60 per cent of the flooding point,

$$G'_{\text{design}} = 0.43\ (0.60) = 0.258 \text{ lb/ft}^2\text{-sec.}$$

$$G_{m_2} = 0.2908 \text{ kg/sec} = 0.64 \text{ lb/sec.}$$

The cross-sectional area of the tower,

$$S_2 = \frac{G_{m_2}}{G'_{\text{design}}} = \frac{0.64}{0.258} = 2.48 \text{ ft}^2 \quad \text{(top)}$$

At the bottom of the tower,

$$\frac{L'}{G'}\sqrt{\frac{\rho_G}{\rho_L}} = \frac{L_{m_1}}{G_{m_1}}\sqrt{\frac{\rho_{G_1}}{\rho_L}} = \frac{0.2457}{0.2996}\sqrt{\frac{0.0736}{62.3}}$$

$$= 0.82\ (0.0344) = 0.0282$$

From Fig. 5.25, the ordinate is again 0.19.

$$G'_{\text{flooding}} = 0.426\ \text{lb/ft}^2\text{-sec.}$$

$$G'_{\text{design}} = 0.6\ (0.426) = 0.2556/\text{ft}^2\text{-sec.}$$

$$G_{m_1} = 0.2996\ \text{kg/sec} = 0.659\ \text{lb/sec.}$$

$$S_1 = \frac{G_{m_1}}{G'_{\text{design}}} = \frac{0.659}{0.2556} = 2.58\ \text{ft}^2\ \text{(bottom)}$$

Selecting the *larger area* for the tower cross-section, we get the diameter of the tower,

$$D = \sqrt{\frac{4(2.58)}{\pi}} = 1.81\ \text{ft} = 55\ \text{cm.}$$

Height of the Tower

$$Z = (H_{OG})\ (N_{OG})$$

$$N_{OG} = \int_{Y_2}^{Y_1} \frac{dY}{Y - Y^*} - \tfrac{1}{2} \ln \frac{1 + Y_1}{1 + Y_2}$$

The equilibrium data are converted into mole ratios:

Y $\dfrac{\text{moles } NH_3}{\text{mole air}}$	0.00	0.0160	0.0245	0.0339	0.0434
X $\dfrac{\text{moles } NH_3}{\text{mole water}}$	0.00	0.0212	0.0319	0.0422	0.0529

The operating and equilibrium lines are constructed as shown in Fig. 5.26 (a) from which $Y - Y^*$ is read and tabulated below:

Y	Y^*	$Y - Y^*$	$\dfrac{1}{Y - Y^*}$
$Y_2 = 0.005$	0	0.005	200
0.01	0.0025	0.0075	133.3
0.02	0.0085	0.0115	87
0.03	0.0145	0.0155	64.5
0.04	0.02	0.02	50
0.05	0.027	0.023	43.5
$Y_1 = 0.0526$	0.0285	0.0241	41.5

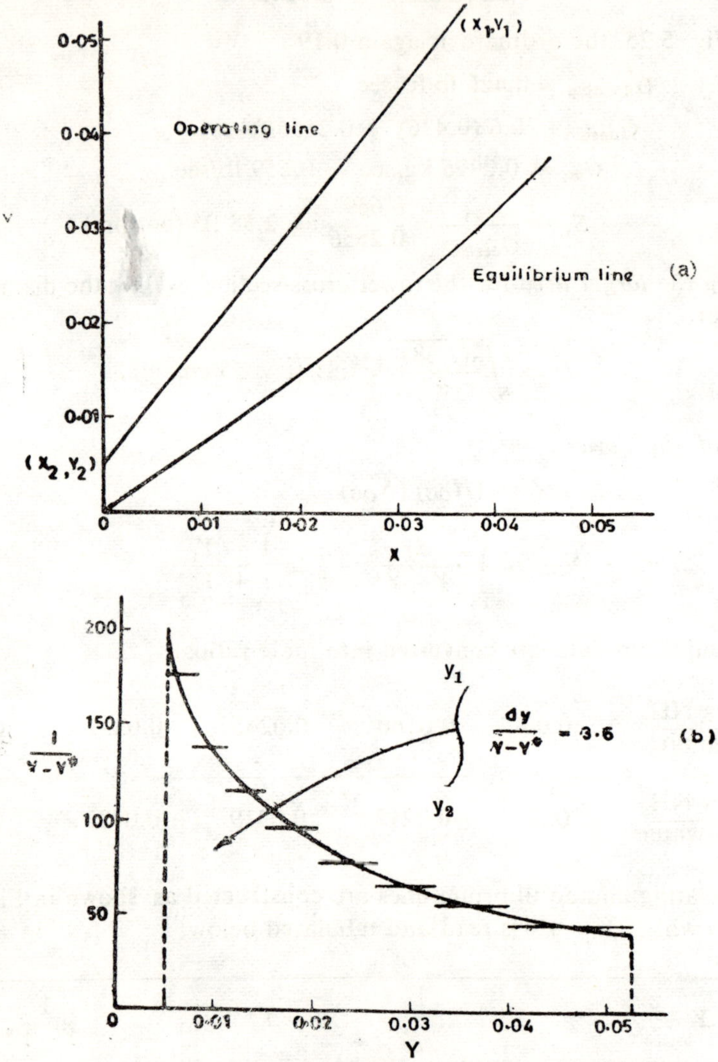

Fig. 5.26 Solution to example 7.

By graphical integration [Fig. 5.26 (b)].

$$\int_{Y_2}^{Y_1} \frac{dY}{Y - Y^*} = 3.6$$

Therefore, $N_{OG} = 3.6 - \frac{1}{2} \ln \frac{(1 + 0.0526)}{(1 + 0.005)}$

$$N_{OG} = 3.6 - 0.02 = 3.58$$

The height of a transfer unit (H_{OG}) can be found through the use of generalized correlations given in Eqs. (5.85), (5.86), and (5.87). The height (H_G) of the gas transfer unit can be calculated from the relation,

$$H_G = \frac{\alpha(G')^\beta}{(L')^\gamma} \sqrt{Sc_G}$$

In the above equation G' and L' are the average values for the entire tower which in this case are 1.254 and 1.007 kg/m².sec, respectively From Table 5.11, for 1 in. Rachig rings, $\alpha = 0.557$, $\beta = 0.32$, and $\gamma = 0.51$. These data are used with some reservation since G' exceeds the limit shown in the table. The Schmidt number $Sc_G = 0.705$,

$$H_G = \frac{0.557\ (1.254)^{0.32}}{(1.007)^{0.51}} \sqrt{0.705} = 0.503 \text{ m}$$

Similarly, H_L is given by

$$H_L = \phi \left(\frac{L'}{\mu_L}\right)^\eta \sqrt{Sc_L}$$

where, from Table 5.12, $\phi = 0.00235$, and $\eta = 0.22$. The Schmidt number $Sc_L = 462$. Substituting these values, we get

$$H_L = 0.00235 \left(\frac{1.007}{9.82 \times 10^{-4}}\right)^{0.22} \sqrt{462}$$

or

$$H_L = 0.232 \text{ m}$$

Employing, Eq. (5.85), we can get the height of the overall transfer unit (H_{OG}). The slope (m) of the equilibrium line can be evaluated to be equal to 0.78. The average values of G and L are 0.0103 and 0.0134 kg moles/sec, respectively. Substituting, these values in Eq. (5.85), we get

$$H_{OG} = 0.503 + 0.78 \left(\frac{0.0103}{0.0134}\right) 0.232$$

or

$$H_{OG} = 0.503 + 0.139 = 0.642 \text{ m}$$

The height of the packing,

$$Z = 3.58\ (0.642) = 2.3 \text{ m}$$

The total height of the tower should be somewhat greater than this value to allow for the distribution space at the top and at the bottom.

Adsorption by Solids

An alternative to absorption by liquids is the adsorption of pollutants on solids. Adsorption is a surface phenomenon by which gas or liquid molecules are captured by and adhere to the surface of a solid adsorbent. The attractive forces holding the molecules at the surface may be either physical (physical adsorption) or chemical (chemisorption) in nature. In physical adsorption the gaseous material condenses upon the surface of the solid, accompanied by an evolution of heat. The adsorbed material can be removed or desorbed by reducing the pressure or by increasing the temperature. Thus, the process is reversible. Chemisorption, which is the result of a chemical interaction between the solid and the adsorbed material, is usually irreversible with the heat of adsorption being much higher than that for physical adsorption. The molecules are held to the solid surface by chemical bonds and the original material undergoes a chemical change. Chemisorption is of particular importance in catalysis, but is not a primary mechanism in air pollution control methods.

Adsorption techniques are widely used in the field of odour control for removing small quantities of pollutants present in a large volume of air. Adsorption techniques are also used for collecting valuable organic substances that cannot be picked up by scrubbing methods. The rate of adsorption of a substance depends on the concentration of the material around the adsorbent, the surface area of the adsorbent, the pore volume of the adsorbent, and some other properties such as temperature, molecular polarity and the chemical nature of the adsorbent surface. Several solids possess adsorptive ability with extremely large specific surface areas. Some of the adsorbents commonly used in air pollution control are activated carbon, activated alumina, silica gel and molecular sieves. Typical data for adsorbent materials and their uses are given in Table 5.14.

Adsorption of a gas on a solid occurs in several stages. The first stage is the diffusion of the pollutant from the bulk gas phase to the external surface of the solid, and this is similar to the diffusion of the gas to the gas-liquid interface in absorption. The second stage is the diffusion of the gas molecules into the pores of the solid, and the third stage is the actual adsorption on the active sites in the pores The third stage is relatively fast compared to the first two stages, and it is these earlier stages that determine the rate of adsorption. It is usually not possible to determine which of the first two stages controls the adsorption rate. In some cases they are of equal significance. For that reason, adsorbers are most often designed using empirical methods based on scale-up from pilot-scale data.

Removal of pollutants by adsorption may be carried out in a batchwise or continuous manner of operation. A typical fixed bed operating on a batch cycle is shown in Fig. 5.27. The polluted gas stream enters the bed at the top, travels downward through the bed, and leaves at the bottom.

When the gas first enters the solid bed, the front portion of the bed adsorbs the pollutant gas very rapidly and the little amount of solute left is taken up by the deeper layers of the bed. Thus, the gas initially leaving the bed is almost completely free of the pollutant. After a period of

Fig. 5.27 Fixed bed adsorber

time, the layers of the solid at the entrance become saturated with the adsorbed gas and the zone of active adsorption moves deeper into the bed. Finally, when the front of the active zone reaches the bottom of the bed the bed starts becoming inactive and the pollutant level in the exit gas stream starts to rise and continues to rise until it approaches the influent concentration. The curve of pollutant concentration as a function of time thus has an *S*-shaped appearance and is commonly called the break-through curve. Fig. 5.28 shows a typical break-through curve. The time at which the break-through curve first begins to rise appreciably is called the break-

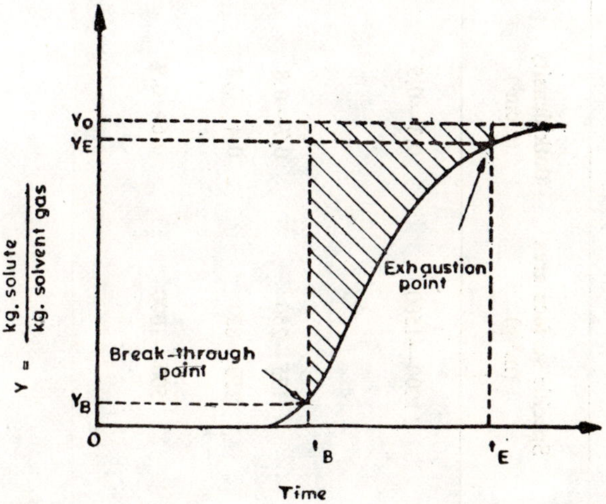

Fig. 5.28 Typical adsorption break-through curve

through point, and the time at which the outlet effluent concentration essentially reaches that of the inlet pollutant concentration is called the exhaustion point.

The adsorber is normally regenerated when the break-through point is

Table 5.14 Properties of adsorbents and their uses

Adsorbent	Specific surface area (m^2/g)	Bulk density (g/cm^3)	Pore volume (cm^3/g)	Average pore size (Å)	Uses
Activated carbon	1000—1500	0.3—0.5	0.6—0.8	20—35	Removal of odours, and trace impurities from gases, purification of industrial gases and hydrocarbons, solvent recovery
Activated alumina	175—250	0.7—0.8	0.4	34	Dehydration of gases and liquids
Silica gel	250—850	0.4—0.8	0.3—0.45	21	Dehydration and purification of gases
Molecular sieves	500—1000	0.6—0.9	0.25—0.3	~4	Selective adsorption of CO_2, NH_3, C_2H_2, H_2S, and SO_2

reached. Regeneration means removing the pollutant from the adsorbent thereby renewing it for further use. For the single-bed adsorber, as shown in Fig. 5.27, the regeneration is accomplished by stopping the gas stream and then allowing steam or an inert gas to flow through the adosrbent bed.

Adsorption Process Design

Adsorption processes are most easily designed through the use of break-through curves obtained in pilot-plant studies. When break-through occurs, the adsorbent in the adsorption zone is practically spent. A major part of the break-through curve analysis is concerned with the determination of the fractional ability of the adsorbent in the adsorption zone and the depth of the adsorption zone.

The mass velocity of the solvent gas stream is G_m kg/(hr) (m^2). The initial concentration of the pollutant (solute) in the inlet gas is Y_0 kg. solute/kg. solvent gas and the concentration at the break-through point is Y_B. The adsorbent has essentially exhausted at some arbitrarily chosen value Y_E close to Y_0. The adsorption zone Z_A in the bed, of constant depth, is that part of the bed in which the concentration is in the process of being changed from Y_B to Y_E. If t_F is the adsorption zone formation time and Z is the height of the adsorbent bed,

$$Z_A = Z \frac{t_E - t_B}{t_E - t_F} \tag{5.88}$$

where t_E = time to exhaustion, t_B = time to break-through point.

The difference $t_E - t_B$ gives the time required for the adsorption zone to move its own depth down the adsorbent bed after the zone has been established.

The amount of solute S, removed from the gas in the adsorption zone from t_B to t_E is given by the shaded area in Fig. 5.28 times G_m. This is given by

$$S = G_m \int_{t_B}^{t_E} (Y_0 - Y)\, dt \tag{5.89}$$

If, however, all adsorbent in the zone were saturated with solute, it would contain $Y_0\, G_m\, (t_E - t_B)$ kg solute/m^2 of bed cross-section. Therefore, the fractional ability of the adsorbent in the adsorption zone still to adsorb solute is,

$$f = \frac{S}{Y_0 G_m (t_E - t_B)} = \frac{\int_{t_B}^{t_E} (Y_0 - Y)\, dt}{Y_0 (t_E - t_B)} \tag{5.90}$$

or

$$f = \int_0^{1.0} \left(1 - \frac{Y}{Y_0}\right) d\left(\frac{t - t_B}{t_E - t_B}\right) \tag{5.91}$$

and
$$t_F = (1 - f)(t_E - t_B) \tag{5.92}$$

Substituting Eq. (5.92) into (5.88), we get

$$Z_A = Z \frac{t_E - t_B}{t_E - (1 - f)(t_E - t_B)} \tag{5.93}$$

In the adsorption column of bed height Z, the amount of adsorbent per unit cross-sectional area of the bed is $Z\rho_b$, where ρ_b is the bulk density (Table 5.14). If the bed was in equilibrium with the entering gas and, therefore, completely saturated at a concentration of X_T (kg adsorbate per kg of adsorbent), the amount of adsorbate would be $Z\rho_b X_T$ kg. At the break-through point the bottom portion of the adsorption zone of depth Z_A is still in the column, but the remaining bed $Z - Z_A$ is saturated. The amount of adsorbate at the break-through point is,

$$(Z - Z_A)\,\rho_b X_T + Z_A \rho_b f X_T$$

The degree of saturation at the break-through point is,

$$\eta = \frac{[(Z - Z_A) + Z_A f]\,\rho_b X_T}{Z\rho_b X_T} = \frac{(Z - Z_A) + Z_A f}{Z} \tag{5.94}$$

or
$$\eta = \frac{Z - (1 - f)\,Z_A}{Z} \tag{5.95}$$

The degree of saturation is strongly dependent upon the fractional capacity f. As f decreases, Z_A increases in size and η decreases in magnitude. Thus, maximizing the fractional ability of the adsorption zone is very beneficial.

For a coordinate system that moves with the adsorption zone, a material balance over the zone is,

$$G_m(Y_0 - 0) = L_m(X_T - 0) \tag{5.96}$$

or
$$\frac{L_m}{G_m} = \frac{Y_0}{X_T} \tag{5.97}$$

This is the operating line in the zone when L_m and G_m are the solid and gas flow rates relative to the zone. Since the operating line passes through the origin, at any level in the zone the solute concentration in the gas (Y) and of the adsorbate upon the solid (X) are related according to the equation,

$$G_m\, Y = L_m\, X \tag{5.98}$$

Over the differential height dZ of the zone, the rate of adsorption is,

$$G_m\, dY = K_Y a\,(Y - Y^*)\, dZ \tag{5.99}$$

where $(Y - Y^*)$ is the difference between the operating line and the equilibrium curve at any given value of X. Fig. 5.29 shows the operating line and the equilibrium curve.

For the adsorption zone.

$$N_{OG} = \int_{Y_B}^{Y_E} \frac{dY}{Y - Y^*} \tag{5.100}$$

and

$$H_{OG} = \frac{G_m}{K_Y a} \tag{5.101}$$

The depth of the zone is,

$$Z = H_{OG}\, N_{OG} \tag{5.102}$$

where N_{OG} is the number of overall gas-side transfer units, and H_{OG} is the height of transfer units in the adsorption zone. For any value of $Z < Z_A$, assuming that H_{OG} is constant, we get

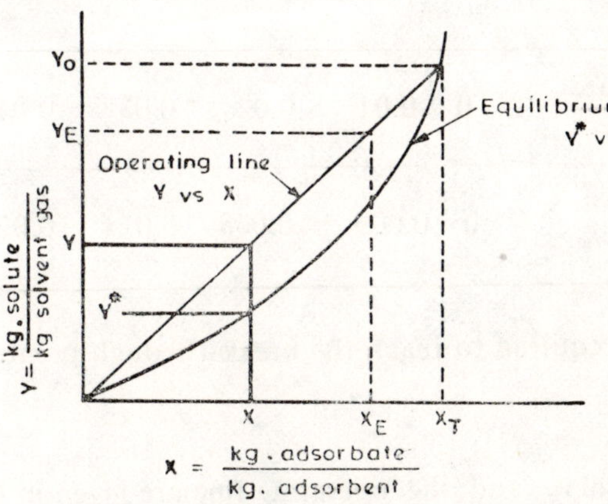

Fig. 5.29 Typical operating line and equilibrium curve in the adsorption zone

$$\frac{Z \text{ at } Y}{Z_A} = \frac{\displaystyle\int_{Y_B}^{Y} \frac{dY}{Y - Y^*}}{\displaystyle\int_{Y_B}^{Y_E} \frac{dY}{Y - Y^*}} = \frac{t - t_B}{t_E - t_B} \tag{5.103}$$

From adsorption equilibrium data, the integrals in Eq. (5.103) can be evaluated graphically to obtain value of $(t - t_B)/(t_E - t_B)$. These values are then plotted against $(1 - Y/Y_0)$ and the area under the curve gives f (Eq. 5.91), the fractional ability of the adsorbent. Also, knowing $K_Y a$ from experimental data, we can estimate the adsorption zone depth Z_A from Eqs. (5.101) and (5.102). The degree of saturation at break-through point can be calculated by Eq. (5.95) and the amount of adsorbate at the break-through point is then equal to $\eta Z \rho_b X_T$.

Example 8

A fixed bed adsorber, packed to the height of 1 m, is employed to reduce the concentration of a pollutant in an air stream. The air enters the adsorber at a rate of 100 kg/(hr) (m^2) and the initial concentration of the pollutant entering with the air is 0.024 kg pollutant/kg air. The adsorption is assumed to be isothermal and the bulk density of the adsorbent is 400 kg/m^3. The break-through point is considered as that time when the effluent air has the pollutant concentration of 0.001 kg pollutant/kg air and the bed is considered exhausted when the concentration reaches 0.022 kg pollutant/kg air. The overall capacity coefficient $K_Y a$ may be assumed to be 7150 kg/(hr) (m^3) ΔY. The adsorption equilibrium data for the system are:

$X\ \dfrac{\text{kg adsorbate}}{\text{kg adsorbent}}$	0	0.01	0.02	0.03	0.04	0.05
$Y\ \dfrac{\text{kg pollutant}}{\text{kg air}}$	0	0.003	0.008	0.014	0.022	0.032

Estimate the time required to reach the break-through point.

Solution

The equilibrium curve and the operating line are given in Fig. 5.30. For $Y_0 = 0.024$ kg pollutant/kg air the value of X_T from the figure is 0.042 kg adsorbate/kg adsorbent. Also, $Y_B = 0.001$, $Y_E = 0.022$ kg pollutant/kg air.

Values of Y on the operating line between Y_B and Y_E and of the equilibrium curve Y^* at the same values of X are given in the first and the second columns of the following table. The values in third and the fourth columns are computed from this information. The total number of transfer units corresponding to the adsorption zone is $N_{OG} = 9.314$ using Eq. (5.100). The values in the fifth column are obtained by dividing the corresponding values in the fourth column by 9.314 in accordance with Eq. (5.103).

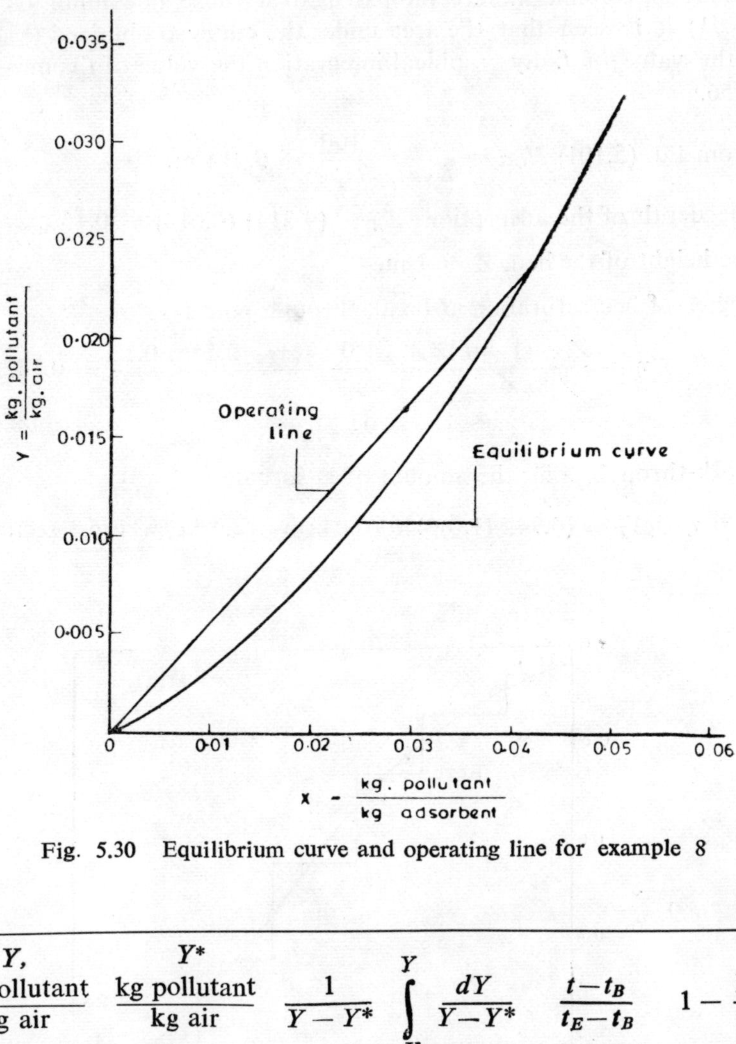

Fig. 5.30 Equilibrium curve and operating line for example 8

Y, $\frac{\text{kg pollutant}}{\text{kg air}}$	Y^* $\frac{\text{kg pollutant}}{\text{kg air}}$	$\frac{1}{Y-Y^*}$	$\int_{Y_B}^{Y} \frac{dY}{Y-Y^*}$	$\frac{t-t_B}{t_E-t_B}$	$1-\frac{Y}{Y_0}$
Y_B = 0.001	0.00037	1587	0	0	0.958
0.002	0.00087	885	1.244	0.134	0.917
0.004	0.002	500	2.484	0.267	0.834
0.008	0.005	333	4.124	0.443	0.667
0.012	0.0085	286	5.364	0.576	0.500
0.016	0.0127	303	6.524	0.700	0.333
0.020	0.0180	500	8.044	0.864	0.167
Y_E = 0.022	0.0207	769	9.314	1.000	0.083

Values in column six are plotted against those in column five; from Eq. (5.91) it is seen that the area under the curve so obtained (Fig. 5.31) gives the value of f. By graphical integration the value of f comes out to be 0.556.

From Eq. (5.101) $H_{OG} = \dfrac{G_m}{K_Y a} = \dfrac{100}{7150} = 0.014$ m.

The depth of the adsorption, $Z_A = (9.314)\ (0.014) = 0.13$ m.

The height of the bed, $Z = 1$ m.

Degree of bed saturation at break-through point,

$$\eta = \frac{Z-(1-f)Z_A}{Z} = \frac{1.0-(1-0.556)\ 0.13}{1.0} = 0.942,$$

$$= 94.2\%$$

At break-through point the amount of adsorbate

$$= \eta Z \rho_b X_T = (0.942)(1.0)(440)(0.042) = 17.4 \text{ kg/m}^2 \text{ cross-section}$$

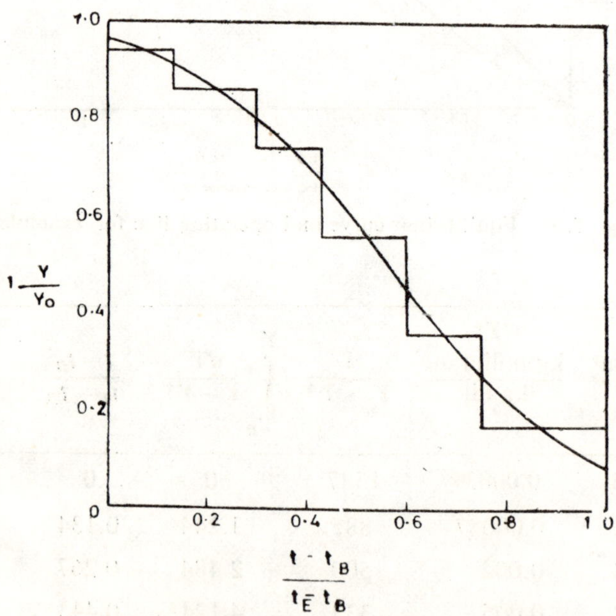

Fig. 5.31 Solution to example 8

The pollutant enters with the air at a rate of

$$(100)(0.024) = 2.4 \text{ kg/(hr)(m}^2\text{)}.$$

The time required to reach the break-through point,

$$t_B = 17.4/2.4 = 7.25 \text{ hr.}$$

Combustion

Many industrial processes produce gas streams that have no recovery value, so absorption or adsorption methods may not be economically feasible. If the waste gas contains sufficient combustible material then incineration may be the simplest route to air pollution control. Combustion applications include the destruction of odours, toxic substances, and reactive materials, the prevention of explosion hazards, and the reduction of pollutants in the oxidised forms. The products of combustion of organic materials are normally odourless, colourless, and harmless giving off carbon dioxide and water vapour. However, high molecular weight hydrocarbons could cause smoke problems. Similarly, certain inorganic materials upon combustion give oxidation products like sulphur dioxide and oxides of nitrogen, and these remain a pollution problem unless they are easily separated from the gas stream.

When a pollutant is to be destroyed, the combustion must be complete; otherwise, intermediate products of combustion will form which may be more noxious than the original contaminant. For complete combustion, the oxygen must come into intimate contact with the combustible material through adequate turbulence at sufficiently high temperature and have a sufficiently long residence time. In fact, time, temperature, and turbulence have so predominant role in combustion reactions that they are often called the "there T's" of combustion.

The normal ranges are: temperature: 375—825°C, residence time: 0.2—0.5 sec and gas velocity: 4.5—7.5 m/sec.

There are three methods of combustion in common use today: direct combustion (or flaring), thermal incineration (also called after-burning or flame combustion), and catalytic oxidation. Highly combustible streams with high heating values can be eliminated by direct flaring. However, flaring is not a satisfactory solution when the gas streams contain excessive amounts of inorganic pollutants like sulphur, chlorine and fluorine. In such cases the gas streams are pretreated before flaring. Often unsaturated hydrocarbons such as olefins and aromatics with their low hydrogen-to-carbon ratios produce smoke unless the design is such that it allows them to burn smokelessly. One such design incorporates steam injection through jets surrounding the flare tip (Fig. 5.32). The steam jets induce turbulence in the periphery of the flame and add heat to the combustion process. In addition, this sets up a water gas reaction giving carbon monoxide and hydrogen which helps to generate smokeless flame. The steam requirements vary from 0.05 to 0.3 kg steam/kg gas, depending on the olefinic and aro-

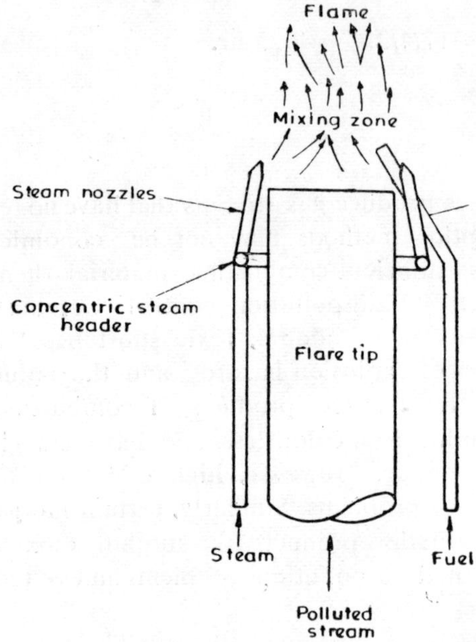

Fig. 5.32 Steam injection type flare

matic content of the gas stream. A pilot flame is also provided to re-ignite the gas stream if flameout occurs.

Preheating of the gas stream is often required if the combustible gases are diluted with inert gases to ensure that the miniuum fume energy concentration is not reached. This is done by using a ring burner in addition to the pilot flame burners. *Thermal incineration* is perhaps the most efficient as well as the most flexible technique for destroying such diluted gas streams. In the thermal incinerator (Fig. 5.33) the waste gas is preheated

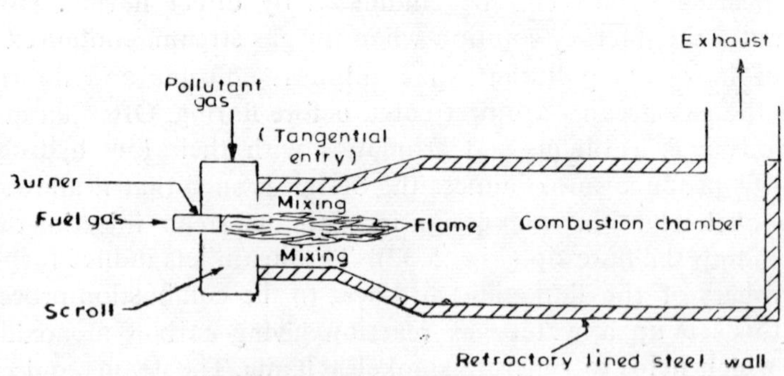

Fig. 5.33 Thermal incinerator

often over an auxiliary fuel-fired burner and passed into a combustion chamber where a temperature of 500° to 800°C is maintained. The gas stream is kept at this elevated temperature long enough (0.3 to 0.7 sec residence time) to allow complete oxidation. The gas stream is introduced in such a geometry and at a velocity that promote turbulence and thorough mixing with the burning fuel. Thermal incinerators require little maintenance if properly engineered. The major operating cost is in the form of fuel required to sustain appropriate temperature levels. The fuel requirements may be reduced by recovering the waste heat either by returning the heat to the process or to preheat the gases.

Catalytic oxidation is an attractive proposition when thermal incineration of very diluted pollutants is not economically feasible due to additional fuel cost or when heat recovery would result in only marginal savings. The combustion takes place on a catalyst usually at temperatures several degrees below those required for thermal incineration. The temperature that is necessary to initiate the reaction depends on the type of pollutant present in the gas stream. Table 5.15 summarizes the catalyst entry temperatures for various industrial systems.

The catalysts used for effective pollution control are the precious metals, primarily platinum and palladium or their alloys These are arranged in such a way as to provide the maximum possible surface area for contact with the gas. The catalyst is coated onto suitable elements such as metal ribbons, ceramic rods, or alumina pellets. These elements are then packed into the catalyst bed. A typical catalytic combustion unit is shown in Fig. 5.34. It consists of a reaction vessel or converter in which the catalyst

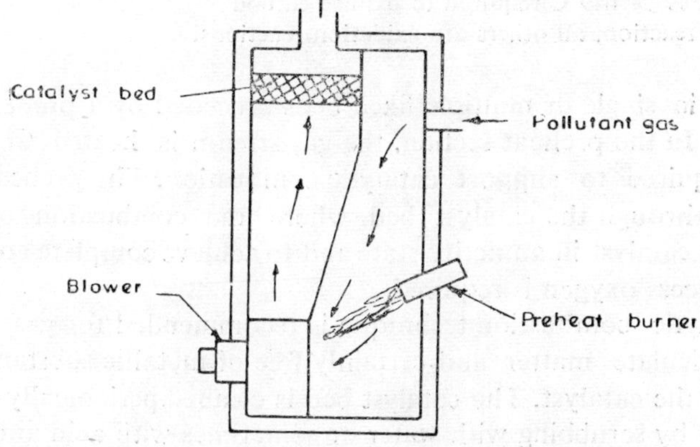

Fig. 5.34 Catalytic combustion unit

Table 5.15 Inlet temperatures for catalyst combustion (ref. 29)

Process	Contaminants in waste gas	Catalytic oxidation temp. (°C)
Asphalt oxidising	Aldehydes, anthracenes, oil vapours, hydrocarbons	315–371
Carbon black mfg.	Hydrogen, carbon monoxide, methane, carbon	649–982*
Catalytic cracking units	Carbon monoxide, hydrocarbons	343–427
Coke ovens	Wax, oil vapours	315–371
Formaldehyde mfg.	Hydrogen, methane, carbon monoxide, formaldehyde	343
Nitric acid mfg.	Nitric oxide, nitrogen dioxide	260–649**
Lithography ovens	Solvents, resins	260–399
Octyl-phenol mfg.	Phenol	315–427
Phthalic anhydride mfg.	Maleic acid. phthalic acid, napthaquinones, carbon monoxide, formaldehyde	315–343
Polyethylene mfg.	Hydrocarbons	260–649
Printing presses	Solvents	315
Varnish cooking	Hydrocarbons	315–371
Wire coating and enameling ovens	Solvents, varnish	315–371

*Temperatures > 649°C required to oxidise carbon.
**Reduction reaction, all others are oxidation reactions.

is arranged in single or multiple fixed beds preceded by a preheat section, if necessary. In the preheat section, the gas stream is heated to the temperature required to support catalytic combustion. The preheated gas is then passed through the catalyst bed where the combustion occurs. To maintain the catalyst in an active state and to achieve complete combustion, about 1% excess oxygen is required.

The catalytic combustion technique is recommended for gases that are free of particulate matter and certainly free of metallic substances which could poison the catalyst. The catalyst bed is cleaned periodically (quarterly or annually) by scrubbing with water or sometimes with acid and by heating to high temperatures.

REFERENCES

1. W. Strauss, "The control of sulfur emissions from combustion processes", in *Air Pollution Control I,* edited by W. Strauss (Wiley-Interscience, New York), 1971.
2. L. Schiller and A. Naumann, Z. Ver. Deut. Ing., Vol. 77 (1933), p. 318.
3. R.H. Perry and C.H. Chilton (Eds.), *Chemical engineers handbook*, 5th edition (McGraw-Hill, New York), 1973.
4. M. Crawford, *Air pollution control theory* (McGraw-Hill, New York), 1976.
5. C.E. Lapple, "Processes use many collection types", *Chem. Engg*, Vol. 58 (1951), p. 145.
6. C.E. Lapple, *Air pollution engr. manual,* edited by J.A. Danielson (U.S. Dept. of Health, Education & Welfare), Pub. No. 999–Ap–40, 1967, p. 95.
7. C.J. Stairmand, "The design and performance of cyclone separators", *Trans., Inst. Chem. Engr.*, Vol. 29 (1951), p. 356.
8. C.B. Shepherd and C.E. Lapple, *Ind. & Engg. Chem.*, Vol. 31 (1939), p. 972.
9. C.N. Davies, *Proc. Phys. Soc.*, Vol. 57 (1945), p. 259.
10. W. Strauss, *Industrial gas cleaning,* Ist Edition (Pergamon Press. London), 1966, p. 222.
11. W.L. Torgeson, *A study of filtration mechanisms* (General Mills Inc. Chicago), 1961.
12. H. Brauer and Y.B.G. Varma, *Air Pollution control equipment* (Springer Verlag, Berlin), 1981, p. 162.
13. J.A. Danielson (Ed.), *Air pollution engr. manual* (National Centre for Air Pollution Control, Pub. No. 999–AP–40, Cincinnati, Ohio), 1967.
14. H.F. Kraemer and H.F. Johnstone, *Ind. Engg. Chem.*, Vol. 47 (1955), p. 2426.
15. C.J. Stairmand, *Trans. Inst. Engr,* (*London*), Vol. 28 (1950), p. 130.
16. H.F. Johnstone and M.H. Roberts, *Ind. Engg. Chem.*, Vol. 41 (1949), p. 2417.
17. H.F. Johnstone, R.S. Field and M.C. Tassler, "Gas absorption and aerosol collection in venturi atomizer" *Ind. Engg. Chem.*, Vol. 46 (1954), pp. 1601-1608.
18. S. Calvert, "Source control by liquid scrubbing", in *Air pollution III,* edited by A.C. Stern (Academic Press, New York), 1968, Chap. 46, p. 475.
19. J.D. Brady and L.K. Legatski, "Venturi scrubbers", in *Air pollution control and design handbook*. part-2, edited by P.N. Chermisinoff and

R.A. Young (Marcel Dekker, Inc., New York), 1977, Chap. 28, p. 747.

20. K.T. Semrau, "Practical process design of particulate scrubbers", *Chem. Engg.*, Vol. 84 (1977). Sept. 26.
21. K.T. Semrau, "Correlation of duct scrubber efficiency", *J. Air Pollu. Control Assoc.*, Vol. 10 (1960), p. 200
22. C.E. Billings, "*Technological source of air pollution*,' in Industrial pollution edited by N. Irving Sax (Van Nostrand Reinhold Co., New York). 1974, Chapter 14, p. 350.
23. W.G. Whitman, *Chem. Met. Engr.*, Vol. 29 (1923), p. 147.
24. R.E. Treybal, *Mass-transfer operations*, 2nd Edition (McGraw-Hill New York), 1968.
25. T.K. Sherwood and R.L. Pigford, *Absorption and extraction*, 2nd Edition (McGraw-Hill, New York), 1952.
26. G.A. Morris and J. Jackson, *Absorption towers* (Butterworth, London), 1953.
27. W.S. Norman, *Absorption, distillation and cooling towers* (Longmans, London), 1961.
28. M. Crawford, *Air pollution control theory*, (McGraw-Hill, New York), 1976, pp. 495-496.
29. W. Strauss, *Industrial gas cleaning* (Pergamon Press, London), 1966, p. 120.

SUGGESTIONS FOR FURTHER READING

1. R.T. Dennis, "Collection efficiency as a function of particle size, shape and density: theory and experience", *J. Air Pollu. Control Assoc.* Vol. 24 (1974), p. 1156.
2. G.J. Celenza, "Designing air pollution control systems", *Chem. Engg. Progr.* Vol. 66 (11) (1970), p. 31.
3. P. Rashmi, "Equipment for controlling gaseous pollutants", *Chem. Engg.*, Vol. 82 (21) (1975), p. 129.
4. R.C. Lasater and J.H. Hopkins, "Removing particulates from stack gases", *Chem. Engg.* Vol. 84 (22) (1977), p. 111.
5. J.H. Abbott and D.C. Drehmel, "Control of fine particulate emissions", *Chem. Engg. Progr.*, Vol. 72 (12) (1976), p. 47.
6. C. Doerschlag and G. Miczek, "How to choose a cyclone dust collector", *Chem. Engg.*, Vol. 84 (4) (1977), p. 64.
7. K. Iinoya and C. Orr, Jr., "Source control by filtration", in *Air Pollution III*, edited by A.C. Stern (Academic Press, New York), 1968, Chap. 44.

8. D.H. Rullman, "Baghouse technology: a perspective", *J. Air Pollu. Control Assoc.*, Vol. 26 (1976), p. 16.
9. S. Oglesby (Jr), "Electrostatic precipitators tackle air pollutants", *Environmental Sci. & Technol.*, Vol. 5 (1971), p. 766.
10. E.B. Hauf, "A guide to scrubber selection". *Environmental Sci. & Technol.*, Vol. 4 (1970), p. 110.
11. J.D. Brady, D.W. Cooper and M.T. Rei, "A wet collector of fine particles", *Chem. Engg. Progr.* Vol. 73 (8) (1977), p. 45.
12. J.R. Fair, "Sorption processes for gas separation", *Chem. Engg.*, Vol. 79 (26) (1972), p. 87.
13. F.A. Zenz, "Designing gas-absorption towers", *Chem. Engg.*, Vol. 79 (25) (1972), p. 120.
14. W.D. Lovett and F.T. Cunniff, "Air pollution control by activated carbon", *Chem. Engg. Progr.*, Vol. 70 (5) (1974), p. 43.
15. W.A. Johnston, "Designing fixed-bed adsorption columns", *Chem. Engg.*, Vol. 79 (26) (1962), p. 86.
16. D.E. Waid, "Controlling pollutants via thermal incineration", *Chem. Engg. Progr.*, Vol. 68 (8) (1972), p. 57.

PROBLEMS

1. A multitray settling chamber handles 6 m³/s of air at 20°C. There are 8 trays including the bottom surface, spaced 0.25 m apart. The chamber is 4 m long and 1 m wide. For particles of density 2000 kg/m³ and sizes (a) 70 and (b) 25 μm, calculate the residence time, the distance settled and the efficiency of collection. Is the tray spacing sufficient to collect all the particles of each size? Assume laminar flow. What will be the efficiency if the flow regime is turbulent in each case? (1.33 s; 0.39 m; 100%; 1.33 s; 0.05 m; 20%; 79.2%; 18.1%).
2. Assuming turbulent flow inside a settling chamber, derive Eq. (5.19).
3. A conventional cyclone with diameter 1.0 m handles 3.0 m³/s of standard air. Using $N_e = 6$, determine the cut size and the efficiency as a function of particle diameter for particles of density (a) 1500, and (b) 2500 kg/m³. (4.2 μm).
4. Solve problem 3 for a high efficiency cyclone.
5. A conventional cyclone with a diameter of 1.0 m is to be used to separate particulate matter of density 1500 kg/m³ from standard air at a flow rate of 2.0 m³/s. An analysis of the particles yielded the following size distribution.

Size, μm	3.0	6.0	10.0	20.0	40.0	80.0	140.0	200.0
Weight fraction less than stated size	0.03	0.08	0.15	0.33	0.60	0.85	0.96	0.99

Using $N_e = 6$ estimate the overall separation efficiency. (86%)

6. A baghouse is to be designed to handle 800 m^3/min of air. The filtration takes place at constant pressure so that the air velocity through each bag decreases during the time between clearings according to the relation,

$$v_{bag} = \frac{1}{0.267 + 0.08\,t}$$

where v_{bag} is the m^3/min of air per square metre of cloth and t is in min The bags are shaken in sequence, row by row on a 15-min cycle. Each bag is 30 cm in diameter and 6 m high. The baghouse is to be square in cross-section, with 30 cm spacing between bags and 30 cm clearance from the walls. Calculate (a) the number of bags required, and (b) the width of the baghouse. (100; 6.3 m)

7. Why a bag filter is not suitable for each of the following applications: particles from lime slaking, flyash from coal burning, radioactive fumes, and heavy mist from a chemical process?

8. Before the installation of an electrostatic precipitator, the stack gas of a power plant contained 6.0 g particulates per m^3 of gas. The gas flowrate is 350 m^3/min and the new precipitator can remove 2500 kg particulates/day.

   (a) What is the emission rate of particulates before and after pollution control in kg/day? (3024 kg/day; 524 kg/day)
   (b) What is the efficiency of the electrostatic precipitator? (82.7%)
   (c) Will the new system meet an emission standard of 0.7 g/m^3?

9. A cylindrical electrostatic precipitator of diameter 0.3 m is used for separating pulverised coal flyash particles from a furnace gas stream. If the volumetric flow rate of the gas is 0.05 m^3/s, what will be the length of the precipitator for obtaining a collection efficiency of 99.90 per cent? What per cent change in electrode collection area is required to increase the collection efficiency from 99.90 to 99.95 per cent. (3.67 m; 10%)

10. Water is used in a spray chamber as the spray fluid to treat standard air containing particles with a density of 1000 kg/m^3. The flow rate

of air is 120 m^3/min and that of water is 0.3 m^3/min. The average drop size is 1000 μm. The spray chamber has a diameter of 1 m and a height of 3 m. What is the overall collection efficiency of the chamber if the average particle size is 2.5 μm? For the above particles the individual drop collection efficiency may be taken as 9.0 per cent. (96%)

11. A venturi scrubber is used to collect open hearth steel furnace fumes. The gas pressure drop is 50 cm of water guage and the scrubber water pressure drop is 1 atm. The gas flow rate is 1500 m^3/min and the water flow rate is 5000 litres/min. Estimate the fume collection efficiency of the scrubber. (92%)

12. In the design of wet scrubbers the disposal of the waste is often a difficult problem. The following articles indicate methods to do this.

    (a) R.W. Rickles: "Waste recovery and pollution abatement," *Chem. Engg.*, 72, pp. 135–152, Sept., 27, 1965.
    (b) A.W. Busch: "Liquid waste disposal system design", *Chem., Engg.*, 72, pp. 83–86, March, 29, 1965.

    Review the above papers and briefly discuss the methods.

13. Work the problem in example 7 using 1 in Berl Saddles instead of Raschig rings.

14. Read the literature available on adsorption and briefly discuss the various types of adsorption isotherms.

15. A process stream containing air and butyric acid ($C_4H_8O_2$), the latter in concentration of 4.0% by volume, is passed through a fixed bed adsorber. The bed is packed to a height of 1 m with activated carbon to recover butyric acid. The adsorption takes place isothermally at 20°C and the adsorption cycle is 1 hr. The gas flow rate is 160 kg/min and the adsorbent retains 0.35 kg butyric acid per kg carbon at 20°C.

    (a) What is the amount of adsorbent in the bed? (3065 kg)
    (b) What is the gas flow area of the bed? (10.22 m^2)

6. Control of Specific Gaseous Pollutants

Among the gaseous air pollutants, the sulphur oxides, the oxides of nitrogen, carbon monoxide and hydrocarbons have attracted major attention because of their common occurrence in the atmosphere and their known harmful effects. In Chapter 5, the general principles, upon which pollution control equipment are based, were examined. This chapter deals primarily with the application of these equipment and certain strategies for the control of major gaseous pollutant emissions from stationary sources such as combustion processes, smelting operations, and chemical and petroleum industries. Mobile sources are also dealt with briefly.

6.1 CONTROL OF SULPHUR DIOXIDE EMISSION

The greatest anthropogenic sources of SO_2 emissions are the fossil-fuel-burning power plants. Probably the most convenient method of solving the sulphur dioxide problem at the source is to disperse the waste gases by stacks of sufficient height. This height can be determined by the parameters of the source and the local geography. The pollutant is diluted through its spread in the atmosphere and thereby its excessive ground level concentrations can be avoided. But this practice is becoming increasingly unacceptable as people are becoming more and more aware of the pollution problems and are showing greater concern about all forms of pollution.

There are three basic procedures for controlling sulphur dioxide emissions from stationary combustion sources: the extraction of sulphur from fuels, sulphur reduction within the combustion chamber, and the treatment of flue gases.

Extraction of Sulphur from Fuels

Removal of sulphur from fuel prior to combustion is theoretically an attractive route to low SO_2 concentrations in the atmosphere. Sulphur in coal is present in both inorganic and organic forms. Practically all inorganic sulphur exists in the coal as iron disulphide (FeS_2), i.e., in the form of pyrites and marcasites. Washing can reduce the pyritic sulphur content by about 30%, but such methods result in a loss of combustible material and may add 20% or more to the cost of coal. Organic sulphur, present in the form of cystin, thiols, sulphides and some cyclic compounds, is intrinsically and almost uniformly bound with most of the carbonaceous material and can be removed only by chemical processing.

Hydrodesulphurization of Coal

Hydrodesulphurization of coal using a solvent extraction process can remove both the organic sulphur as well as its inorganic forms. In this process, finely ground coal is slurrried with anthracene oil; a small amount of hydrogen (about ½%) is added to avoid repolymerization, and the slurry is heated at a high temperature (450°C) in order to dissolve the coal (Fig. 6.1). The ash residue consisting of pyritic sulphur and other minerals is eliminated by pressure filtration. The coal solution filtrate is sent to a flash evaporator where the light fraction is removed; subsequent operation in a

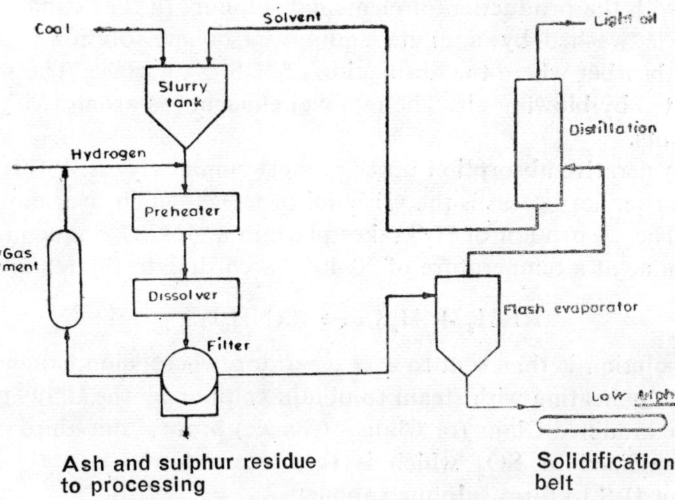

Fig. 6.1 Solvent refining of coal (ref. 1)

distillation unit recovers the solvent. The hot liquid evaporator residue is cooled to a brittle solid fuel which can be pulverized. The product can be liquified at about 250°C, has a higher heating value than raw coal, and contains less than 1% sulphur even when the starting material is coal of high sulphur content.

Coal Gasification

Coal gasification appears to be an attractive method for reducing sulphur dioxide pollution and is finding better use in India's vast coal reserves. Several processes for turning solid coal into gas are in use, or are under development throughout the world. The new processes are more sophisticated than the early water-gas and producer-gas processes and give a better quality gas and simultaneously remove both organic and inorganic types of sulphur as well.

The process of gasification is affected with partial oxidation of coal in the presence of oxygen and steam and it yields mainly carbon monoxide and hydrogen:

$$C + H_2O \rightarrow CO + H_2^* \quad (6.1)$$

The oxygen added with the coal and steam burns some of the coal to attain the high temperature needed for the reaction to occur at a reasonable rate. The sulphur content of the coal is converted in the process mainly into hydrogen sulphide and must be separated from the coal gas. This is usually accomplished by either absorption or adsorption. In absorption processes the gases are scrubbed with an alkaline reagent such as sodium carbonate or ethanolamine, followed by the regeneration of the reagent with the production of elemental sulphur. In the Seaboard process[2] the gas is washed by a dilute sodium carbonate solution in a countercurrent absorber where the absorption of H_2S takes place. The solution is regenerated by blowing air. The removal efficiency is around 90% in single stage plants.

A regenerative absorption process, most commonly used for the purification of refinery gases is the Girbotol process[2] which uses monoethanolamine. The absorption of H_2S takes place in a 15-20% aqueous solution of the amine at a temperature of 30-40°C according to the reaction:

$$RNH_2 + H_2S \rightleftharpoons RNH_3HS \quad (6.2)$$

The solution is then sent to a regenerator where monoethanolamine is separated by heating with steam to obtain sulphur by the Claus process.

In the modified Claus (or Claus—Chance) process, one-third of the H_2S feed is oxidized to SO_2 which is then reacted over a catalyst with the remaining H_2S to form sulphur vapour:

$$1/3\ H_2S + 1/2\ O_2 \rightarrow 1/3\ SO_2 + 1/3\ H_2O \quad (6.3)$$

$$1/3\ SO_2 + 2/3\ H_2S \rightarrow S + 2/3\ H_2O \quad (6.4)$$

The sulphur is condensed and sent to a sulphur storage tank. Fig. 6.2 shows schematically the process diagram of Claus process.

The absorption processes are simple and effective, but require cooling of the gas which involves considerable heat losses. In this respect, dry methods of purification, which can be carried out at high temperatures, are rather more promising. One such process which has been successfully

*Under certain conditions some methane is also formed during gasification.

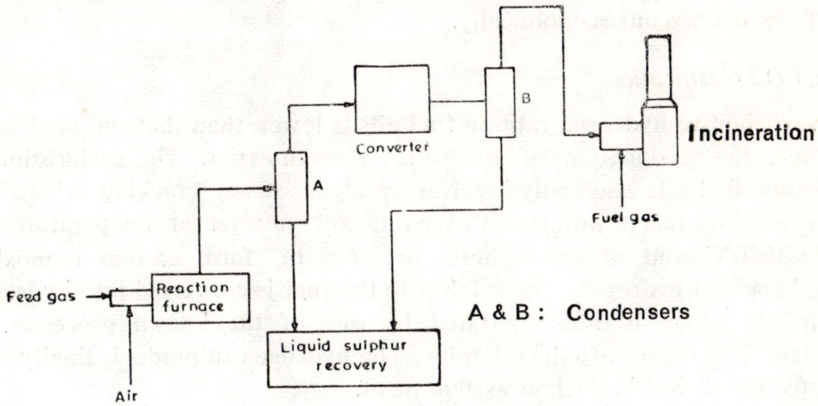

Fig. 6.2 Claus sulphur recovery plant

used on a commercial scale involves the adsorption of H_2S on ferric oxide in a fluidized bed at 400°C. The ferric oxide is then regenerated by roasting it in air at 800°C producing SO_2 for sulphuric acid manufacture.

Desulphurization of Fuel Oils

During atmospheric distillation of crude oil, only a small quantity of sulphur in the crude passes to the light fractions, and the bulk of sulphur is concentrated in the residue. The residue commonly referred to as the heavy fuel oil, with a boiling point above 350°C, can be directly used as a commercial fuel in power plants or further processed, e.g., by vacuum distillation. Sulphur can be removed from the fuel oil by several different processes but by far the most effective method is the hydrodesulphurization process.

Hydrodesulphurization process, now well established on a large commercial scale, treats fuel oil with hydrogen over a catalyst at temperatures of 300–500°C and pressures of 40–150 atm. Hydrogen reacts with organic sulphur compounds and forms hydrogen sulphide which is collected and can be used for the production of sulphur. The catalyst (cobalt-molybdenum based) is susceptible to poisoning by organometallic compounds present in the oil. Asphaltenes, which contain sulphur in a molecular structure fairly resistant to hydrogenation, may also cause catalyst fouling. Further, the volumes of hydrogen used are considerable and the process necessitates the construction of expensive plants for generation of hydrogen. All these factors substantially increase the cost of desulphurization of fuel oil. Estimated costs for reducing the sulphur content of the fuel oil by 50% would, with available technology, increase the cost of the fuel oil by 25%

or more. Hence, new methods of desulphurizing are badly needed if this is to be carried out economically.

Fuel Oil Gasification

The carbon to hydrogen ratio in fuel oils is lower than that in coal and, hence, the gasification of fuel oils is much easier. The gasification of residual fuel oils essentially involves catalytic steam cracking where the hydrocarbon reacts with steam over a nickel catalyst at temperatures of 700–1000°C and at atmospheric pressure to form carbon monoxide, methane and hydrogen. The sulphur in the fuel is converted into hydrogen sulphide which is then separated by one of the known processes, for instance, by means of ethanolamine. The hydrogen sulphide is finally converted to sulphur in a Claus system plant.

Sulphur Reduction During Combustion

In this method, which is applicable to conventional combustion chambers, finely ground limestone ($CaCO_3$) is injected directly into the combustion chamber. The limestone is calcined to CaO by the heat of combustion and it reacts with SO_2 contained in the flue gas to form sulphites and sulphates.

$$CaCO_3 \rightarrow CaO + CO_2 \tag{6.5}$$

$$CaO + SO_2 \rightarrow CaSO_3 \tag{6.6}$$

$$CaO + SO_2 + 1/2\ O_2 \rightarrow CaSO_4 \tag{6.7}$$

The solid reaction products, unreacted materials and fly ash are removed either by dry collectors or by wet scrubbing. Reaction (6.7) is the most favourable at temperatures above 1000°C; however, limestone does not completely react with SO_2 to produce a stoichiometric yield of $CaSO_4$ because of insufficient gas residence time in the combustion zone. Furthermore, at temperatures above 1200°C, $CaSO_4$ is unstable and SO_2 removal either by sulphite or sulphate formation is inhibited. As a result, the process removes less than half of the sulphur oxides.

Due to the low SO_2 removal efficiencies, the dry limestone injection process, although relatively simple and easy to operate, is not very attractive from the point of view of control technology. Attempts to increase the desulphurization efficiencies have been fairly successful and one such improved process is the fluidized bed combustion process in which limestone and crushed coal together form the fluidized bed and air is used as the fluidizing medium. The process chemistry is basically the same as that in the case of the dry limestone technique but the operating temperatures are of the order of 700–1000°C within the bed due to high heat transfer capability of the fluidized beds. These temperatures are much lower than those in the conventional combustion chamber (~1500°C). At ~900°C, $CaCO_3$ dissociates into CO_2 and CaO and the latter reacts with SO_2 to form $CaSO_4$. The degree of desulphurization in this process may amount to 90% which is more than twice as high as that in the dry limestone technique.

In addition to high SO_2 removal efficiencies, the fluidized bed combustion process, due to the low operating temperatures, prevents the onset of ash fusion, and as a result the fouling and corrosion of boiler tubes associated with the molten slag are considerably reduced. Also, the formation of nitrogen oxides by the nitrogen fixation reaction is reduced. Along with these advantages, the method has some drawbacks. It requires novel designs of boilers and additional installations for the preparation of limestone, collection of solid particles carried off from the fluidized bed and regeneration of $CaSO_4$.

Desulphurization of Flue Gases

Flue gas desulphurization has probably been the subject of more research than any other gas purification operation. Nearly fifty flue gas desulphurization processes have been proposed, some have been tested on a large scale, but as such no ideal process exists. These processes, in general, are classified in terms of whether they are wet or dry. Wet scrubbing processes are effective but the equipment may be very costly and difficult to add to many existing plants. In addition, these processes in the absence of reheating tend to cause localized pollution problems because of lost plume buoyancy. Dry, elevated temperature removal processes are, therefore, preferable as a cold plume is not formed and the problem of handling large amounts of slurry in flue gas scrubbing can be avoided. However, the technical problems encountered in dry processes have now caused a shift towards wet processes, although a few dry processes are being used on a commercial scale. The reheating of the desulphurized flue gas is a common design feature of all modern processes. Here a few promising processes available or under development are discussed.

Dry Methods

The dry techniques for the desulphurization of flue gases may be classified as (a) adsorption of SO_2 by metal oxides to form stable sulphites or sulphates with subsequent regeneration of the oxide and recovery of sulphur and (b) adsorption on activated carbon followed by regeneration and conversion of concentrated SO_2 to sulphuric acid or elemental sulphur.

Processes Using Metal Oxides

Several metal oxides and oxide mixtures have been identified[3] as adsorbents or catalysts for SO_2 in flue gases. Among them aluminium sodium oxide (alkalized alumina) and manganese oxide have been found to be better adsorbents; bulk oxides of Co and Cu are also active. Metal oxides on support exhibit an enhanced reactivity compared with that of bulk oxides. Na(or K), Sr, Cr, and Cu impregnated oxides supported on alumina are rated as ·' nerior catalysts.

Alkalized Alumina Process

A cyclic adsorption process called the alkalized alumina process[4] has been developed by the U.S. Bureau of Mines. In this process, the dust-free flue gas is fed to a reactor wherein the adsorbent, a porous form of sodium aluminate ($Na_2O.\ Al_2O_3$), adsorbs SO_2 at a temperature of ~315°C. During the adsorption, the SO_2 and O_2 in the flue gas react with the adsorbent:

$$Na_2O.Al_2O_3 + SO_2 + \tfrac{1}{2}O_2 \rightarrow Na_2SO_4 + Al_2O_3 \qquad (6.8)$$

The spent material is then contacted with a reducing gas such as hydrogen in a regenerator at about 680°C to produce hydrogen sulphide:

$$Na_2SO_4 + Al_2O_3 + 4H_2 \rightarrow Na_2O.Al_2O_3 + H_2S + 3H_2O \qquad (6.9)$$

The sodium aluminate pellets are recycled and the H_2S gas is sent to a Claus unit for the production of sulphur. A simplified flow sheet for the process is shown in Fig. 6.3.

The major problem with the alkalized alumina process is related to maintenance of the grannular strength of the sorbent. Exposure of the sorbent to continued sorption—regeneration cycle results in uneconomically high attrition rates.

Manganese Oxide Process

Mitsubishi Heavy Industries of Japan have developed an activated manganese oxide process[5], the DAP-Mn process, where the powdered adsorbent is fed into an entrainment reactor. In the reactor, the activated manganese oxide reacts with SO_2 to produce manganese sulphate:

$$MnO_x.yH_2O + SO_2 + 1/2(2-x)\ O_2 \rightarrow MnSO_4 + yH_2O \qquad (6.10)$$

where x is between 1.5 and 1.8 and y between 0.1 and 1. The regeneration of the spent sorbent is carried out by reacting the manganese sulphate with air and ammonia to produce ammonium sulphate in accordance with the reaction:

$$MnSO_4 + 2NH_3 + 2H_2O + (y-1)\ H_2O + 1/2\ (x-1)\ O_2 \rightarrow$$

$$(NH_4)_2SO_4 + MnO_x\ yH_2O \qquad (6.11)$$

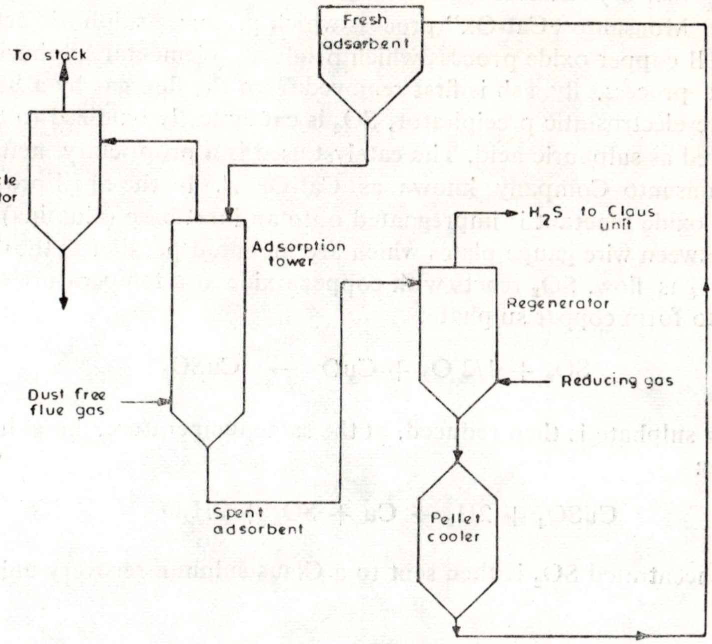

Fig. 6.3 Alkalized alumina process (ref. 4)

The activated manganese oxide is filtered from the $(NH_4)_2SO_4$ solution and the filtrate is passed through a crystallizer where ammonium sulphate is recovered. The ammonium sulphate can be used as a fertilizer and the degree of SO_2 removal by this process is around 90%. Fig. 6.4 shows the process.

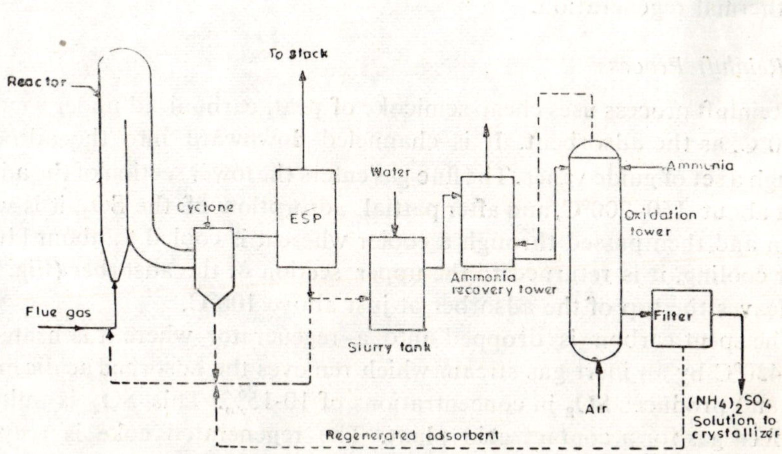

Fig. 6.4 Mitsubishi manganese oxide process (ref. 5)

Two other dry removal systems which have attained industrial importance are the Monsanto "Cat-Ox" process which produces sulphuric acid[6], and the Shell copper oxide process which produces elemental sulphur[7]. In the Cat-Ox process, fly ash is first removed from the flue gas by a high temperature electrostatic precipitator, SO_2 is catalytically oxidized to SO_3 and recovered as sulphuric acid. The catalyst used is a proprietary material of the Monsanto Company known as Cat-Ox *A*. In the Shell process, the copper oxide reactant is impregnated onto an inert base (alumina) and is held between wire gauge plates which are mounted parallel to the direction of flue gas flow. SO_2 reacts with copper oxide at a temperature of about 400°C to form copper sulphate.

$$SO_2 + 1/2\ O_2 + CuO \rightarrow CuSO_4 \qquad (6.12)$$

Copper sulphate is then reduced, at the same temperature, in a hydrogen rich gas:

$$CuSO_4 + 2H_2 \rightarrow Cu + SO_2 + 2H_2O \qquad (6.13)$$

The concentrated SO_2 is then sent to a Claus sulphur recovery unit.

Processes Using Activated Carbon

Adsorption on a material like activated carbon provides an attractive method for continuous removal of SO_2 from flue gases because of the high surface area and low cost of activated carbon. It can be produced from a variety of sources such as wood, coconut or other types of nut shells, or from petroleum feed stocks. The main limitations of activated carbon as an adsorbent in flue gas desulphurization processes are its attrition, a tendency towards spontaneous ignition, and problems associated with thermal regeneration.

The Reinluft Process

The Reinluft process uses cheap semicoke of peat, carbonized under vacuum at 600°C, as the adsorbent. It is channeled downward into the adsorber through a set of guide vanes. The flue gas enters the lower section of the adsorber at about 150–200°C, and after partial adsorption of the SO_2, it is withdrawn and then passed through a cooler where it is cooled to about 110°C. After cooling, it is returned to the upper section of the adsorber (Fig. 6.5) and leaves the top of the adsorber at just above 100°C.

The spent carbon is dropped into a regenerator where it is heated to 380-450°C by an inert gas stream which removes the adsorbed acidic materials and produces SO_2 in concentrations of 10-15%. This SO_2 is suitable as a feed gas for a contact acid plant. The regenerated coke is recycled after the fines have been removed.

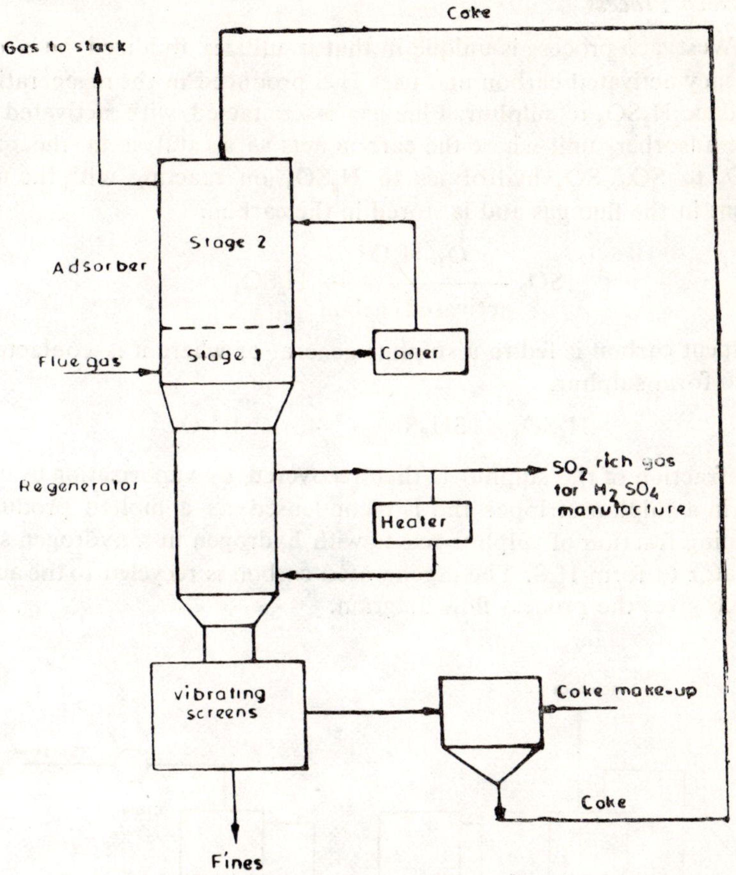

Fig. 6.5 The Reinluft process

One of the main difficulties of the Reinluft process is the instability of carbon in the presence of flue gas oxygen which leads to carbon combustion. In addition, attrition of carbon and corrosion are major problems. Cost of recirculating very large amounts of carbon and costs of heat and heat reducing agent are other major handicaps. These problems led to the development of similar processes which use harder, less combustible, activated carbons. The BF-process[8] is an improvement over the Reinluft process. It uses a specially developed activated carbon having high resistivity against ignition and attrition, and high SO_2 adsorption capacity in a circuit similar to that of Reinluft process. The adsorbent can be regenerated either thermally (with concentrated SO_2 as the byproduct) or by washing with water (with dilute H_2SO_4 as the byproduct).

Westvaco Process

The Westvaco process is unique in that it utilizes fluidized beds of high-efficiency activated carbon and uses H_2S produced in the regeneration train to reduce H_2SO_4 to sulphur. Flue gas is contacted with activated carbon in the adsorber unit where the carbon acts as a catalyst in the oxidation of SO_2 to SO_3. SO_3 hydrolyses to H_2SO_4 on reacting with the moisture present in the flue gas and is stored in the carbon.

$$SO_2 \xrightarrow[\text{activated carbon}]{O_2,\ H_2O} H_2SO_4$$

The spent carbon is fed to a sulphur generat or where it is contacted with H_2S to form sulphur.

$$H_2SO_4 + 3H_2S \rightarrow 4S + 4H_2O \qquad (6.14)$$

A fraction of the sulphur is then recovered by vaporization by hot inert gases in a sulphur stripper and is recondensed as a molten product. The remaining fraction of sulphur reacts with hydrogen in a hydrogen sulphide generator to form H_2S. The regenerated carbon is recycled to the adsorber. Fig. 6.6 gives the process flow diagram.

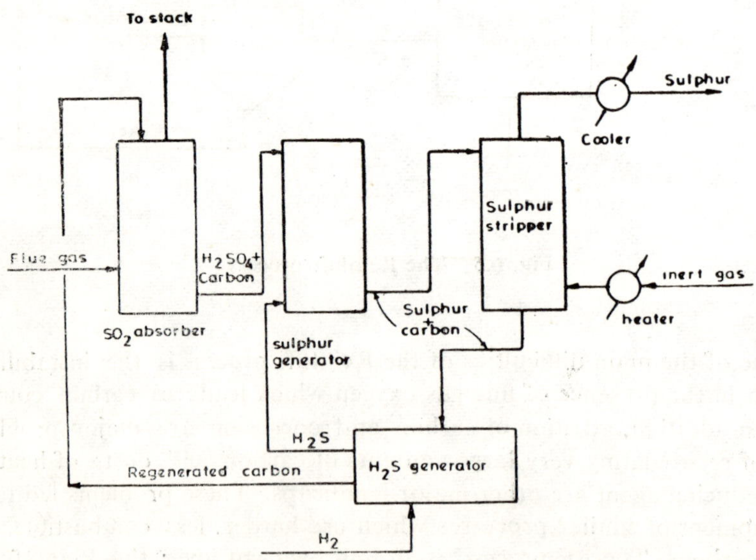

Fig. 6.6 Westvaco process

Wet Scrubbing Methods

Several wet flue gas desulphurization methods are available which use solutions and slurries of compounds of calcium, magnesium and sodium.

Some of these processes require the removal of flyash prior to scrubbing in a prescrubber while others are capable of removing SO_2 and the ash simultaneously in the scrubber. The main disadvantage of all the wet methods is that the temperature of the gases is reduced to 30-50°C, thus impairing their dispersion in the atmosphere sharply. To rectify this shortcoming, it is necessary to reheat the effluent gases, and the consumption of heat for this purpose is usually around 3% of the heat of the fuel used in the combustion chamber.

Lime-Limestone Scrubbing

These systems use a slurry of lime or limestone in water to absorb SO_2 from the flue gas in a gas-liquid scrubber. Waste gases are freed of ash in an ash collector and then move to a wet scrubber sprayed by a slurry of fine-ground limestone or hydrated lime. The sulphur dioxide is absorbed to produce calcium sulphite and calcium sulphate.

$$2\ CaCO_3 + 2SO_2 + 1/2O_2 \rightarrow CaSO_3 + CaSO_4 + 2CO_2 \quad (6.15)$$

$$2\ Ca(OH)_2 + 2SO_2 + 1/2O_2 \rightarrow CaSO_3 + CaSO_4 + 2H_2O \quad (6.16)$$

The spent liquor from the scrubber is neutralized by a fresh lime-limestone solution and sent to a settler. The clear solution is then recycled to the scrubber.

Various types of scrubbers are employed commercially: spray towers, venturies, packed beds, and turbulent-contact absorbers (lightweight hollow plastic spheres held between restraining grids in a countercurrent scrubbing tower). These scrubbing systems can be designed for almost any desired level of SO_2 removal. Commercial power generation systems are generally designed for 80% to 95% SO_2 removal; however, some systems have at times achieved more than 99% removal. Limestone processes have received considerable attention more recently, because of their relative simplicity, low cost and the easy availability of limestone; however, sludge disposal remains a major problem in this process.

Magnesium Oxide Scrubbing

The magnesium oxide process developed in the U.S.A. by the Chemical Construction Corporation (Chemico) is a "regenerative" process rather than a "throwaway" operation like the limestone scrubbing. In other words, it does not produce waste material—SO_2 removed from the flue gas is concentrated and used to make marketable H_2SO_4 or elemental sulphur.

In the chemico process the flue gas is scrubbed with a slurry of magnesium oxide in a venturi scrubber which absorbs SO_2 and yields magnesium sulphite and sulphate according to the following reactions.

$$MgO + SO_2 \rightarrow MgSO_3 \quad (6.17)$$

$$MgSO_3 + SO_2 + H_2O \rightarrow Mg(HSO_3)_2 \quad (6.18)$$

$$Mg(HSO_3)_2 + MgO \rightarrow 2MgSO_3 + H_2O \quad (6.19)$$

$$MgSO_3 + 1/2O_2 \rightarrow MgSO_4 \quad (6.20)$$

The mixed sulphite/sulphate slurry along with unreacted MgO is separated from the liquid phase in a centrifuge and the mother liquor is recycled to the absorber. The centrifuged wet cake is dried and regenerated in a calciner. Carbon is added in the calcining step to reduce any $MgSO_4$ to MgO and SO_2:

$$MgSO_3 \xrightarrow[800°C]{heat} MgO + SO_2 \quad (6.21)$$

$$MgSO_4 + 1/2\ C \rightarrow MgO + SO_2 + 1/2\ CO_2 \quad (6.22)$$

MgO is returned to the process, while the concentrated SO_2 (10 to 15%) can be reprocessed into H_2SO_4 or elemental sulphur. Fig. 6.7 schematically illustrates the process.

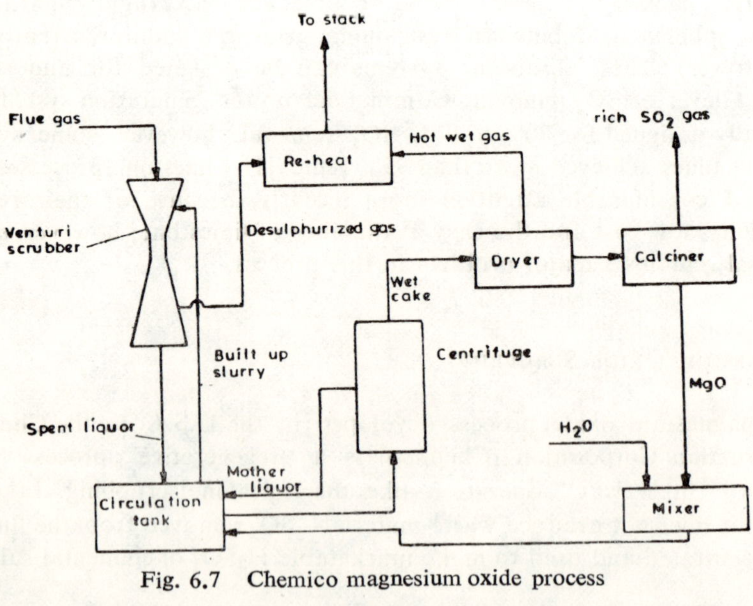

Fig. 6.7 Chemico magnesium oxide process

WELMAN-LORD PROCESS

The Welman-Lord process (Fig. 6.8) employs a saturated solution of sodium sulphite to absorb SO_2 from flue gases in a scrubber; the sulphite is converted to bisulphite:

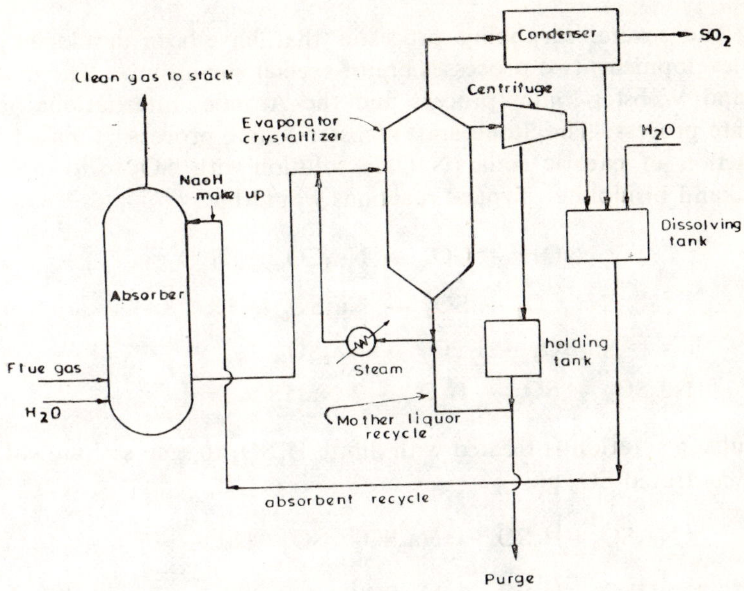

Fig. 6.8 Wellman-Lord process

$$Na_2SO_3 + SO_2 + H_2O \rightarrow 2\,NaHSO_3 \tag{6.23}$$

Fly ash removal prior to SO_2 scrubbing is necessary to keep the rate of its accumulation in the scrubbing liquid low. The sodium sulphite/bisulphite solution thus formed is fed to a double effect evaporator-crystallizer where the bisulphite decomposes into sodium sulphite around 110°C, releasing concentrated SO_2 and steam:

$$2\,NaHSO_3 \xrightarrow[110^\circ C]{heat} Na_2SO_3 + SO_2 + H_2O \tag{6.24}$$

The outlet from the evaporator contains crystals of sodium sulphite, which are separated from the mother liquor in a centrifuge. The mixture of SO_2 and water vapour, which is released from the evaporator, is sent to a condenser and the condensate goes to the dissolving tank for dissolution of Na_2SO_3 crystals; the rich SO_2 gas is further processed.

During the process, SO_2 oxidation leads to the formation of sodium sulphate which cannot be regenerated and, therefore, it is purged from the system. An equivalent amount of NaOH is added to maintain the sodium balance. The degree of desulphurization of flue gases in this process is around 90%.

Other Flue Gas Scrubbing Processes

Among the several scrubbing processes that have been developed or are under development, two processes are of special significance. These are the Stone and Webster-Ionics process and the Atomics International molten carbonate process. The Stone and Webster-Ionics process is based upon the reaction of caustic soda (NaOH) solution with SO_2 to form sodium sulphite and bisulphite. Typical reactions would be

$$2\,NaOH + CO_2 \rightarrow Na_2CO_3 + H_2O \tag{6.25}$$

$$Na_2CO_3 + SO_2 \rightarrow Na_2SO_3 + CO_2 \tag{6.26}$$

$$Na_2SO_3 + 1/2O_2 \rightarrow Na_2SO_4 \tag{6.27}$$

$$Na_2SO_3 + SO_2 + H_2O \rightarrow 2\,NaHSO_3 \tag{6.28}$$

The resulting solution is treated with dilute H_2SO_4 to get sodium sulphate and concentrated SO_2 gas.

$$Na_2SO_3+H_2SO_4 \rightarrow Na_2SO_4+SO_2+H_2O \tag{6.29}$$

$$2NaHSO_3+H_2SO_4 \rightarrow Na_2SO_4+2SO_2+H_2O \tag{6.30}$$

The heart of the process is the electrolytic regeneration cell which converts Na_2SO_4 into caustic soda (NaOH) and H_2SO_4. The overall reaction is

$$Na_2SO_4+3H_2O \rightarrow 2NaOH+H_2SO_4+H_2+1/2O_2 \tag{6.31}$$

The caustic soda is returned to the absorber, and the acid is used in the desorption tower.

The Atomics International molten salt process is an unconventional scrubbing process which uses a mixture of lithium, sodium, and potassium carbonates in a melt at about 425°C. SO_2 and SO_3 react with the metal carbonates to yield corresponding sulphites and sulphates respectively.

$$M_2CO_3+SO_2 \rightarrow M_2SO_3+CO_2 \tag{6.32}$$

$$M_2CO_3+SO_3 \rightarrow M_2SO_4+CO_2 \tag{6.33}$$

(M stands for alkali metal ions)

The mixture is regenerated in a two-stage process. First the sulphite and sulphate are reduced to sulphide by using producer gas ($CO+H_2$).

$$M_2SO_3+2CO+H_2 \rightarrow M_2S+2CO_2+H_2O \tag{6.34}$$

$$M_2SO_4+2CO+2H_2 \rightarrow M_2S+2CO_2+2H_2O \qquad (6.35)$$

The reduced melt is then reacted with steam and CO_2 to produce H_2S and M_2CO_3.

$$M_2S+H_2O+CO_2 \rightarrow M_2CO_3+H_2S \qquad (6.36)$$

Metal Smelting Operations

The metal smelting processes—mainly copper, lead, and zinc smelting—are second only to combustion processes as sources of sulphur dioxide emissions. The concentration of SO_2 emissions from smelting is often high enough ($>3.5\%$ by volume) to produce either sulphuric acid or sulphur. Liquid sulphur dioxide can be produced using the ASARCO process by absorbing SO_2 in dimethylamine. Direct production of ammonium sulphate is possible by COMINCO process where aqueous ammonia is used as an absorbent for concentrating sulphur dioxide from lean streams.

Byproduct Sulphuric Acid Manufacture

The manufacture of sulphuric acid from smelter gas streams involves gas conditioning, drying, catalytic conversion, and absorption[10]. The gas conditioning step removes impurities like arsenic, chlorides, fluorides, fumes of metal sulphide or oxide, mercury, etc., and this step is necessary to avoid rapid deactivation of the catalyst. Gas leaving the smelters has a temperature in excess of 600°C and it is generally cooled to about 55°C by injecting water into scrubbing tower. This prevents acid contamination and minimizes catalyst fouling. The gas is then dried in a tower by using 93% H_2SO_4 to absorb the moisture. After gas-conditioning and drying, the smelter gas is sent to a catalytic converter. In the conventional single absorption contact plant, the gas stream is passed through a four stage catalytic converter where SO_2 is converted to SO_3. The gas is cooled and is then absorbed with high efficiency in a circulating stream of 98 to 99% sulphuric acid. SO_3 unites with the small excess of water in the acid to form more sulphuric acid. This results in an effluent gas with SO_2 concentration of about 2000 ppm, which is usually well above the permitted standards of 500 ppm. Hence, tail gas scrubbing is further necessary if the effluent SO_2 concentration is to be brought within the acceptable limits.

A modern modification of the contact plant, which was developed to reduce the emission of SO_2 in exit gases, includes two stages of absorption. The process often referred to as the double-contact, double-absorption (DCDA) method provides higher plant efficiencies (99.5%) and SO_2 emissions of below 500 ppm. The gases from the converter, after 90% of SO_2 has been converted to SO_3, are interrupted and passed at an intermediate stage to an absorber (Fig. 6.9) to remove the SO_3. The gases are then reheated and returned to the converter for further conversion. They then pass through additional catalyst, are cooled, and flow through a second absorber and then to the atmosphere. The DCDA plants usually require higher SO_2 concentration (> 6%) for autothermal operation.

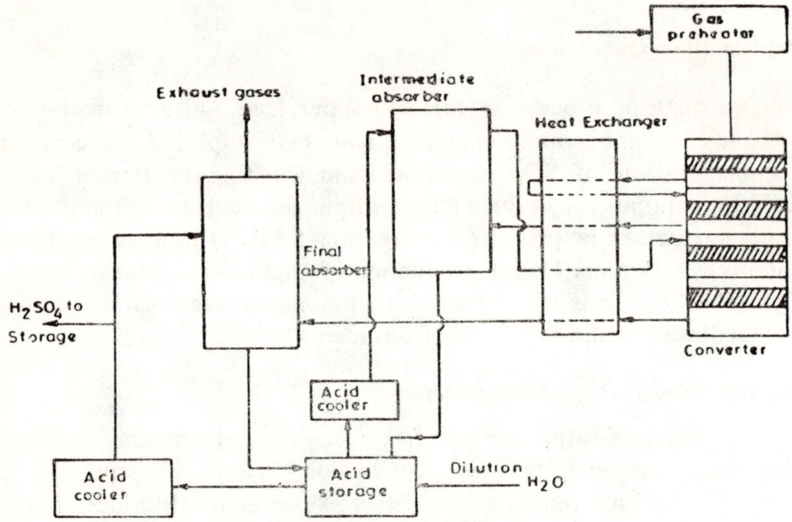

Fig. 6.9 Schematic diagram of a DCDA process

Absorption in Dimethylaniline

The ASARCO (American Smelting & Refining Co.) — DMA process is a regenerative absorption process where the waste gases containing SO_2 are absorbed in dimethylaniline in the lower stages of an absorber. The desulphurized gas containing dimethylaniline is further scrubbed with sodium carbonate solution and then with dilute H_2SO_4 to remove traces of SO_2 and dimethylaniline before it is sent to the stack. The SO_2-rich liquor is fed to a steam distillation column which strips SO_2 from the liquor and dimethylaniline is recycled. The resultant SO_2 is dried with 98% H_2SO_4 in a drying tower, liquified and sent to storage for further processing. Dimethylaniline is an efficient absorbent for SO_2 concentrations above 3.5%, and the tail gas leaving the stack after scrubbing contains about 500 ppm SO_2. A simplified flow diagram of the process is given in Fig. 6.10. The ASARCO process is economical only for SO_2 gas concentrations of above 2%.

Absorption in Ammonia

Aqueous ammonia is used in the COMINCO (Consolidated Mining and Smelting Co.) process for reducing the SO_2 concentration in zinc roaster plant

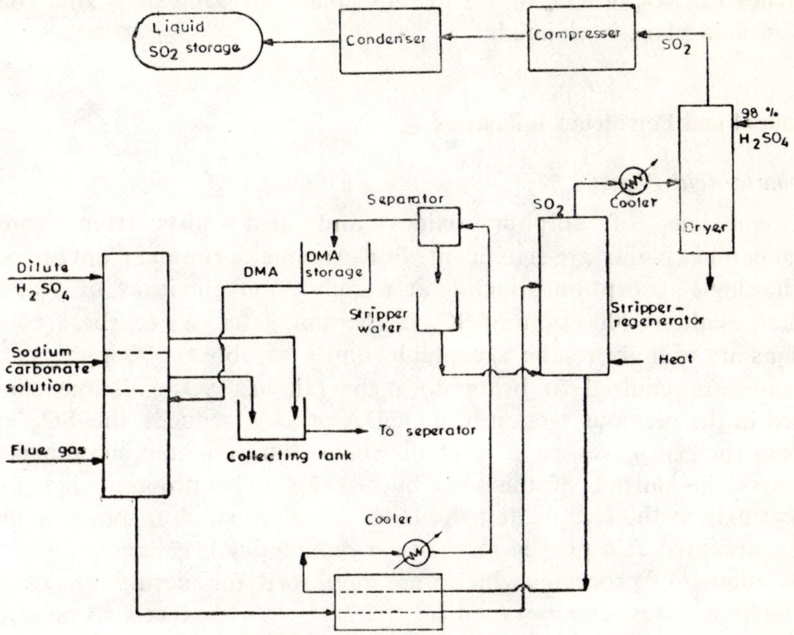

Fig. 6.10 ASARCO process

gases from 5.5% to less than 0.2%. A 30% solution of aqueous ammonia absorbs SO_2 producing ammonium bisulphite (NH_4HSO_3) in solution. SO_2 is then stripped from this solution by adding 93% H_2SO_4 in a stripping column (Fig. 6.11) when ammonium sulphate is produced. A removal

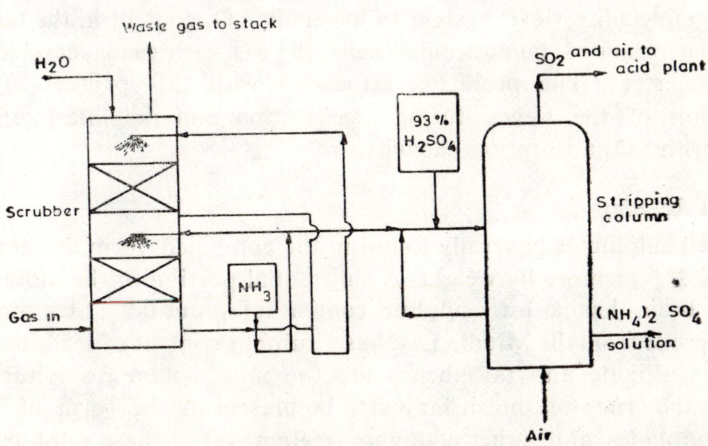

Fig. 6.11 COMINCO process

efficiency for SO_2 of 85% in a lead sinter plant to 97% in a zinc roaster and an acid plant is claimed.

Chemical and Petroleum Industries

Sulphuric Acid Plants

The emissions of sulphur oxides and acid mists from conventional contact plants are significant. For example, a contact plant operating on the single absorption principle at a conversion efficiency of 98-98.5% discharges about 2000 ppm of SO_2 in the tail gases. Since these concentrations are well above the acceptable limits of about 500 ppm, control measures are required to bring down the tail gas SO_2 emissions. As discussed in the previous section, the DCDA process reduces the SO_2 emissions in the exit gases to a level of about 500 ppm and simultaneously increases the output of the acid by $1-1\frac{1}{2}\%$. This process is being used increasingly in the U.S.A. and the U.K. for new sulphur burning plants and is accepted as a proven method for reducing SO_2 emissions.

A number of processes which were developed for desulphurization of combustion gases can be used for tail gases in a contact acid plant. The sodium (Welman-Lord) and magnesium (Chemico) scrubbing systems[11] have been used to obtain an exit gas containing less than 300 ppm SO_2. Another process which has been used commercially to treat the tail gases is the ammonia scrubbing process. Here the ammonium sulphite-bisulphite formed is reacted with nitric or orthophosphoric acid to produce the appropriate ammonium salt, and SO_2 is recovered for further processing.

The most significant advance in adsorption technique over the last 25 years has been the development and utilization of microporous molecular sieve adsorbents. These are based upon the well-known crystalline aluminosilicate zeolites, primarily of the synthetic types. The Pura Siv process[12] utilizes a molecular sieve system to lower the SO_2 content in the tail gas. Upon regeneration of the molecular sieve, the SO_2-rich gas is recycled back to the converters. The problems associated with this process are the deterioration of the sieves during regeneration and the interference of moisture with SO_2 adsorption capacity.

Petroleum Refining

In crude oil sulphur is generally found in the combined form in amounts as high as 2% or more by weight. A substantial portion of the indigenous crude in India has a low sulphur content of about 0.2%; by contrast, imported crude from the Middle East has a sulphur content of 1.5% to 2.5%. Hydrogen sulphide and thiophenes are the most common sulphur compounds in the crude oil, but it may also be present in the form of mercaptans, sulphides and other compounds. Removal of these sulphur compounds from the basic crude is not technically or economically feasible and, hence, desulphurization of products is usually carried out, not only for the sake of pollution control but also for improving the product quality and minimising corrosion problems in the processing plant.

An effective means of removing simpler sulphur compounds like hydrogen sulphide and the mercaptans is by washing the oil with aqueous soda solution. More complex sulphur compounds found in heavy oil fractions have to be broken down catalytically to remove the sulphur. Hydrodesulphurization is the most common method in which hydrogen reacts with the organic sulphur compounds and forms hydrogen sulphide, which is collected and can be converted to elemental sulphur. Extension of hydrodesulphurization technology for residual fuel oil is costly although certain commercial processes are available.

Sulphur dioxide is produced during the regeneration of catalysts used in catalytic cracking and desulphurization processes. Amounts of SO_2 released in flue gases during regeneration in catalytic cracking vary from about 300 to 2200 ppm. These gases are usually disposed of along with other gases using tall stacks.

6.2 CONTROL OF NITROGEN OXIDES

There are two sources of nitrogen which contribute to the formation of nitrogen oxides during combustion. One is the atmospheric nitrogen in the combustion air and the other is the bound nitrogen in the fuel itself. This normally non-reactive nitrogen manages, without the presence of any known catalyst, to combine with oxygen during combustion to form the oxides of nitrogen. The primary-oxide is NO, which rapidly converts by reaction with O_2 or ozone in the urban atmosphere to NO_2. The reason for the formation of NO_x by chemical reactions that are uncatalysed is still not clearly understood. However, there is considerable evidence that water vapour present in fuel-air combinations could act as a catalyst and allow the N_2-O_2 reaction to occur at energy levels typical of fossil fuel flames[13].

Nitrogen oxides are formed in combustion processes according to the reactions:

$$1/2\ N_2 + 1/2\ O_2 \rightleftharpoons NO \tag{6.37}$$

and

$$NO + 1/2\ O_2 \rightleftharpoons NO_2 \tag{6.38}$$

The conditions responsible for the formation of NO and NO_2 can be found out by examining the thermodynamic equilibrium. Table 6.1 gives the equilibrium constant K for the reaction (6.37) at various temperatures. The equilibrium constant $K(T)$ is related to the mole fractions and the stoichiometric coefficients as:

Table 6.1 Equilibrium constant (*K*) for the reaction (6.37)

Temp (K)	*K*
1000	8.7×10^{-5}
1400	1.92×10^{-3}
2000	2.0×10^{-2}
3000	0.121

$$K(T) = \frac{Y_{NO}^{\nu_{NO}}}{Y_{O_2}^{\nu_{O_2}}\, Y_{N_2}^{\nu_{N_2}}} \left(\frac{P}{P_0}\right)^{\nu_{NO}-\nu_{O_2}-\nu_{N_2}} \tag{6.39}$$

where Y is the mole fraction, ν is the stoichiometric coefficient, P_0 is the reference pressure (1 atm.) and P is any pressure.

The equilibrium concentration of NO can be calculated easily by picking a ratio of nitrogen to oxygen. For the purpose of calculation two ratios N/O=4/1 and 40/1 are chosen as the former closely represents the composition of air (3.75 : 1) and the latter represents the products of combustion from a hydrocarbon burned in 10% excess air. The stoichiometric coefficients ν for the reaction in Eq. (6.37) are:

$$\nu_{NO} = 1,\ \nu_{O_2} = 1/2 \quad \text{and} \quad \nu_{N_2} = 1/2.$$

For $P = P_0 = 1$ atm, Eq. (6.39) becomes

$$K(T) = \frac{Y_{NO}}{Y_{O_2}^{\frac{1}{2}}\, Y_{N_2}^{\frac{1}{2}}} \tag{6.40}$$

Furthermore, the sum of the mole fractions

$$Y_{NO} + Y_{N_2} + Y_{O_2} = 1 \tag{6.41}$$

For a specified ratio of N/O = 4/1, a third condition is

$$\frac{Y_{NO} + 2Y_{O_2}}{Y_{NO} + 2Y_{N_2}} = 1/4 \tag{6.42}$$

From Eqs. (6.40), (6.41) and (6.42), Y_{NO} can be calculated at various temperatures using Table 6.1. These results are shown in Table 6.2 It is clear from the Table 6.2 that the equilibrium NO formation increases rapidly with increasing temperature and sufficient amount of NO could be formed in the vicinity of 2000K, which is the adiabatic flame temperature for most of the combustion reactions. In addition, the amount of NO formed depends on the available oxygen (and nitrogen). These results clearly demonstrate that by minimizing the peak combustion temperature and

Table 6.2 Equilibrium NO concentration at various temperatures as a function of initial N: O ratio

Temp. (K)	N: O=4 Y_{NO} (ppm)	N: O=40 Y_{NO} (ppm)
1000	210	80
1400	770	298
2000	8000	2950
3000	45000	15000

operating with as little excess air (and therefore oxygen) as possible, it is possible to control the formation of NO_x. These, in fact, are the two basic NO_x control techniques.

Fortunately, in a real combustion process the time the gas remains in the combustion chamber at a high temperature is lower than the actual time required to reach equilibrium. As a result, the NO concentrations formed are much below the equilibrium concentrations. Table 6.3 lists the

Table 6.3 Time period required for the formation of 500 ppm of NO (gas contains 75% N_2; 3% O_2)

Temp. (K)	Time (sec)	Equilibrium conc. (ppm)	Equilibrium (%)
1588	1370	550	91
1801	16.2	1380	36
2033	1.1	2600	19
2255	0.11	4150	12

time period required to reach 500 ppm NO concentration at different temperatures. Thus, the residence time of gas in the high temperature zone is another important control parameter. The lower the residence time, the lesser will be the formation of NO. Actual residence times in power plant boilers are around 2 seconds at or near their maximum temperatures.

So far only the formation of NO was considered and not its further oxidation to NO_2. In actual combustion processes, the reactions (6.37) and (6.38) occur simultaneously. Table 6.4 shows the predicted equilibrium concentrations that are valid for both the reactions. At room temperature the ratio NO_2/NO becomes large, but little NO or NO_2 is formed. Upto 800K, the amounts of NO and NO_2 present are still negligible, although

Table 6.4 Calculated equilibrium concentrations of NO and NO_2 at various temperatures

Temp. (K)	NO (ppm)	NO_2 (ppm)	NO_2/NO
300	1.1×10^{-10}	3.3×10^{-5}	3.0×10^{5}
800	0.77	0.11	0.14
1400	250	0.87	0.0035
1873	2000	1.8	0.0009

the NO concentration exceeds that of NO_2. However, NO can form in significant concentrations at the conventional combustion temperature (> 1500K) but negligible amounts of NO_2 are present. The reason for this is that the gases do not leave the power station at high temperatures but are cooled and their sensible heat recovered. The decreasing temperature theoretically, would shift the equilibrium concentrations of NO and NO_2. Unfortunately, as soon as the gas leaving the heat zone is quenched, the reaction rate drops several orders of magnitude and the time for the reaction to reach equilibrium is longer than the residence time available so that only 5% to 10% of the available NO is oxidized to NO_2.

Another possibility is the decomposition of NO into elemental nitrogen and oxygen. This is a slow reaction as it requires appreciable activation energy (~ 80 kcal/g-mole) and, hence, its rate is very small. This means that once NO is formed it is quite stable, neither decomposing nor oxidizing to NO_2. The above discussion suggests three ways to reduce the NO_x emissions from stationary sources:

1. By minimizing the residence time at peak temperature.
2. By reducing the peak temperature.
3. By minimizing the availability of O_2 for reaction with N_2.

Modification of Operating Conditions

(*A*) *Low excess-air combustion*

Normal boiler units operate in the presence of 10 to 20% excess air to ensure complete combustion of the fuel, but this excess air provides sufficient oxygen for reacting with nitrogen. The effect of excess air on nitric oxide formation for pulverized coal firing is shown in Fig. 6.12. In this case, NO_x decreased from about 600 ppm at 25% excess air to as low as 175 ppm at 1.4% excess air. However, this overall decrease in NO_x results in lowering of carbon combustion efficiency. For example, it was observed[14] that the combustion efficiency fell from 99.5% at 25% excess air to around 92% at 2% excess air.

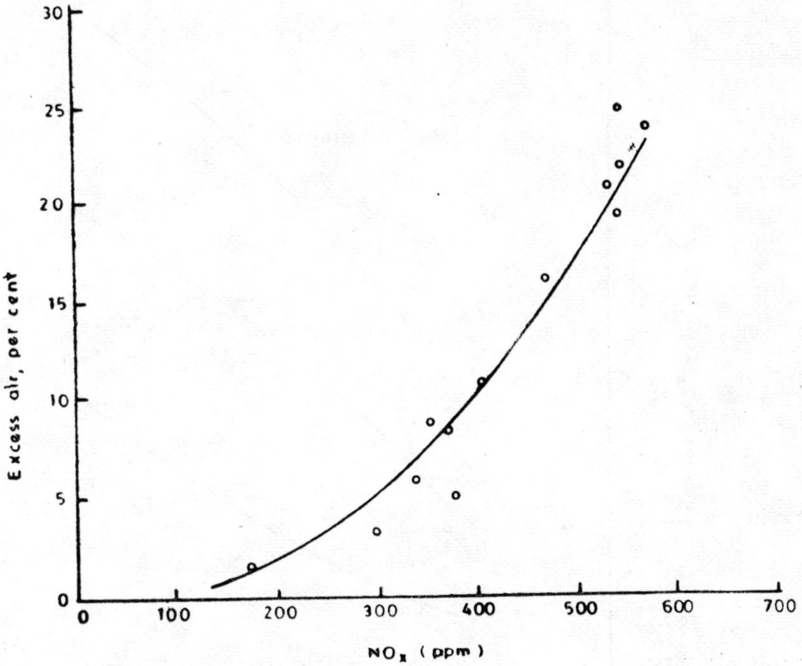

Fig. 6.12 NO_x formation as a function of excess air (ref. 14).

(B) Two stage combustion

This is one of the most effective control methods for the suppression of the formation of nitrogen oxides. In this method all the fuel is fired with only 90–95% of the total air required at the bottom of the furnace followed by secondary air injection higher up in the furnace to complete the combustion. Thus, in the primary zone, incomplete combustion of the fuel takes place resulting in reduced concentration of nitrogen oxides in furnace gases. Heat removal and gas dilution between zones cause the temperature of the gases after the primary zone to decrease so that the final stage of the combustion process occurs at a lower temperature. With this arrangement, the air requirement is the same as that needed for normal complete combustion (115-130% of the stoichiometric amount). Reduction in the emission of NO_x by ~ 38% for coal and oil firing and by 50% for natural gas has been observed under these conditions.

(C) Flue-gas recirculation

The recirculation of cool flue gas (about 10 to 20%) into the combustion chamber is found to be quite effective in reducing NO_x formation. This

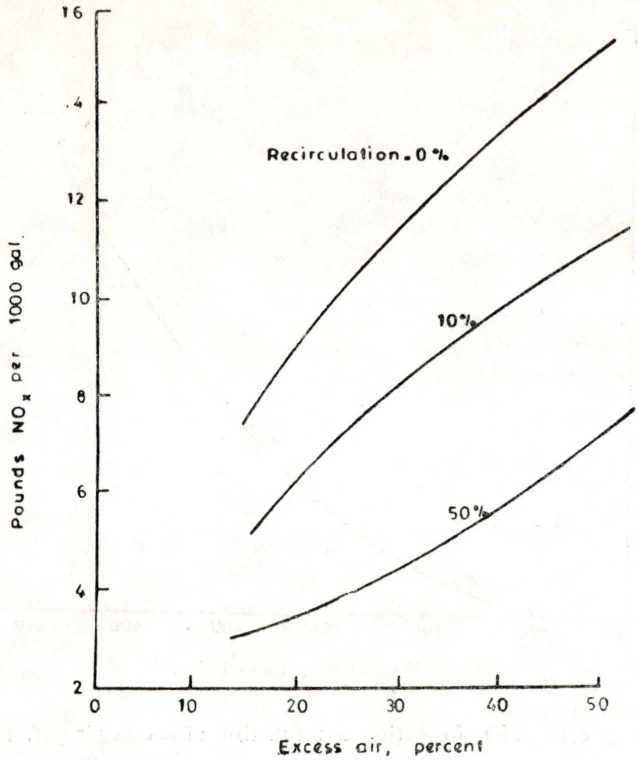

Fig. 6.13 Combined effect of recirculation and excess air on NO_x emissions

method is effective only when the gas is mixed with the primary air prior to combustion[15]. By recirculating the flue gas, the peak flame temperature is reduced because of deficiency of oxygen. Results of the effect of recirculation in oil-fired furnaces are shown in Fig. 6.13. A 10% recirculation, for example, at 15% excess air leads to a potential reduction in NO_x emission of approximately 30%. Similar reductions occur at high percentages of excess air.

(*D*) *Injection of water and steam*

Injecting water or steam into the combustion zone is also effective in reducing emissions of nitrogen oxides. At present, there are still no reliable quantitative data on the effect of this method in boilers, though certain positive results have been achieved in gas turbines and internal combustion engines. Injection of water or steam to the extent of 5-10% of the total combustion air can somewhat reduce the temperature level in the furnace, an effect similar to that occurring in flue gas recirculation.

The best approach to the control of nitrogen oxides by combustion modifications appears to be a combination of a number of methods discussed above. For example, two stage combustion used in conjunction with low excess air firing in natural gas fired boilers results in as high as 90% reduction in NO_x emission. Similarly, low excess air firing with flue gas recirculation has been successfully used in gas and oil firing to achieve significant reductions in NO_x levels. The methods mentioned for use in coal-fired boilers are associated with a number of limitations. For instance, reduction of the excess air and temperature in the furnace may favour high concentrations of carbon monoxide. So any modification of the combustion systems, which is proposed to control oxides of nitrogen, must be examined carefully for its simultaneous effect on carbon monoxide emissions.

Modification of Design Conditions

Burner configuration in combustion chamber can have a substantial influence on the formation of nitrogen oxides. The two basic furnace designs are the tangential and the horizontal methods of firing. In the tangential firing, the furnace itself is used as the burner so that the flame and combustion products rotate in an upward spiral around the walls of the furnace. This results in a low peak flame temperature and consequently reduction in NO_x emissions. In the horizontal firing, the flame is at right angles to the walls of the firebox. This tends to concentrate the hot gases thereby leading to higher flame temperatures and more NO_x emissions. Fig. 6.14 shows the effect of two different forms of boiler firing on NO_x as a function of excess air. Thus, burner configurations that produce more intense combustion and higher temperatures lead to more NO_x emissions than those which permit combustion to occur over a wider area.

Fluidized bed combustion offers great promise for controlling emissions of NO_x as it operates at a temperature which is less than 1050°C. Fluidized beds offer very high heat transfer rates with resultant low combustion temperatures. Also, because of good gas-solid contact within the bed, the excess air rate for combustion can be lowered.

Effluent Gas Treatment Methods

The aim of modification in design and combustion techniques, in general, is to prevent or suppress the emission of nitrogen oxides in the combustion chambers; but, if such measures are not enough or too costly, stack gas treatment is necessary. Stack gas treatment provides potential alternative methods for the control of NO_x emissions, but these methods suffer from problems of treating low concentrations of NO_x in large volumes of effluent gases.

The practical methods for removal of nitrogen oxides can be grouped into two categories: (i) scrubbing and (ii) catalytic decomposition and

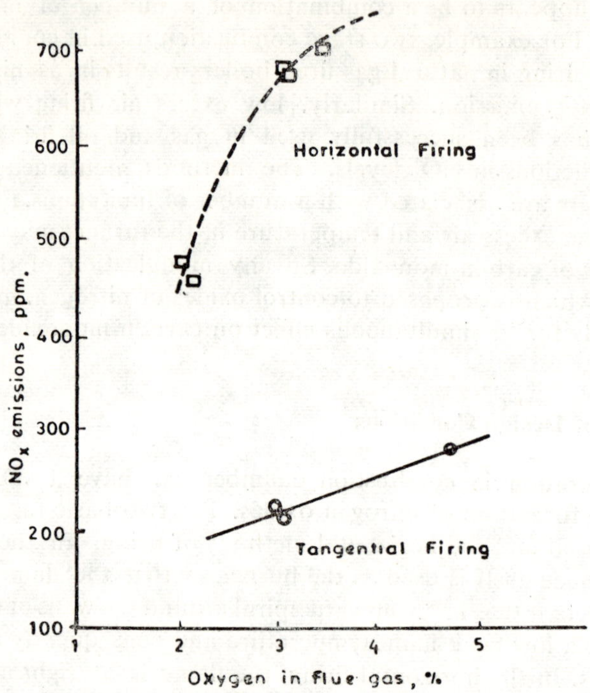

Fig. 6.14 Effect of excess air on NO_x emission for two types of boiler firing (ref. 14).

reduction. The scrubbing methods may be subdivided into absorption by liquids and adsorption by solids. The catalytic reactions involve decomposition of nitric oxide into N_2 and O_2 or reaction with another gas such as carbon monoxide.

Absorption by Liquids

For controlling NO_x emissions from power plants (200 to 1500 ppm), three processes have been proposed which partly improve the economics of SO_2 recovery also. These processes are listed below:

1. Treatment with lime slurry where nitric acid and gypsum are recovered as the byproducts[16].
2. Scrubbing with magnesium hydroxide liquor where concentrated NO is recovered as the byproduct[17].
3. Absorption in sulphuric acid which produces both nitric and sulphuric acids[18].

All the three processes are complex and require attainment of equimolar concentrations of NO and NO_2 in the gas, since the absorption of the combined oxide, N_2O_3 is most favourable. A typical flow diagram of the NO_x absorption by magnesium hydroxide is shown in Fig. 6.15.

In the above process the oxides of nitrogen are absorbed by magnesium hydroxide liquor in an absorption tower. The magnesium nitrate/nitrite

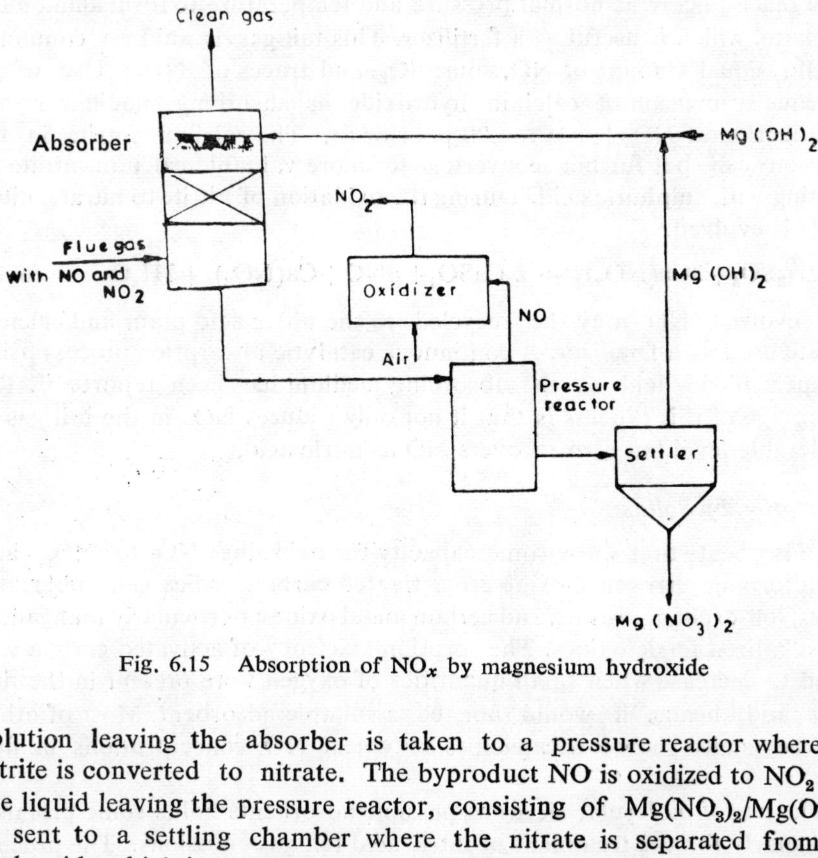

Fig. 6.15 Absorption of NO_x by magnesium hydroxide

solution leaving the absorber is taken to a pressure reactor where the nitrite is converted to nitrate. The byproduct NO is oxidized to NO_2 and the liquid leaving the pressure reactor, consisting of $Mg(NO_3)_2/Mg(OH)_2$, is sent to a settling chamber where the nitrate is separated from the hydroxide which is recycled to the absorption tower. Part of the NO_2 from the oxidizer is sent to the absorber to maintain equimolar concentrations of NO and NO_2 while the rest of NO_2 is used for nitric acid production.

Tail gas emissions from nitric acid plants may range from 0.1 to 0.69 volume per cent with an average of 0.37%. Several scrubbing techniques were developed which initially used solutions of sodium and calcium hydroxide; some major techniques[19, 20] are summarized below:

1. Two stage absorption, first in water and then in sodium hydroxide, yielding nitrite and nitrate salts.
2. Absorption in various types of ammonical solutions such as waste caustic ammonia liquor, ammonium bicarbonate and ammonium bisulphite.

3. Absorption with an aqueous suspension of lime, where calcium nitrate and nitrite can be recovered for use as fertilizers.

Plant performance data with waste caustic ammonia liquor as the absorbing medium showed that the NO_x concentrations can be reduced to 100 ppm. Ammonium bisulphite can be used as a scrubbing liquor when NO_x react quickly at normal pressure and temperature to form ammonium sulphate, which is useful as a fertilizer. This tail gas is said to contain a small residual amount of NO, some SO_2, and traces of NH_3. Use of an aqueous suspension of calcium hydroxide as absorbing medium results in reduction of NO_x levels to 200 ppm or so. The calcium nitrite in the solution can be further converted to more valuable calcium nitrate by treating with sulphuric acid. During the oxidation of nitrite to nitrate, nitric oxide is evolved.

$$2H_2SO_4+3Ca(NO_2)_2 \rightarrow 2\,CaSO_4+4NO+Ca(NO_3)_2+2H_2O \qquad (6.43)$$

The evolved NO may be recycled to the nitric acid plant and calcium nitrate used as a fertilizer. A continuous catalytic absorption process using stripped nitric acid as the absorbing medium has been reported[21]. The advantage of this process is that it not only reduces NO_x in the tail gas to a tolerable level but also recovers NO as nitric acid.

Adsorption by Solids

The adsorbents that show some capacity for oxidizing NO to NO_2, and for adsorbing nitrogen dioxide are activated carbon, silica gel, molecular sieves, ion-exchange resins, and certain metal oxides, particularly manganese and alkalized ferric oxides. The sorption efficiency of activated carbon was found to decrease when small quantities of oxygen were present in the flue gases and, hence, it would not be a suitable adsorbent. Most of other adsorbents also show low capacities at typical NO concentrations in flue gas.

The use of molecular sieves as possible adsorbents holds some promise, particularly for the treatment of nitric acid tail gas emissions. The process developed by Union Carbide[22] generates dry tail gas containing less than 50 ppm NO_x. In this process two beds operate batchwise: one adsorbs NO_x from the tail gas whilst the other is regenerated. The main disadvantage of using molecular sieves is the simultaneous adsorption of water vapour in the tail gas which decreases the efficiency of the bed.

The most suitable adsorbent for NO_x is the one which can be regenerated and at the same time which does not preferentially react with water vapour or with CO_2 in the flue gas. The most promising adsorbent is a ferrous salt. Gidespow and Onischak[23] have studied the adsorption of nitric oxide on silica gel containing ferrous sulphate in the concentration range 50-1000 ppm. Some conversion (2-4%) of ferrous sulphate was observed, but it was felt that this low conversion would not be a practical limitation in a regenerative process since the main aim was to remove NO and not preparing a nitric oxide complex. The practical applicability of the above method to large scale removal of NO is yet to be explored.

Catalytic Decomposition

The homogeneous decomposition of NO is extremely slow and it has also been noted that the heterogeneous decomposition as well does not proceed rapidly. Heterogeneous decomposition using a variety of catalysts (Al_2O_3, Cr_2O_3, Fe_2O_3, ZrO_2, Co_3O_4, $CuO{:}\,SiO_2$, $CuO{:}\,Al_2O_3$, copper chromite) has been tried, but no catalyst has been found which may provide sufficient activity at reasonable temperatures.

Catalytic Reduction

Because of the decomposition reaction being slow and also because of the presence of reducing agents in the flue gas, the catalytic reduction is a much more attractive method of controlling nitrogen oxide emissions. This method has been successfully applied to the control of nitric acid emissions in tail gas. As a first control step, catalytic decolourizers were developed to reduce NO_x to NO, but NO reduction to N_2 was limited. Next came the development of catalytic abatement process where the decolourization product NO could be reduced to molecular nitrogen in the presence of reducing agents like hydrogen, natural gas, coke oven gas and carbon monoxide. Palladium and platinum metals were found to be most effective catalysts. When natural gas is used, it is necessary to preheat the tail gas to about 500°C to ensure ignition, and with hydrogen and carbon monoxide the preheat temperatures are as low as 150°C. However, the formation of some ammonia and carbon monoxide with hydrogen and carbon-containing fuels, respectively, in the treated gas is undesirable.

For application to flue gas emission control, the noble metal catalysts used in nitric acid plants would be prohibitively expensive and subject to sulphur poisoning in the case of coal and oil fired equipment. This has led to the study of the reduction reaction over a variety of commercial catalysts. Table 6.5 gives a list of the variety of catalysts used, the reducing agents and the temperature range. An examination of the results with Ba-promoted Cu-chromite catalyst indicates that it is a reasonable choice as a catalyst for temperatures around 250 to 450°C because of its high activity for NO reduction with a variety of reducing agents.

There are two types of reduction process, which are termed selective and non-selective. In selective reduction, the added reactant preferentially reduces NO_x and in non-selective reduction, the excess oxygen must be consumed first. Selective reduction is to be preferred since it minimizes the

Table 6.5 Commercial catalysts and reducing agents used in catalytic reduction of NO (ref. 24)

Catalyst	Temp. Range (°C)	Reducing agent
Cr-promoted Fe-oxide	250–340	CO or H_2
Fe-chromite	300–450	CO or H_2
Supported platinum	350	CO
	300	C_2H_6
Supported Cu-chromite	120–300	CO
Zn-promoted Cu-chromite	375–425	H_2
	370–425	CH_4
Ba-promoted Cu-chromite	225–525	C_1 to C_8-Hydrocarbons

amount of reactant required. Selective reduction can be carried out with H_2, CO, or NH_3 as the reactant gas. Typical reactions are:

$$2NO + 2H_2 \rightarrow N_2 + 2H_2O \tag{6.44}$$

$$2NO + 2CO \rightarrow N_2 + 2CO_2 \tag{6.45}$$

$$6NO + 4NH_3 \rightarrow 5N_2 + 6H_2O \tag{6.46}$$

In non-selective reduction, two types of reactions take place. The first type involves the reaction of the fuel with oxygen and nitrogen dioxide, the latter being reduced to nitric oxide.

$$CH_4 + 2O_2 \rightarrow CO_2 + 2H_2O \tag{6.47}$$

$$CH_4 + 4NO_2 \rightarrow CO_2 + 2H_2O + 4NO \tag{6.48}$$

The second reaction reduces the NO but the reduction does not take place until all the oxygen in the flue gas has been reacted with the fuel and reducing conditions have been achieved.

$$CH_4 + 4NO \rightarrow CO_2 + 2H_2O + 2N_2 \tag{6.49}$$

6.3 CARBON MONOXIDE CONTROL

Carbon monoxide is formed as an intermediate product of chemical reaction between fossil fuels and oxygen. The formation is due to two reasons: (1) in rich mixtures, CO is formed because of insufficient quantity of oxygen; (2) in lean mixtures, CO is formed due to poor turbulence of the fuel and air in the reaction chamber or due to dissociation of CO_2 to CO in high temperature zones. Hence, the effects of fuel-air ratio, degree of turbulence, and temperature may be significant for CO formation in the hot combustion zone.

Carbon monoxide, unlike some other major gaseous pollutants, does not lend itself to removal by exhaust gas techniques. Hence, the best approach is the control of its formation. However, this is not easy because the control strategies for carbon monoxide and nitrogen oxides are basically in conflict. As a result, merits of each type of combustion equipment need to be analyzed such that both the pollutants could be reduced simultaneously. The most practical method of reducing the CO emissions from stationary combustion sources is by proper design, installation, operation, and maintenance of the combustion equipment. Carbon monoxide and hydrogen are widely used in the chemical industry for the manufacture of methanol, ammonia, and various organic acids and aldehydes. Although significant amounts of CO are generated and handled, the emissions from the chemical and petroleum industry are minor because of good design codes and careful operating practices.

6.4 CONTROL OF HYDROCARBONS

Four techniques are used to control hydrocarbon emissions from stationary sources. These techniques are incineration, adsorption, absorption, and condensation. The incineration devices were discussed in the previous chapter. Hydrocarbon removal efficiencies are high in the flame afterburner which is simple to operate and maintain. Some disadvantages of the flame after burner are that it emits NO_x, may require expensive heat recovery equipment, and may emit CO. In the catalytic afterburner, NO_x emissions are less than those in the flame unit and the fuel costs are also lower. However, the hydrocarbon removal efficiencies are lower than those of the flame unit, the catalyst is susceptible to poisoning and the catalytic unit requires regular maintenance.

In the adsorption method, the polluted gas stream is passed through one or several adsorbers operated in parallel. The adsorbers are generally operated in a temperature range between 30 and 60°C. The preferred adsorbent is granular activated carbon with a mean diameter of 2 to 4 mm. The adsorbed vapours are usually removed by passing steam through the system. The mixture of steam and hydrocarbons is liquified in a condenser and cooled down to ambient temperature in a cooler. The hydrocarbons are separated from the water and recovered for further use. Adsorption units are fairly expensive in terms of capital cost but need little in the way of maintenance or operational manpower. Probably the most effective way of removing hydrocarbons from a polluted stream is by contact with a liquid, or a series of different liquids, in one or more absorption towers. Water is a satisfactory liquid if the hydrocarbons have sufficiently high solubility; however, for many hydrocarbons this is not the case and appropriate chemical reagents must be used to achieve a good performance. The condensation method makes use of the fact that at sufficiently low temperatures gaseous hydrocarbons condense to liquids. The effluent gas stream is passed over low temperature surfaces, and the condensed hydrocarbons are left behind and collected.

6.5 MOBILE SOURCES

The principal pollutants of interest emitted from mobile sources such as the automobiles are carbon monoxide, nitrogen oxides, and hydrocarbons. The types of pollutants emitted by petrol engines and diesel engines are similar but vary their proportions due to difference in the mode of operation of the two types of engines.

The major source of air pollution from automobiles is the exhaust pipe which accounts for about 70 per cent of the pollution whereas the crank case emissions account for about 20 per cent. Evaporations from fuel tank and carburettor account for the remaining part of pollution percentage. The exhaust from petrol engines is almost colourless due to homogeneity of the fuel-air mixture and high temperatures involved during combustion. On the other hand, the diesel engine exhaust is usually dirty due to improper mixing of fuel and air and lower combustion temperatures. The diesel engine smoke may be white, blue or black.

Control methods applied to one pollutant often influence the output of other pollutants. For example, the amount of CO formed during the combustion period is related to the ratio of air and fuel in the cylinder. If the mixture is rich, i.e., there is an excess of fuel, then the products of combustion contain a high level of CO and hydrocarbons and low levels of NO_x. The stoichiometric mixture gives low CO and hydrocarbon values but a large amount of NO_x. Fig. 6.16 compares the exhaust NO_x, CO and hydrocarbon concentrations at various mixture strengths for a typical automobile. According to the figure the apparent solution to the problem lies in using high air-fuel ratios, but this in turn presents difficult engine design problems and if the air-fuel ratio is raised too much, hydrocarbon emissions again increase.

The control of pollutant emissions from automobiles is based on the following four approaches:

(1) Reduction of the amount of pollutants formed during combustion by suitable modification of the internal combustion engine.

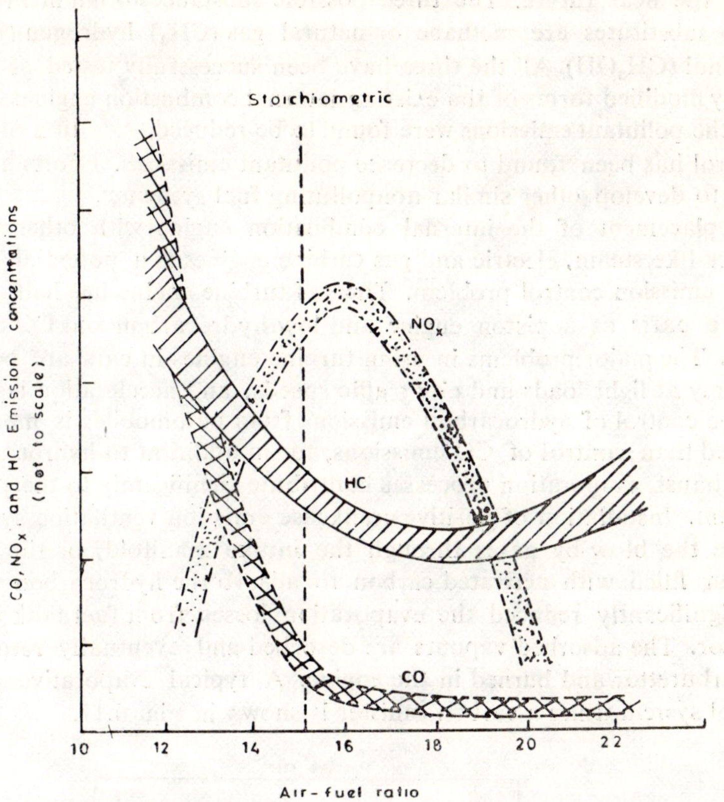

Fig. 6.16 Effect of air-fuel ratio on exhaust composition

(2) Development of exhaust system reactors that will complete the combustion process and change potential pollutants into more acceptable materials.
(3) Development of substitute fuels for petrol that will produce low levels of pollutants upon combustion.
(4) Replacement of internal combustion engine with low pollution-producing engines.

Engine modifications are directed toward improved control over the combustion cycle. These modifications include electronic fuel injection, quick-heating intake manifolds, new ignition systems that do not use spark plugs, and cylinder-head modifications to reduce quenching zones and improve combustion. Exhaust system reactors, both thermal and catalytic, are now being used to reduce CO and hydrocarbon emissions. Another possible solution to the CO problem involves changing the fuel either by mixing petrol with other substances or by substituting another fuel for petrol. Alternative fuels are now being explored, particularly because of shortage of petroleum products and this approach could be successful in the near future. The three possible substances often mentioned as petrol substitutes are methane or natural gas (CH_4), hydrogen (H_2), and methanol (CH_3OH). All the three have been successfully tested as fuels in slightly modified forms of the existing internal combustion engines. In each case, the pollutant emissions were found to be reduced. Addition of alcohol to petrol has been found to decrease pollutant emissions. Efforts are being made to develop other similar nonpolluting fuel systems.

Replacement of the internal combustion engine with other types of engines like steam, electric and gas turbine engines is a potential solution to the emission control problem. The gas turbine engine has half as many moving parts as a piston engine and low hydrocarbon and CO emission levels. The major problems in using turbine engines in cars are poor fuel economy at light loads and city traffic speeds, and acceleration lag.

The control of hydrocarbon emissions from automobiles is more complicated than control of CO emissions, for in addition to hydrocarbons in the exhaust, evaporation processes contribute significantly to the pollution problem. Installation of positive crankcase emission ventilation systems to recycle the blow-by gases through the intake manifold, or the use of a canister filled with activated carbon to adsorb the hydrocarbon vapours has significantly reduced the evaporation losses from fuel tank and carburettor. The adsorbed vapours are desorbed and eventually returned to the carburettor and burned in the engine. A typical evaporative emission control system using a carbon canister is shown in Fig. 6.17.

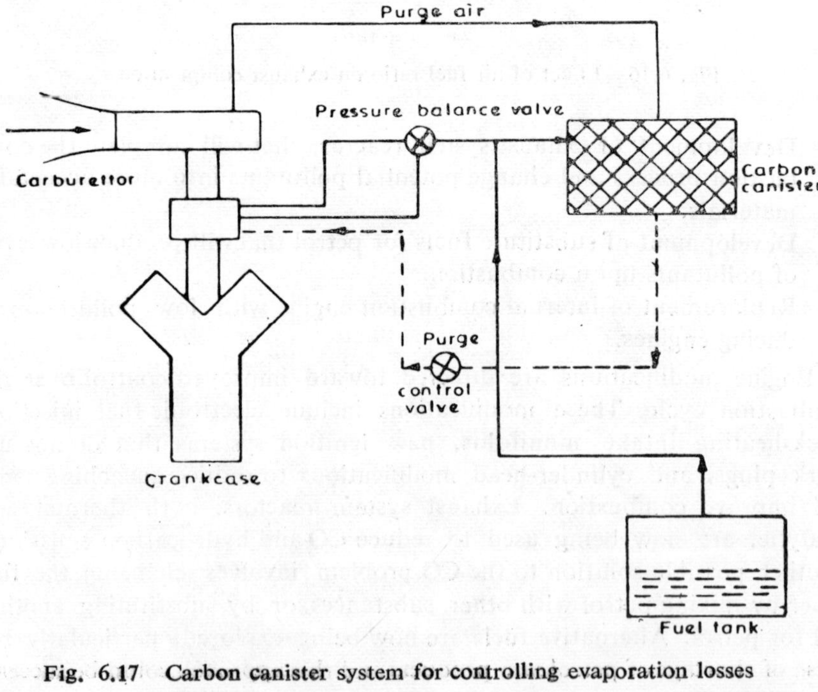

Fig. 6.17 Carbon canister system for controlling evaporation losses

The control of NO_x emissions from automobiles can be accomplished by two methods: exhaust gas recirculation (EGR) and catalytic reduction.

In EGR method, a portion of the exhaust gas, which is inert, is continuously recirculated through the engine so that the burning air-fuel mixture is diluted. The dilution decreases the oxygen concentration in the burning mixture thereby lowering the peak combustion temperature. The exhaust gas recirculated is regulated by an EGR valve, and is generally 15 to 22% by the volume of the air-fuel mixture.

The catalytic reduction method makes use of a dual-catalyst system in which NO is reduced to N_2 while hydrocarbons and CO are oxidized to CO_2 and H_2O simultaneously. The exhaust gases are initially passed over a reductive catalyst where NO reacts with CO (also in the exhaust) to form N_2 and CO_2. Air is then injected and the mixture is passed over an oxidative catalyst where the remaining hydrocarbons and CO are oxidized to CO_2 and H_2O. In the reduction catalyst chamber some amonia may form which will subsequently be converted to NO or N_2O in the oxidizing catalyst chamber. Fig. 6.18 shows the intricate series of interdependent reactions that go on in a dual catalyst exhaust emission control system.

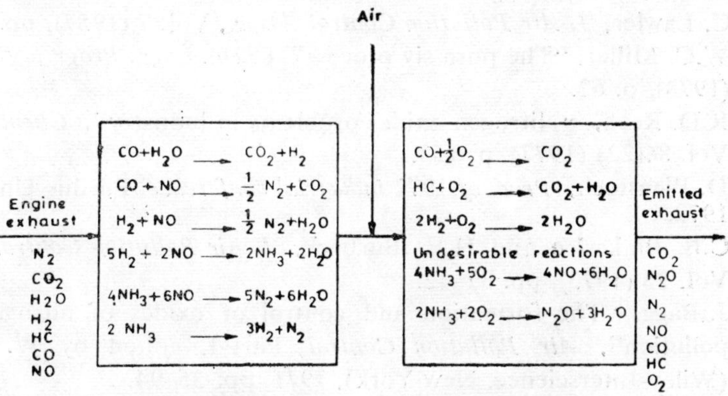

Fig. 6.18 Typical reactions in a dual catalyst exhaust control system (ref. 25, reproduced with permission of the author.)

REFERENCES

1. R.M. Jimeson, "Utilizing solvent refined coal in power plants", *Chem. Engg. Progr.*, Vol. 62 (10) (1966), p. 53.
2. A.L. Kohl and F.C. Riesenfeld, *Gas purification* (McGraw-Hill, New York), 1960.

3. D. Bienstock and J.H. Field, *J. Air Pollution Control Assoc.*, Vol. 10(2) (1960), p. 121.
4. S. Katell, "Removing sulfur dioxide from flue gases", *Chem. Engg., Progr.*, Vol. 62(10) (1966), p. 67.
5. A.V. Slack, *Sulfur dioxide removal from waste gases,* PCR–4 (Noyes Data Corp., U.S.A.), 1971.
6. W.E. Miller, "The Cat-Ox process at illinois power", *Chem. Engg. Progr.*, Vol. 70(6) (1974), p. 49.
7. F.M. Dautzenberg, J.E. Nadar and A.J.J. Van Ginneken, *Chem. Engg. Progr.*, Vol. 67 (1971), p. C–122.
8. H. Juntgen, K. Knoblauch and W. Peters, "SO_2 removal from flue gases by special carbon", in *Proc. of the second inter. clean air congress*, edited by H.M. Englund and W.T. Beery (Academic Press, New York), 1971, p. 390.
9. R.D. Oldenkamp and E.D. Margolin, "The molten carbonate process for sulfur oxide emissions", *Chem. Engg. Progr.*, Vol. 65(11) (1969), pp. 73–76.
10. N. Kaplan and M.A. Maxwell, "Removal of SO_2 from industrial waste", *Chem. Engg.*, Vol. 84 (1977), p. 127.
11. C. Lawler, *J. Air Pollution Control Assoc.*, Vol. 7 (1957), pp. 29–31.
12. W.C. Miller, "The pura siv process", *Chem. Engg. Progr.*, Vol. 69(8) (1973), p. 62.
13. R.D. Reed, "Nitrogen oxides problems in industry", *Chem. Engg.*, Vol. 84(22) (1977), p. 153.
14. D. Bienstock, *Proc. of* 1972 *industrial conference*, Purdue Univ., Oct. 1972.
15. C.E. Blakeslee and H.E. Burback, *J. Air Pollution Control Assoc.*, Vol. 23 (1973), pp. 37–42.
16. J. Bagg, "The formation and control of oxides of nitrogen in air pollution", *Air Pollution Control*, Part–I, edited by W. Strauss (Wiley-Interscience, New York), 1971, pp. 35–94.
17. H. Brauer and Y.B.G. Varma, *Air pollution control equipment* (Springer-Verlag, Berlin), 1981, p. 248.
18. W. Bartok, "Stationary sources and control of nitrogen oxide emissions", in *Proc. of the second international clean air congress*, edited by H.M. Englund and W.T. Beery (Academic Press, New York), 1971, p. 801.
19. *Atmospheric emissions from nitric acid manufacturing processes,* U.S.P.H.S. Publication, No. 99–Ap–27, 1966.
20. H.R.L. Streight, *Can. J. Chem. Engg.*, Vol. 36 (1958), p. 3.
21. B.J. Mayland and R.C. Heinze, *AIChE Symposium Series,* Vol, 70 (137) (1974), p. 83.
22. L.L. Fornoff, "A new molecular sieve process for NO_x removal and recovery of nitric acid plant tail gas", *AIChE Symposium Series,* Vol. 68(126) (1972), p. 111.

23. D. Gidaspow and M. Onischak, *Can. J. Chem. Engg.*, Vol. 51 (1973), p. 337.
24. C.S. Rao, "Control of oxides of nitrogen from effluent gases", *Chem. Engg. World*, Vol. 13(4) (1978), pp. 102–108.
25. John E. Seinfeld, *Air Pollution* (Mc-Graw-Hill, New York), 1975.

SUGGESTIONS FOR FURTHER READING

1. R.E. Harrington, "SO_2 removal from waste gases: a status report United States: recovery processes", in *Pollution engineering and scientific solutions,* edited by F.S. Barrekatte (Plenum Press, New York), 1973, p. 224.
2. J. Ando and S. Hori, "SO_2 removal from waste gases: a status report Japan", in *Pollution engineering and scientific solutions,* edited by F.S. Barrekatte (Plenum Press, New York), 1973, p. 233.
3. J.B. Hyne, "Desulphurization of effluent gas streams—a review and comparison of techniques", *Proc. of 51st NGPA Annual Convention, New Orleans,* U.S.A., April 10–12, 1972.
4. G.T. Rochelle, "Economics of flue gas desulphurization", *Proc. of flue gas desulphurization symposium,* New Orleans, U.S. Environ. Protection Agency, May 14–17, 1973.
5. R.G. Temple, "Control of internal combustion engines", in *Air pollution control,* Part I, edited by W. Strauss (Wiley Interscience, New York), 1971, pp. 177-225.

PROBLEMS

1. At 610°C the equilibrium constant for the reaction $SO_2+\frac{1}{2}O_2 \rightarrow SO_3$ is found to be K = 8.5, when pure gases at 1 atm. pressure are used as the reference state. The off gases from a typical ore-roasting oven contain 12% SO_2, 8% O_2, and 80% N_2 on a molal basis. If this gas is passed into a catalytic reactor for conversion to SO_3 at a temperature of 610°C and 1 atm. pressure, what is the equilibrium fractional conversion of SO_2 to SO_3? (63.5%)
2. It is desired to control SO_2 emissions from thermal power plants. There are two alternatives available—one is the dry process and the other is the wet process. Draw generalized flow diagrams for the two processes and identify the control devices used in each case.
3. Major pollutants from pulp and paper mills are particulate matter, SO_2, odourous sulphur compounds like H_2S, methyl mercaptan and dimethyl sulphide Prepare a scheme to control the above pollutant emissions.

4. Fluorine contaminants may be emitted to the atmosphere by a wide variety of industrial processes in which fluorine compounds are manufactured, utilized as catalysts, or are present as impurities in the process materials. Identify the specific fluorine compounds and their sources of emission, and suggest methods for their control.
5. For the reaction $\frac{1}{2}N_2 + \frac{1}{2}O_2 \rightarrow NO$, the equilibrium compositions of NO and O_2 at 4000 K and 1 atm. pressure are 100,000 ppm and 150,000 ppm respectively.

   (a) To what N:O ratio this corresponds to? (4)
   (b) What is the value of the equilibrium constant K? (0.3)
   (c) If the N : O ratio is now changed to 5 : 1 and the equilibrium concentration of O_2 obtained is the same as before, what will be the concentration of NO in ppm? (33,000)

6. Gasoline is a mixture of hydrocarbons and impurities. Let us suppose its empirical chemical composition to be C_8H_{16}.
   Calculate the amount of oxygen required for combustion if the reaction is stoichiometric according to the equation,

   $$C_8H_{16} + 12O_2 \rightarrow 8CO_2 + 8H_2O$$

   Air contains 23.2 per cent oxygen by weight. How much air is required per kilogram of fuel? Calculate the air to fuel ratio, as kilogram of air per kilogram of fuel. (1655 kg; 14.77)
7. The quantities of various exhaust products of internal combustion can be estimated on the basis of chemical equilibrium considerations. Compute the equilibrium mole fraction of CO in a system of CO, CO_2, and O_2 at 1 atm. pressure and at 2000, 3000, and 4000 K according to the equation,

   $$CO_2 \rightleftharpoons CO + \tfrac{1}{2}O_2$$

   under conditions in which the ratio of the number of oxygen atoms to the number of carbon atoms is 2, 3.125 (stoichiometric combustion of octane), and 5. The equilibrium constant is

   $$K = 3 \times 10^4 \exp(-67{,}000/RT)$$

8. An automobile is equipped with a catalytic converter to control CO and hydrocarbon emissions in the exhaust gas. The catalyst temperature must be controlled at 300°C by adjusting the air intake to the converter. The sensible heat of the product gas after combustion in the converter is 1.8×10^5 J/m³ at 300°C. During driving, the exhaust gas rate is 2 m³/min at the gas temperature of 425°C and its molecular weight is 30. If the air intake to the converter occurs at 25°C, what is the amount of air required? Take the average specific heat of the exhaust gas and air as 1050 J/kg. K, and neglect the heat loss from the converter to the surroundings. (1.5 kg/min)

7. Sources and Classification of Water Pollutants

Water is a vital natural resource which is essential for a multiplicity of purposes. Its many uses include drinking and other domestic uses, industrial cooling, power generation, agriculture (irrigation), transportation and waste disposal. In the chemical process industry water is used as a reaction medium, a solvent, a scrubbing medium, and a heat transfer agent. As a source of life for man, plants and other forms of life it cannot be replaced. This chapter begins with an outline of India's water budget. Since pollution of our aquatic system continues to grow, the main emphasis in the chapter is on the discussion of the origin of wastewater, the classification of water pollutants and their effects.

7.1 WATER RESOURCES

The main source of water in India is rain. However, the time-distribution of precipitation is quite uneven as can be seen from Table 7.1. As a result, there are floods in many parts of the country during the June-September period; the same regions might experience acute water scarcity during the January-May period. The annual yield from rainfall is estimated to be about 400 million hectare metres (MHM) and out of this, 70 MHM of

Table 7.1 Season-wise distribution of rainfall in India (ref. 1)

Rainfall season	Duration	Approximate percentage of the annual rainfall
South-west monsoon	June—Sept.	73.7
Post-monsoon	Oct.—Dec.	13.3
Winter or North-East monsoon	Jan.—Feb.	2.6
Pre-monsoon	March—May	10.4
	Total	100.0

water evaporates immediately, 115 MHM runs off into surface water bodies and the remaining percolates into the soil[2]. Based on rainfall records and run-off coefficients, the total annual flow through rivers in India is estimated to be around 168 MHM. Table 7.2 gives a detailed breakdown of the surface water resources. In practice, however, most of this water flows back to the sea or is lost through evaporation. A part percolates into the ground. The utilizable portion of water for irrigation, power generation and other uses is around 70 MHM.

Table 7.2 Surface water resources in India (ref. 3)

Region	Water potential (Million Hectare Metres)
1. West flowing rivers including the Indus system	39.00
2. East flowing rivers other than Ganga and Brahmaputra systems	41.19
3. Ganga system	48.96
4. Brahmaputra system	38.08
Total	167.23

Data regarding ground water resources, that is, water obtained from precipitation and stored in aquifers after percolating through porous rocks and intake areas, are more limited than those on the surface water resources. It is estimated that the net annual recharge in India is 67 MHM and out of this only 35 MHM may be assumed to be available for utilization. Combined with surface water, it gives the total utilizable potential of fresh water to be around 100 MHM.

Two potential sources of water for human use are: (i) desalinated seawater or brackish ground water and (ii) reclaimed wastewater. With the rapid growth of population, many countries in the world are now thinking of desalinated sea water or brackish water as a potential source of supply of potable water in scarcity hit regions where natural water resources are either non-existent or are inadequate to meet the increasing demand. Desalination may be accomplished by such processes as distillation, electrodialysis, freezing and reverse osmosis. Even though well over 2 million cubic metres of desalted water are being produced daily, the high cost of production at present restricts the use of desalination technology mainly to those places where conventional sources are practically non-existent.

In India desalination technology has great potential. The need of potable water in arid and semi-arid regions is so great that though the desalination technology is somewhat costlier, it will still be of immense benefit to those regions. Development work has been going on for a number of years in India and in other countries and it is quite possible that desalted water may be produced at reasonable costs so that it can be used not only for conventional human needs but even for agriculture.

Wastewater reclamation has often been overlooked in the study of water resources. Treated wastewater may be used for such municipal purposes as firefighting, irrigation of parks, and for sanitary needs. One of the most effective and economical way of using wastewater in industry is the inter-plant re-use of treated and recycled industrial effluents. Reclaimed process water, however, will require dual supply systems within a manufacturing plant and care must be taken to prevent possible cross-connections and the contamination of potable water used in the same plant.

Utilisation of Water

Water resources are developed primarily for irrigation, domestic use, power generation, and industrial use. For the efficient development of any water resource an understanding of the nature of the various water uses is essential. A brief summary of the most important uses follows.

Irrigation

Water requirements for irrigation are by far the highest among various uses of fresh water. The amount of water required for irrigation purposes varies with the climate of the region and the type of crops that are to be raised. Surface waters used for irrigation may be diverted through canals which afford gravity-aided flows to the irrigation area. Where the lands to be irrigated lie above the river, lift irrigation may be required. In 1968–69 about 20.5 MHM of surface water was utilized to irrigate a gross area of about 25 million hectares, and about 8.0 MHM of ground water was utilised to irrigate 13.2 million hectares of land. The ultimate irrigation potential of the country is not known with certainty. Current estimate puts the potential irrigated area around 106 million hectares—72 million hectares irrigated by surface water resources and 34 million hectares by ground water resources.

Domestic Water Supply

Domestic water requirements vary from season to season and from rural to urban areas. In summer, the need for water is greater than that in winter. Water consumption in cities is greater than that in rural areas. Moreover, it depends on factors like the standard of living and habits of people in the community.

An accurate assessment of water required for domestic needs is not possible as several factors complicate the estimation in India. The per capita consumption of water in both urban and rural areas is generally not known because people augment municipal supplies with other sources such as private wells, rivers, and lakes. On the basis of certain gross assumptions, the daily per capita use for both urban and rural population has been estimated for the year 1968–69 and the projected demand has been shown for the year 2000-01 in Table 7.3.

Table 7.3 Norms of water use for domestic purpose (ref. 4)

	Daily per capita consumption (litres)	
	1968–69	2000–01
Urban population		
Served through organized supplies		
(a) with water-borne sanitation	135	180
(b) without water-borne sanitation	90	115
No organized supplies	45	45
Rural population		
Served through organized supplies	45	60
No organized supplies	25	25

The per capita consumption of water has been assumed to increase with the standard of living; however, those who live in major cities of India are painfully familiar with intermittent water services, frequently limited to 2 to 3 hours per day. The per capita water use has declined in many cities as a result of tremendous growth in their population.

Power Generation

Cooling is the principal water use in thermal power generation and more than 99 per cent of water used is required for condenser cooling. The water carrying waste heat is directly discharged into the natural aquatic system or sometimes it is passed through a cooling tower and then recirculated back into the condenser. In both cases, losses due to evaporation are roughly of the same order of magnitude. It has been estimated that in a 500 MW power plant about 40 million litres of water would evaporate per day. With increasing capacity for thermal power generation, the demand for water will increase significantly. On the basis of the assumption that a moderate one-third of total intake water is recirculated, the total annual withdrawal requirement in the year 2000 is estimated to vary from 3.6 MHM for a low economic growth and high population scenario to 9.3 MHM for a high economic growth and low population scenario.

Industrial Water Use

Industry is much dependent on adequate water supplies. The enormous demand that industry has for water is indicated by the water requirement of some key industrial processes. Table 7.4 presents some typical industrial water requirements.

Table 7.4 Water requirements for selected industries per unit of product (refs. 5, 6)

Industry	Unit of product	Water requirement (litres)
Bread	ton	2100–4200
Beer	kilolitre	15000
Milk products	ton	20000
Wood pulp	ton of pulp and paper	236000
Cotton bleaching	ton	300000
Chemicals—acetic acid	ton	417000–1000000
Sulphuric acid	ton of 100% H_2SO_4	10400
Gasoline	kilolitre	7000–10000
Steel	ton	260000

As can be seen from the table, the quantities of water used by industry vary widely. Major water users are steel, pulp and paper, textiles, chemicals and petroleum refining; they account for nearly 80 per cent of industrial water demand. Accurate projected requirements are not easily available for various industries in India, but a rough estimate indicates that the annual intake of water by all industries will be between 1.7 and 3.9 MHM in the year 2000.

With a growing population, the demand for fresh water is steadily increasing in India, but as with all other resources, there is a limit to fresh water supply. In addition, the availability of high-quality water is dwindling because of misuse, waste and pollution.

7.2 ORIGIN OF WASTEWATER

Wastewaters can be classified by their origin as domestic wastewater and industrial wastewater. Any combination of wastewaters that is collected in municipal sewers is termed as municipal sewage. Domestic wastewater is that which is discharged from residential and commercial establishments, whereas industrial wastewater is that which is discharged from manufacturing plants. The pollutants in domestic wastewater arise from residential and commercial cleaning operations, laundry, food preparation, body cleaning functions, and body excretions. The composition of domestic wastewater is relatively constant.

Industrial wastewater is formed at industrial plants where water is used for various processes, and also for washing and rinsing of equipment, rooms, etc. These operations result in the pollution of the nearby aquatic systems because some of the products and byproducts are discharged, either deliberately or unintentionally into them.

Normally, wastewaters are conducted to treatment plants for removing undesirable components which include both organic and inorganic matter as well as soluble and insoluble material. These pollutants, if discharged directly or with improper treatment, can interfere with the self-cleaning mechanisms of water bodies. The capacity for self-cleaning is due to the presence of relatively small numbers of different types of micro-organisms in the water bodies. These micro-organisms use as food much of the organic pollutants and break them down into simple compounds such as CO_2 or methane, and the micro-organisms produce new cells also. But often either a pollutant does not degrade naturally or the sheer volume of the pollutant discharged is sufficient to overwhelm the self-cleaning process. Also, the microbial population can be destroyed by toxic wastes discharged into the waterway. If that happens, the pollutant concentrations will build up and reach high enough levels that will prevent re-establishment of a microbial population. The water quality thus becomes permanently degraded.

Various constituents of wastewater are potentially harmful to the environment and to human health. In the environment, the pollutants may cause destruction of animal and plant life, and aesthetic nuisance. Drinking water sources are often threatened by increasing concentration of pathogenic organisms as well as by many of the new toxic chemicals disposed of by industry and agriculture. Thus, the treatment of these wastes is of paramount importance.

Wastewater Composition

Depending on the amounts of physical, chemical, and biological constituents of wastewaters, they may be classified as strong, medium, or weak.

Table 7.5 Typical composition of domestic waste water (ref. 7*)

Constituent	Concentration†		
	Strong	Medium	Weak
Solids, total	1200	700	350
Dissolved solid, total	850	500	250
fixed	525	300	145
volatile	325	200	105
Suspended solids, total	350	200	100
fixed	75	50	30
volatile	275	150	70
Settleable solids (ml/1)	20	10	5
Biochemical oxygen demand, 5-days, 20°C	300	200	100
Total organic carbon (TOC)	200	135	65
Chemical oxygen demand (COD)	1000	500	250
Nitrogen (total as N)	85	40	20
Organic	35	15	8
Free ammonia	50	25	12
Nitrites	0	0	0
Nitrates	0	0	0
Phosphorus (total as P)	20	10	6
Organic	5	3	2
Inorganic	15	7	4
Chlorides††	100	50	30
Alkalinity (as $CaCO_3$)††	200	100	50
Grease	150	100	50

†All values except for settleable solids are expressed in mg/l.
††Values should be increased by amount in carriage water.
*Reprinted with permission of McGraw-Hill, N.Y., 1972.

Typical compositions for domestic wastewater are given in Table 7.5. The compositions and concentrations are highly variable, and hence, the table is intended only to serve as a guide and not as a basis for design. Data on typical characteristics of municipal sewage in some urban cities of India are given in Table 7.6.

Table 7.6 Composition of sewage in some urban cities of India (refs. 8, 9)

Constituent	Delhi	Kanpur	Madras	Hyderabad	Ahmedabad
Suspended solids (mg/l)	354	560	550	985	148
TDS (mg/l)	803	1000	1170	723	1600
BOD_5 at 20°C (mg/l)	203	255	352	339	133
COD (mg/l)	377	532	—	—	—
Nitrogen (total as N) (mg/l)	28.5	73.9	59.2	50.9	27.3
PO_4 (mg/l)	13.7	2.5	25.3	—	—
Chlorides (mg/l)	147	114	259	104	300
pH	7.5	7.0	7.4	7.4	7.5

TDS=total dissolved solids; BOD=biochemical oxygen demand; COD=chemical oxygen demand

The composition of industrial wastewater is quite varied, its constituents ranging from organic solvents, oils, suspended solids to dissolved chemical compounds. Table 7.7 lists a number of potentially polluting chemical substances released by different industries. Most of the chemicals are toxic and some are even suspected of causing cancer. Such "cancer-suspect agents" include[11] 4-nitrobiphenyl, 4-aminodiphenyl, α- and β-naphthylamine, methyl chloromethyl ether, benzidine, 3, 3′-dichlorobenzidine, β-propiolactone, 4-dimethylaminoazo-benzene, N-nitrosodimethylamine and vinyl chloride. Another important aspect of industrial wastewaters is that some of the chemicals are valuable enough to warrant recovery. This aspect of recovery has taken on a new perspective in recent years and the recovered substance can either be recycled within the plant, or offered to other companies for use as basic or intermediate chemicals.

Table 7.7 Major pollutants in various Industries (ref. 10)*

Pollutants	Industry type†
Organic	
Proteins	(1), (2), (3), (4), (13)
Carbohydrates	(1), (2), (3), (5), (6), (13)
Fats and oils	(1), (2), (3), (4), (7), (9), (10), (11), (12), (13)
Dyestuffs	(3), (4), (5), (6)
Organic acids	(6), (13)
Phenols	(3), (6), (7), (8), (9), (10), (11)
Detergents	(1), (3), (12), (13)
Organo-pesticides	(2), (13)
Inorganic	
Acids	(3), (6), (10), (11)
Alkalies	(3), (6), (7), (8), (10), (11), (12)
Metals	(3), (5), (6), (10), (11), (13)
Metallic salts	(4), (5), (10), (11), (13)
Other salts	(1), (3), (4), (5), (6)
Phosphates, nitrates	(1), (3), (4), (5), (6)
Bleaches	(3), (5), (12)
Sulphides	(3), (4), (7), (8)
Cyanides, cyanates	(8), (10), (11)
Chromates	(4), (7), (10), (11)
Minerals (china clay and soil)	(1), (5), (13)

*Reprinted with permission of John Wiley and Sons, Inc., N.Y., 1981.

†Industry type: (1) dairy, (2) food processing, (3) textiles, (4) tanning, (5) paper making, (6) chemical, (7) petro-chemical, (8) coke ovens, (9) industrial oil production, (10) engineering, (11) metallurgy, (12) laundry and (13) agriculture.

7.3 TYPES OF WATER POLLUTANTS AND THEIR EFFECTS

The problem of water pollution due to discharge of domestic and industrial wastes into aquatic systems has already become a serious problem in the country. Nearly 75 to 80% of India's population is exposed to unsafe drinking water. As a result, enteric diseases, often reaching epidemic proportions, devastate several parts of the country. The rivers and lakes near urban centres emit disgusting odours and fish are being killed in millions along sea coasts. The meat of some of them is tainted and unsafe to eat because of excessive levels of mercury and pesticides in their bodies. The origin of these problems must be attributed to many sources and types of pollutants. Some pollutants may have indirect effects whilst substances normally not considered as pollutants may become so under special circumstances. To aid in a systematic discussion of water pollutants, they have been classified into nine categories as described below.

1. Oxygen demanding wastes
2. Disease-causing agents
3. Synthetic organic compounds
4. Plant nutrients
5. Inorganic chemicals and minerals
6. Sediments
7. Radioactive substances
8. Thermal discharges
9. Oil

Pollution of the waterways is often caused by a combination of the above categories which can severely compound the problem. Because of their vital importance in water quality control programmes, these categories are discussed briefly in the following sections.

Oxygen Demanding Wastes

Dissolved oxygen (DO) is essential for sustaining the plant and animal life in any aquatic system. For example, warm-water fish requires a minimum DO level of at least 5 mg/l (5 ppm). If the DO level drops below the level necessary to sustain normal life, then the aquatic system is classified as polluted.

There are four processes which actually affect the DO content in the water: reaeration, photosynthesis, respiration and the oxidation of wastes. Reaeration is the process by which oxygen transfer takes place from the atmosphere to water. The solubility of oxygen in fresh water at saturation point decreases with an increase in temperature and the typical variation at sea level is shown in Fig. 7.1. When the DO level falls below the saturation value given in the figure, oxygen diffuses from the atmosphere at a rate which is proportional to the deficit.

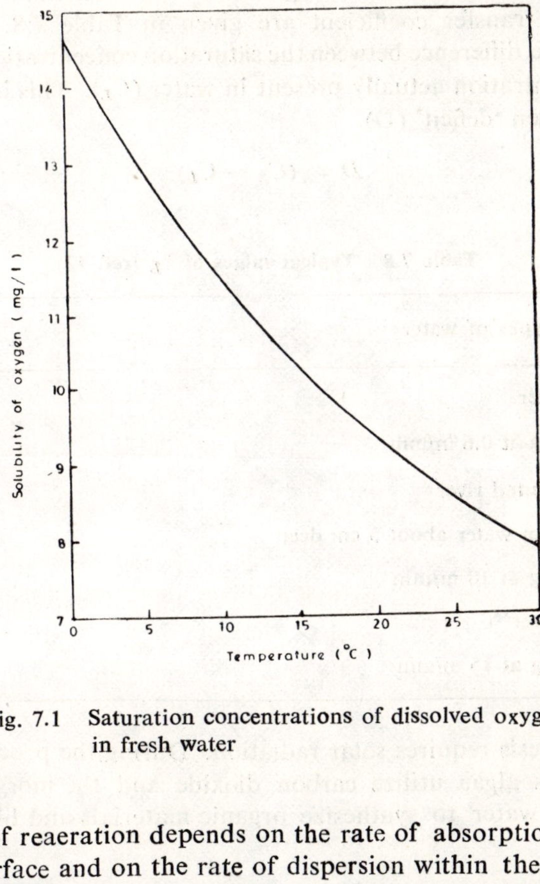

Fig. 7.1 Saturation concentrations of dissolved oxygen in fresh water

The rate of reaeration depends on the rate of absorption through the air-water interface and on the rate of dispersion within the body of water beneath the surface. The driving force for oxygen transfer is the difference between the DO level of saturation (C_S) and the actual concentration (C_L) present in the water. Since the solubility of oxygen in water is very low, the main resistance to mass transfer is on the liquid side of the interface so that the absorption is liquid-film controlled. The transfer rate can be expressed as the 'mass flux' of oxygen across unit area of the surface in unit time, N, so that

$$N = k_L (C_S - C_L) \tag{7.1}$$

where k_L is the liquid-phase mass transfer coefficient.

If oxygen is transferred into a volume of water, V, having a total interfacial area, A, in contact with the gas phase, then the rate of oxygen transfer per unit volume of water is given by,

$$\frac{dC_L}{dt} = \frac{NA}{V} = k_L \frac{A}{V}(C_S - C_L) = r(C_S - C_L) \quad (7.2)$$

where dC_L/dt is the oxygen transfer rate, $r = k_L A/V$, and is known as the reaeration rate constant or reoxygenation rate constant. Typical values for the mass transfer coefficient are given in Table 7.8. In Eq. (7.2), $(C_S - C_L)$ is the difference between the saturation concentration of DO (C_S) and the concentration actually present in water (C_L). This is often referred to as the oxygen 'deficit' (D).

$$D = (C_S - C_L) \quad (7.3)$$

Table 7.8 Typical values of k_L (ref. 12)

Types of water	k_L (m/day)
Stagnant water	0.10–0.15
Water flowing at 0.6 m/min	0.25
Sluggish polluted river	0.5
Sluggish clean water about 5 cm deep	1.0
Water flowing at 10 m/min	2.0
Open sea water	3.0
Water flowing at 15 m/min	7.0

Photosynthesis requires solar radiation. During the process, the green plants such as algae utilize carbon dioxide and the inorganic nutrients present in the water to synthesize organic materials and liberate oxygen.

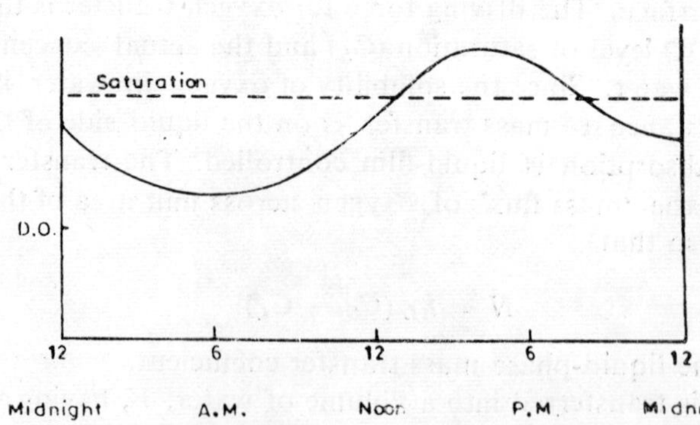

Fig. 7.2 Diurnal variation of dissolved oxygen

Since the process occurs only in the presence of sun-light, the DO level in the water increases during the day. At night the algae and the micro-organisms compete with each other for both dissolved oxygen and organic compounds. This bacterial and algal respiration is responsible for the production of carbon dioxide and subsequent depletion of dissolved oxygen. Combination of the three effects of respiration, photosynthesis, and reaeration results in a diurnal variation in dissolved oxygen concentration (Fig. 7.2) from values that may far exceed saturation during the day to much lower values or depletion at night.

The primary cause of deoxygenation of aquatic systems is the presence of organic substances collectively called oxygen demanding wastes. When these substances enter a waterway, dissolved oxygen is consumed in their breakdown by micro-organisms, so the organic substances can be said to exert a demand on the availability of dissolved oxygen. The more the oxygen is required for the breakdown of the substance, the greater will be the deoxygenation of the waterway. Pollution results when the oxygen demand exceeds the available oxygen.

Although some inorganic substances are found in the category of oxygen demanding wastes, most of them are organic in nature. For example, pollutants in this category typically come from such sources as sewage, food processing plants, pulp and paper mills, tanning operations, and other organic waste producing activities. These wastes, in addition to depleting the DO levels, produce annoying odours, impair domestic and livestock water supplies by affecting its taste, odour and colour.

Biochemical Oxygen Demand

The biochemical oxygen demand (BOD) is a measure of the oxygen utilised by micro-organisms during the oxidation of organic materials. It is the most widely known measure for assessing the water pollution potential of a given organic waste. On an average, the demand for oxygen is directly proportional to the amount of organic waste which has to be broken down. Hence, BOD is a direct measure of oxygen requirements and an indirect measure of biodegradable organic matter.

If a given amount of organic matter is introduced into a water sample and its decomposition is monitored, one would observe that the rate of oxidation of organic matter (rate of decline of BOD) can be approximated as a first-order chemical reaction whose kinetics may be expressed as,

$$\frac{dL}{dt} = -k_1 L \tag{7.4}$$

where L = the BOD remaining in time t — the concentration of the remaining oxidisable material (mg/l), and k_1 = the reaction rate constant or deoxygenation constant, day^{-1}.

Assuming that at time $t = 0$, $L = L_u$, where L_u represents the ultimate

BOD (the original concentration of the organic material before any biological action has occurred), we can integrate Eq. (7.4) to get

$$\frac{L}{L_u} = \exp(-k_1 t) = 10^{-k_1' t} \qquad (7.5)$$

where $k_1' = \dfrac{k_1}{2.303}$

The amount of BOD remaining at any time t, equals

$$L = L_u\,(10^{-k_1' t}) \qquad (7.6)$$

If Y represents the amount of BOD utilised at any time, t, then

$$L_u = L + Y \qquad (7.7)$$

or

$$Y = L_u - L = L_u\,(1 - 10^{-k_1' t}) \qquad (7.8)$$

A standard way to measure BOD is to determine the amount of oxygen required by the bacteria during the first five days of decomposition at 20°C. The result is known as the 5-day BOD at 20°C, commonly abbreviated as BOD_5.

Note that BOD_5 equals

$$Y_5 = L_u - L_5 = L_u\,(1 - 10^{-5k_1'}) \qquad (7.9)$$

The relationship among L_u, L and Y is shown in Fig. 7.3 and 7.4.

The value of k_1' varies significantly with the type of waste and the temperature. Several methods are available for determining k_1' and L_u which include: (1) The least squares method[13]. (2) the method of moments[14], (3) the daily difference method[15], (4) the rapid rate method[16], and (5) the Thomas method[17]. The Thomas method is probably the simplest to use and it is the one illustrated here.

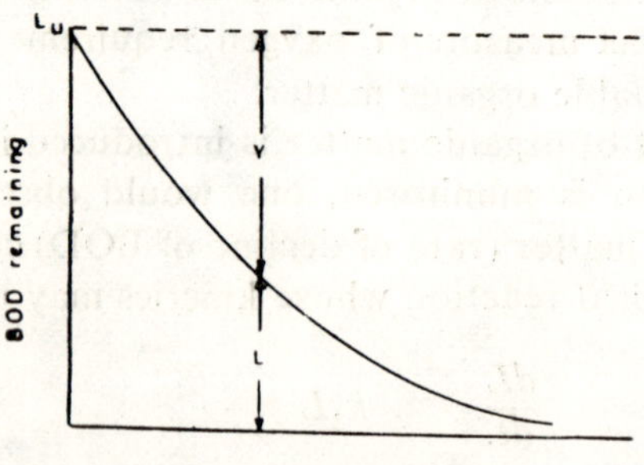

Fig. 7.3 BOD remaining versus time

The method is based on the similarity of two functions:

$$(1 - 10^{-k_1't})$$

and

$$2.3k_1't\,[1 + (2.3/6)\,k_1't]^{-3}$$

Their respective series expansions give:

$$(1 - 10^{-k_1't}) = 2.3k_1't\left[1 - \tfrac{1}{2}(2.3k_1't) + 1/6\,(2.3k_1't)^2 - \frac{1}{24}(2.3k_1't)^3 + \ldots\right] \quad (7.10)$$

and

$$2.3k_1't\,[1 + (2.3/6)\,k_1't]^{-3} = 2.3k_1't\left[1 - \tfrac{1}{2}(2.3k_1't) + \frac{1}{6}(2.3k_1't)^2 - \frac{1}{21.6}(2.3k_1't)^3 + \ldots\right] \quad (7.11)$$

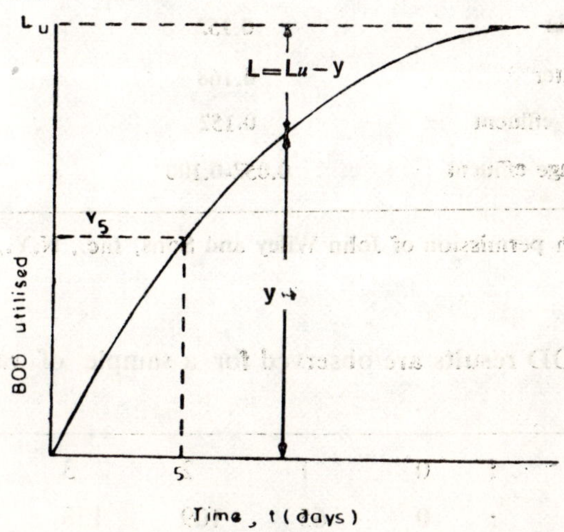

Fig. 7.4 BOD utilised versus time

As can be seen, the first three terms within the brackets in the two series are identical, and the difference between the fourth terms is small. Hence, the Eq. (7.8) can be written as

$$Y = L_u(2.3k_1't)\,[1 + (2.3/6)\,k_1't]^{-3} \quad (7.12)$$

Taking the inverse and the cube root, and rearranging, we get

$$\left(\frac{t}{Y}\right)^{1/3} = (2.3k_1'L_u)^{-1/3} + \left[\frac{(k_1')^{2/3}}{3.43\,(L_u)^{1/3}}\right]t \quad (7.13)$$

A plot of $(t/Y)^{1/3}$ versus time, t, will give a straight-line whose slope, b, and intercept, a, can be used to calculate k_1' and L_u,

$$k_1' = 2.61\frac{b}{a} \tag{7.14}$$

$$L_u = \frac{1}{2.3k_1'a^3} \tag{7.15}$$

Several observations of Y as a function of time are required for a meaningful determination of k_1' and L_u. The observation data are, however, limited to the first 10 days because of the probable interference from nitrogenous matter. Some typical values of k_1' and L_u are given in Table 7.9.

Table 7.9 Some typical values of k_1' and L_u (ref. 18)*

Type of wastewater	k_1' (day^{-1})	L_u (mg/l)
Weak wastewater	0.152	150
Strong wastewater	0.168	250
Primary sewage effluent	0.152	75–150
Secondary sewage effluent	0.052-0.100	15–75

*Reprinted with permission of John Wiley and Sons, Inc., N.Y., 1971.

Example 1

The following BOD results are observed for a sample of raw sewage at 20°C:

t (days)	:	0	1	2	3	4	5
y (BOD, mg/1)	:	0	65	109	138	158	172

Calculate the reaction-rate constant k_1' and the ultimate BOD, L_u.

Solution

1. Construct the following table for plotting:

$(t/y)^{1/3}$	0.249	0.264	0.279	0.294	0.307
t	1	2	3	4	5

2. Plot the values $(t/y)^{1/3}$ versus t (see Fig. 7.5).
3. From the plot the values obtained are slope $b = 0.012$, and the intercept $a = 0.235$.

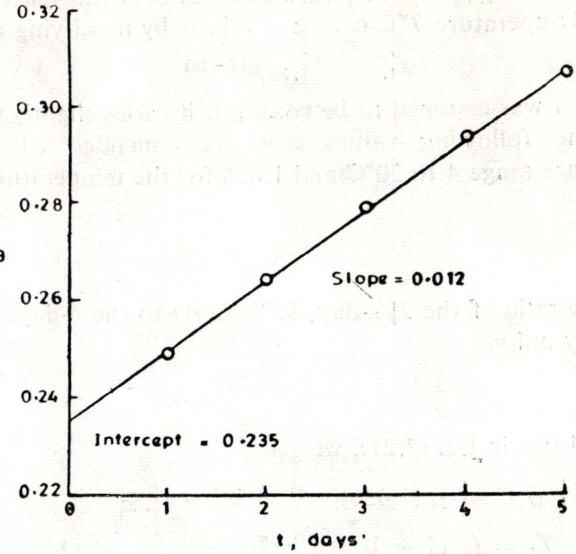

Fig. 7.5 Application of Thomas method

4. Compute k_1' and L_u

$$k_1' = 2.61 \frac{0.012}{0.235} = 0.133 \text{ day}^{-1}$$

$$L_u = \frac{1}{2.3\ (0.133)\ (0.235)^3} = 252 \text{ mg/l}$$

The reaction-rate constant k_1' depends on temperature (T) and may be described by van't Hoff-Arrhenius equation:

$$\frac{d\,(\ln k_1')}{dT} = \frac{E}{R_u T^2} \tag{7.16}$$

where E = activation energy for the reaction (cal/mol), T = reaction temperature (K) and R_u = universal gas constant.

Integrating between the limits of T_0 and T, we get

$$\ln\left(\frac{k'_{1,T}}{k'_{1,T_0}}\right) = \frac{E(T-T_0)}{R_u T T_0} \tag{7.17}$$

For most cases of the wastewater treatment processes, the term $E/R_u T T_0$ is nearly constant. Thus, Eq. (7.17) may be approximated by the expression,

$$\ln\left(\frac{k'_{1,T}}{k'_{1,T_0}}\right) = \text{constant } (T-T_0) \tag{7.18}$$

Therefore $$k'_{1,T} = k'_{1,T_0} \exp\ (\text{constant } (T-T_0)) \tag{7.19}$$

Let exp (constant) = θ = temperature coefficient. Then, Eq. (7.19) becomes

$$k'_{1,T} = k'_{1,T_0}\ \theta^{(T-T_0)} \tag{7.20}$$

If the value of k'_{1,T_0} is determined at 20°C, then the reaction-rate constant $k'_{1,T}$ at temperature T°C can be obtained by modifying Eq. (7.20):

$$k'_{1,T} = k'_{1,20}\,\theta^{(T-20)} \tag{7.21}$$

Although θ was assumed to be constant, it varies slightly with temperature and the following values are recommended[19]: $\theta = 1.135$ for the temperature range 4 to 20°C and 1.056 for the temperature range 20 to 30°C.

Example 2

Show that the ratio of the $2\frac{1}{4}$—day, 35°C BOD to the 5-day, 20°C BOD is approximately unity.

Solution

Putting $\theta = 1.056$ in Eq. (7.21), we get

$$k'_{1,35} = k'_{1,20}\,(1.056)^{35-20} = 2.264\,k'_{1,20}$$

Now $$Y_5 = L_u\,(1 - 10^{-5\,k'_{1,20}})$$

and $$Y_{2\frac{1}{4}} = L_u\,(1 - 10^{-(2.25)\,k'_{1,35}})$$

or $$Y_{2\frac{1}{4}} = L_u\,(1 - 10^{-(2.25)\,(2.264)\,k'_{1,20}})$$

$$= L_u(1 - 10^{-5.09\,k'_{1,20}}) \approx Y_5$$

Thus, the 5-day BOD of wastewater can be obtained in $2\frac{1}{4}$ days if the temperature is 35°C rather than 20°C.

The assumption of a first-order reaction for the BOD oxidation is an over simplification since the model does not include bacterial cell concentration as a parameter and does not correspond to established microbial growth relationships. However, the model is adequate for most practical purposes.

The importance of the concept of BOD in water quality must be emphasized. Drinking water usually has a BOD of less than 1 mg/1, and water is considered fairly pure with a BOD of 3 mg/1. But, when the BOD value reaches 5 mg/1, the water is of doubtful purity. A comparison of these BOD levels with the range of values encountered in industrial wastewaters (Table 7.10) indicates the seriousness of the problem.

The curve shown in Fig. 7.4 is typical of what is known as the first-stage BOD where carbonaceous matter is degraded by bacterial oxidation. But when nitrogenous material is also present, nitrifying bacteria exert an additional oxygen demand. This causes a second hump in the curve called the second stage BOD or nitrification. A typical BOD curve showing carbonaceous oxidation and nitrification phases for wastewater is given in Fig. 7.6.

Table 7.10 Range of BOD levels of some industrial effluents

Source	BOD range (mg/l)
Dairy wastes	400—2000
Food processing wastes	500—4000
Mixed chrome tanning	500—5000
Wool scouring	500—10000
Pharmaceutical manufacture	400—10000
Paper pulping	1500—25000

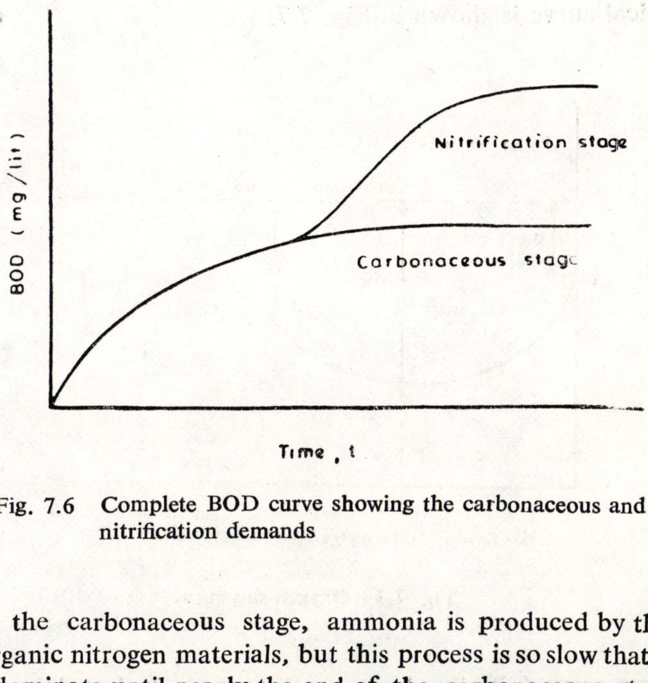

Fig. 7.6 Complete BOD curve showing the carbonaceous and nitrification demands

During the carbonaceous stage, ammonia is produced by the breakdown of organic nitrogen materials, but this process is so slow that nitrifiers do not predominate until nearly the end of the carbonaceous stage. Ammonia (or ammonium ion) is then oxidized to nitrite and then to nitrate by *Nitrosomonas* and *Nitrobacter* microorganisms respectively.

$$2NH_4^+ + 3O_2 \xrightarrow{\text{Nitrosomonas}} 2NO_2^- + 2H_2O + 4H^+ \quad (7.22)$$

$$2NO_2^- + O_2 \xrightarrow{\text{Nitrobacter}} 2NO_3^- \quad (7.23)$$

The oxidation process speeds up towards the end of the first stage and slows down again as the ammonia is oxidized. Ammonia exerts a very high oxygen demand, requiring over 4½ times its own weight of oxygen for complete oxidation. Thus, if nitrification is allowed to occur in the receiving stream, second-stage BOD exerts its presence and a further decrease in the oxygen resource will be experienced.

Oxygen Sag Curve

The discharge of wastes into a body of water results in the depletion of dissolved oxygen level as the wastes are oxidized by the bacteria. Opposing this drop in dissolved oxygen is reaeration which replaces oxygen through the surface, at a rate which is proportional to the depletion of oxygen below the saturation value. The simultaneous action of deoxygenation and reaeration produces a typical pattern in the dissolved oxygen concentration of the aquatic system. This pattern is known as the dissolved-oxygen sag and a typical curve is shown in Fig. 7.7.

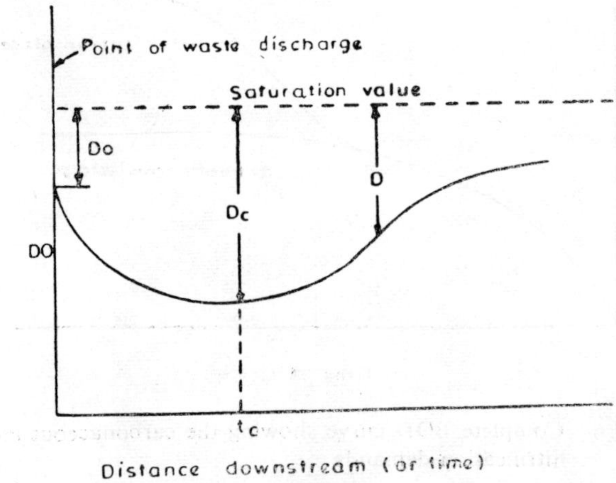

Fig 7.7 Oxygen sag curve

The sag curve initially drops as the wastes deplete the oxygen faster than it can be replaced. At the point where the dissolved oxygen (DO) is a minimum, the rate of reaeration becomes equal to the rate of deoxygenation. Beyond this point, the rate of reaeration exceeds the rate of deoxygenation and the DO level begins to increase and eventually returns to normal.

The simultaneous action of deoxygenation and reaeration can be described by the equation :

$$\frac{dD}{dt} = k_1L - k_2D \qquad (7.24)$$

where, D = dissolved oxygen deficit, L = concentration of the organic material—the BOD remaining in time t, k_1 = deoxygenation constant and k_2 = reaeration constant.

The concentration of the organic material can be expressed in terms of the ultimate BOD, L_u, as (see Eq. 7.5)

$$L = L_u e^{-k_1 t} \tag{7.25}$$

Substituting Eq. (7.25) in Eq. (7.24) and integrating, we get

$$D = \frac{k_1 L_u}{k_2 - k_1} (e^{-k_1 t} - e^{-k_2 t}) + D_0 e^{-k_2 t} \tag{7.26}$$

where D = oxygen deficit in time t, (mg/1) and D_0 = initial oxygen deficit at the point of waste discharge (mg/1).

Since $$e^{-kt} = 10^{-k't},$$

where $$k' = \frac{k}{2.303}$$

$$D = \frac{k_1' L_u}{k_2' - k_1'} (10^{-k_1' t} - 10^{-k_2' t}) + D_0\, 10^{-k_2' t} \tag{7.27}$$

Both the constants k_1' and k_2' are temperature dependent; the temperature effect on k_2' can be evaluated as follows[20]:

$$k'_{2,T} = k'_{2,20}\, (1.047)^{T-20} \tag{7.28}$$

where $k'_{2,T}$ = reaeration constant at temperature T°C and $k'_{2,20}$ = reaeration constant at 20°C.

The critical dissolved oxygen deficit, D_c, where the rate of oxygen utilized for waste material decomposition equals the rate of atmospheric reaeration, is of engineering significance and can be evaluated by simply setting $\frac{dD}{dt} = 0$ in Eq. (7.24):

$$k_2 D_c = k_1 L = k_1 L_u \exp(-k_1 t_c)$$

$$D_c = \frac{k_1}{k_2} L_u \exp(-k_1 t_c) \tag{7.29}$$

The critical time (t_c) can be obtained by differentiating Eq. (7.26) with respect to t and setting $dD/dt = 0$.

$$t_c = \frac{1}{k_2 - k_1} \ln \left[\frac{k_2}{k_1}\left(1 - D_0 \frac{k_2 - k_1}{k_1 L_u}\right)\right] \tag{7.30}$$

The constant k_1 can be evaluated from the BOD test in the laboratory while k_2 must be determined from field information as it depends on the characteristics of a stream and varies from reach to reach in most streams.

Langbein and Durum[21] summarized the results of several investigators and gave the following empirical equation for determining k_2 at 20 °C.

$$k_2 = 3\cdot 3 \frac{V}{H^{1\cdot 33}} \tag{7.31}$$

where k_2 = reaeration co-efficient (base *e*), days^{-1}; V = velocity of flow, ft/sec, and H = depth of flow, ft.

The time required to reach the critical point is significant because at this point the conditions in the stream will be at their worst. The DO level will be minimum and if it reaches zero, then anaerobic conditions may prevail leading to the death of aerobic aquatic inhabitants. This is illustrated in Fig. 7.8 for three different levels of BOD discharge from a point source.

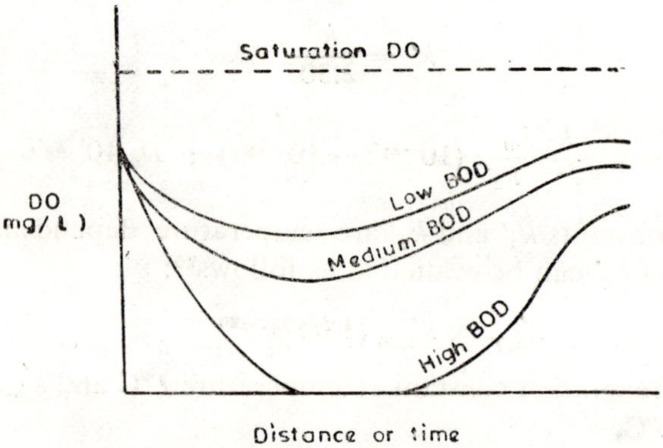

Fig. 7.8 Effect of BOD on the oxygen sag curve

Disease Causing Agents

Water is a potential carrier of pathogenic micro-organisms. These pathogens are carried into the water bodies by sewage and wastes from farms and various industries, specially tanning and meat packaging industries. Contact with the pathogens can be made by drinking the water or through other activities involving contact with water.

Some bacteria are water-borne, and these include those responsible for causing cholera, typhoid, amoebic dysentery and gastroenteritis. Viruses are also found in water, including strains which are responsible for polio, infectious hepatitis, and coxsackie fever. Water may also contain some animal parasites including the round worm, and the pork tape worm. All these types of organisms occur in faeces, and so are present in the sewage. Water also plays an indirect but vitally important role in the transmission of diseases like malaria, yellow fever, filariasis, and schistomiasis.

Although techniques are available for identifying and enumerating many types of pathogens commonly transmitted by water, these methods are time consuming, complex and expensive and, hence, are not ordinarily used to evaluate the bacteriological quality of water. The coliform group of bacteria are considered reliable indicators of the hygienic quality of water. *Escherichia coli* (*E. coli*), belonging to the coliform group, is essentially harmless bacteria found in large numbers in human feces. Since pathogens leave the human body through body wastes and since the coliforms commonly travel together with the pathogens, a large concentration of *E. coli* in the water indicates fecal contamination and a presumptive evidence of the presence of pathogens.

Synthetic Organic Compounds

These include pesticides, synthetic organic chemicals, and detergents. These compounds, in contrast to the organic wastes, are not biodegradable and may persist for long periods. These are of great concern to environmentalists because most of the synthetic organic compounds are accumulative toxic poisons and ultimately may reach objectionable levels in water or in aquatic life. Analyses of polluted waters show the presence of a wide variety of these compounds and many others are probably not being detected. Of these, two (pesticides and detergents) have been subjected to intensive study.

Pesticides

Pesticides can be classified into the three main classes of insecticides, herbicides, and fungicides. There are also other types like rodenticides, molluscicides, and acaricides. Pesticides are not common constituents of sewage but result primarily from surface runoff from agricultural lands, waste discharge by pesticide manufacturers and by other means. Because of their world-wide usage, nearly all the rivers and oceans of the world contain pesticide residues. Pesticide residues in water may reach humans through drinking water, but the concentrations in most cases are far below the harmful levels. Limiting standards for some common pesticides in drinking water are given in Table 7.11.

The chlorinated organic pesticides like DDT, dieldrin and aldrin are hazardous mainly due to their concentration in the food chain. They have high stability, low vapour pressure and very low solubility in water, but have substantial solubility in oils and fats. For example, the concentration of DDT in fresh water is about 0.00001 ppm. It tends to become more concentrated from one trophic level to another due to the process of biological magnification. Typical values for a food chain concentration pattern are given in Fig. 7.9.

Table 7.11 Limiting standards for pesticides in drinking water (ref. 22)*

Pesticides	Objective and acceptable limit	Maximum permissible limit (mg/l)
Aldrin	Absent†	0.017
Chlordane	Absent	0.003
DDT	Absent	0.042
Dieldrin	Absent	0.017
Endrin	Absent	0.001
Heptachlor	Absent	0.018
Heptachlor epoxide	Absent	0.018
Lindane	Absent	0.056
Methoxychlor	Absent	0.035
Organic phosphates and carbamates	Absent	0.100
Toxaphene	Absent	0.005
Heribicides (e.g. 2, 4-D, 2, 4, 5-T, 2, 4, 5-TP)	Absent	0.100

*Reprinted with permission of Marcel Dekker, Inc., N.Y., 1971.

†Not detectable by an acceptable method of analysis.

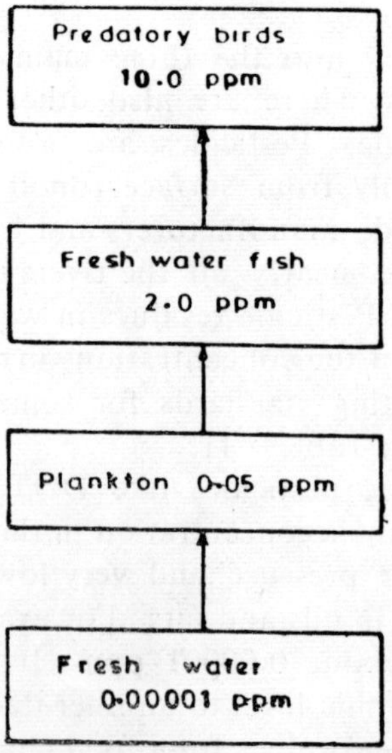

Fig. 7.9 Typical food-chain concentration of DDT

As a result of biological magnification fish and predatory birds are very susceptible to chlorinated hydrocarbons like DDT, dieldrin and aldrin. Dieldrin has been found to affect the calcium metabolism in predatory birds, and this has resulted in the birds laying eggs which have very thin shells and consequent reproductive failure. Man stands at the top of the food-chain in which he is involved and accumulation of these pesticides should, therefore, occur in his system. Table 7.12 gives the average concentration of DDT in the body fats of man in several countries of the world.

Table 7.12 Average DDT levels in man (ref. 23)*

Country	Concentration (ppm)
India	6
Israel	8.5
Hungary	5.7
Czechoslovakia	5.5
U.S.A.	5.3
France	1.7
Canada	1.6

*Reprinted with permission of Elsevier Applied Science Publishers Ltd., London, 1981.

Evidently, these concentrations have not caused any ill-effects in man so far, but there is experimental evidence of harmful longer term effects in mammals. Several pesticides have been shown to cause cancer in mice and rats.

Detergents

The basic active ingredient in detergents is the surfactant or surface-active agent. It lowers the surface tension and allows dirt particles to become linked to water. The dirt is subsequently lifted or floated from the soiled garment during the washing process. Although reducing the surface tension is an essential feature of cleaning agents, detergants usually contain only 10–30% surfactant. The remainder comprises polyphosphate salts (the builder) and a number of other ingredints such as optical whiteners in small amounts to increase the cleaning efficiency of the final formulated product.

Surfactant concentrations as low as 1 ppm produce foam in rivers and in sewage treatment plants. Although these concentrations are non-toxic to humans, the presence of surfactants gives an off-taste to drinking water. The acceptable concentration for surfactants in water supplies has been set at 0.5 ppm. The presence of surfactants reduces the rate of oxygen absorption in water; the rate could be cut down to nearly half for concentrations as low as 0.1 ppm. It has been found that some fish, like the trout, are very sensitive to concentration levels of surfactants as low as 1 ppm and show symptoms similar to asphyxia.

Till the early 1960s the surfactant present in synthetic detergents was alkylbenzene sulphonate (ABS). This showed remarkable resistance to biodegradation (the so-called hard detergents) and has been subsequently replaced by a new surfactant called linear alkyl sulphonate (LAS) which is of comparable cost and cleaning potential but is rapidly biodegradable. At present, one of the most pressing problems is not the effects of the surfactant itself but the release of polyphosphate builders into natural waters.

These detergent builders contribute significantly, approximately 50%, to the phosphate present in sewage effluents. When these phosphates are released into streams and lakes, they act as plant nutrients, thus supporting eutrophic conditions. In the U.S., a legislation restricting the phosphate levels in detergents was proposed and became a law in some states. Nitrilotriacelate (NTA) was considered as a replacement but it proved to be hazardous to human health. At present there is no acceptable substitute available for polyphosphate builders in detergents. The best alternative is to minimize the use of phosphates in detergents wherever possible. May be we should go back to the use of plain soap.

Plant Nutrients

Nitrogen and phosphorus are essential elements which are required by plants and animals for maintaining their growth and metabolism. Small amounts of nitrates and phosphates occur in all aquatic systems and these are sufficient to maintain a balanced biological growth. In wastewaters, these nutrients are present in abundance as phosphates, nitrates, ammonia or combined organic nitrogen. These compounds may enter the water bodies directly from the manufacture and use of fertilizers, and from the processing of biological materials such as food and textiles, or via domestic sewage treatment plants. Thus, when unusually large concentrations of nutrients are present in water bodies, an excess growth of algae, known as an algal bloom, appears. This produces an unsightly green slime layer over the surface of the water body and subsequently leads to eutrophication. The slime layer reduces light penetration and restricts atmospheric reoxygenation of the water The dense algal growth eventually dies and the subsequent biodegradation produces an oxygen deficit which can result in foul-smelling anaerobic conditions.

The anaerobic conditions created by rotting algae can present a health hazard. Birds feeding on the algae are killed by the powerful toxin produced by the anaerobic bacterium *Clostridium botulinum* which fluorishes in the environment. Other problems resulting from heavy algal growth are the clogging of filters in water treatment plants and slowing down of stream flow by filamentous weeds which can grow up to 12 m long. These weeds can trap solid particles carried by a stream and if they are organic, a large decaying mass accumulates which exerts a high oxygen demand.

High concentrations of nitrate in drinking water are also of concern. In the intestine, the nitrate is reduced to nitrite by bacterial action. When it enters the blood stream it becomes attached to haemoglobin, forming a complex known as methaemoglobin. As a result, the oxygen carrying capacity of the blood is reduced producing a condition called methaemoglobin anemia or the blue baby disease. In young babies, it is frequently fatal. The nitrates can be further converted to amines and nitrosoamines in the human body leading to a possible cause of gastric cancer.

Inorganic Chemicals and Minerals

This category of water pollutants includes inorganic salts, mineral acids, finely divided metals and metal compounds. These pollutants enter the water bodies from municipal and industrial wastewaters and mine runoff. Most of these substances are toxic and are capable of killing living organisms in the water bodies.

Mine drainage is an important source of increased acidity in natural waters, the other source being acidic rainfall which has been recognized only recently. Acid mine drainage can originate from the mining of sulphur bearing ores containing lead, zinc and copper, but it is associated primarily with coal deposits which contain varying amounts of iron sulphide (pyrite). The main constituents of mine drainage are sulphuric acid and iron compounds which are formed as a result of reactions between air, water and pyrite, and are carried into water bodies. Acids cause corrosion of metals and concrete and can be fatal to fish. Alkalies discharged by industries such as textiles, tanneries and coke-oven operations can also destroy aquatic life.

The sources of fresh water salinity are varied. They include industrial effluents, minerals dissolved by water used in irrigation, salt brines from mines or oil wells, and ocean salt. Salts cause the 'hardness' of water. Hard water can cause scale formation in pipes, boiler tubes and other industrial equipment. Chloride concentrations, if present in excess of 400 ppm, can be fatal to fresh water fish.

The toxic properties of numerous inorganic compounds, particularly heavy metals, have been known for years. The detection of these substances in natural waters in concentrations approaching toxic levels has created a great amount of concern. Most of these substances produce physiological poisoning by becoming attached to the tissues of aquatic organisms and accumulate. Consequently, their increasing concentrations can build up in food chains. The metals of particular concern in industrial wastewaters are cadmium, chromium, lead, mercury and silver.

Cadmium is generated in waste streams from pigment works, textiles, electroplating, chemical plants, etc. In rivers polluted by cadmium, the metal may be undetectable in filtered water, particularly if the *p*H is neutral or alkaline, although it is present in high concentrations in the particulate fraction. The itai-itai disease in Japan was probably due to the transport of cadmium-containing particulates in water to the irrigation fields. Shellfish can accumulate cadmium from bottom sediment and concentrations of around 0.01 mg/l can retard the growth of aquatic plants. It has been found that 0.1 mg/l of *Cd* can be lethal to both oysters and aquatic plants. It can combine synergistically with other metals like copper and zinc which increases its toxicity. The permissible level for cadmium in drinking water supplies is 0.01 mg/l.

Chromium is found in two forms in wastewaters—hexavalent and trivalent. Hexavalent chromium is present in the wastewaters of plating operations, aluminium anodizing, paint and dye operations, and other industries. Trivalent chromium is present in wastewaters from textile dyeing, the ceramic and glass industry, and photography. For domestic water supplies chromium in the hexavalent form should not exceed 0.05 mg/l. Many plant species are adversely affected by chromium (vi) concentrations of 5 mg/l.

Lead may arise as a contaminant from various industrial and mining effluents. It is present in industrial effluents arising from battery manufacture, printing, painting and dyeing. Lead is toxic to aquatic organisms, but its effects are less pronounced than those of cadmium and mercury. It is usually particle-bound and does not inhibit the production of phytoplankton. Lead is a cumulative poison and concentrates primarily in the bones. The permissible limit of lead concentration in drinking water supplies is 0.05 mg/l, but if lead pipes are used, tap water may exceed the tolerance limit.

Mercury is discharged into natural waters from various industrial effluents such as those from the manufacture of paint and paper, chlorine and caustic soda, fertilizers and pesticides. Mercury has proved to be the most toxic aquatic pollutant beause of its rapid methylation in the aquatic environment and in this form, it gets accumulated in the food chains, culminating in levels viewed as unacceptable in organisms at the top of the trophic structure. The 'enrichment factor' can be of the order of hundreds of thousands. The recommended drinking water standard for mercury is 0.002 mg/l. One aspect of mercury pollution that is causing concern is its effect on the rate of photosynthesis of phytoplanktons. There is some evidence that the oxygen producing activities of phytoplankton can be inhibited at mercury levels as low as 0.001 mg/l.

The principal sources of silver in wastewaters are electroplating and photographic industries. Marine animals show a marked diversity in tolerance to silver. Fish and lower organisms are susceptible to silver poisoning and silver concentrations of 0.0004 to 0.4 mg/l have proved toxic. It is also a cumulative poison and chronic injestion can cause a permanent discolouration of skin in humans. The allowable limit of silver in drinking water has been set at 0.05 mg/l.

Sediments

Sediments include soil, sand and mineral particles washed into the aquatic environment by storms and flood waters. In addition, large deposits of sewage sludge, pulverized coal ash, and various industrial solids are disposed off into rivers and marine waters. The sediments erode pumping equipment and power turbines, and have the effect of increasing turbidity and consequently of reducing the amount of sunlight available to green, water plants. In suspension, the solids may cause thickening of fish gills which may lead to eventual asphyxiation of the fish. On the river bed, the sediment may interfere with spawning sites and decrease the amount of food available for fish.

Radioactive Substances

Very little is known about the threshold of radiation damage to aquatic environment from wastes of uranium and thorium mining and refining, from nuclear power plants, and from industrial use of radioactive materials. The refining of uranium ore is an important source of radioactive waste producing radionuclides of radium, bismuth, etc. Radium is the most significant waste product and is considered to be a hazard in drinking water. Water supplies must not contain more than 3 picocuries per litre of radium-226, nor more than 10 picocuries per litre of strontium-90.

Certain marine organisms have the capacity for accumulating radionuclides from water. This "biomagnification" may cause objectionable radioactivity in living organisms, although the radiation level in water may be low enough to be considered safe. Phytoplankton and fish may concentrate metal radionuclides by factors of 10^2 to 10^5.

Radioactive substances can enter humans with food and water, and get accumulated in blood and certain vital organs like the thyroid gland, the liver and bone and muscular tissues. At present some low and medium level wastes are sealed in containers and dumped into the ocean. If there is any damage to, or leakage from the containers, the nuclear wastes could escape and enter the marine system. Then they could be circulated by storms and ocean currents and eventually reach coastal waters.

Thermal Discharges

Power plants and industry use large quantities of water for cooling purposes. Used coolant water is usually discharged directly into water bodies. This could result in increase in temperature of the water bodies with deleterious consequences for aquatic inhabitants. An increase in water temperature decreases the oxygen saturation percentage (Fig. 7.1) and at the same time accelerates the lowering of DO levels. This is because the hot water tends to form a separate layer above the cool water due to density differences between the two. The hot layer, which itself holds less oxygen than the cooler layer below, prevents the replacement of oxygen in the cooler layer as it is denied contact with the atmosphere. The DO level falls rapidly due to normal biological functions in the lower layer and may lead to anaerobic conditions.

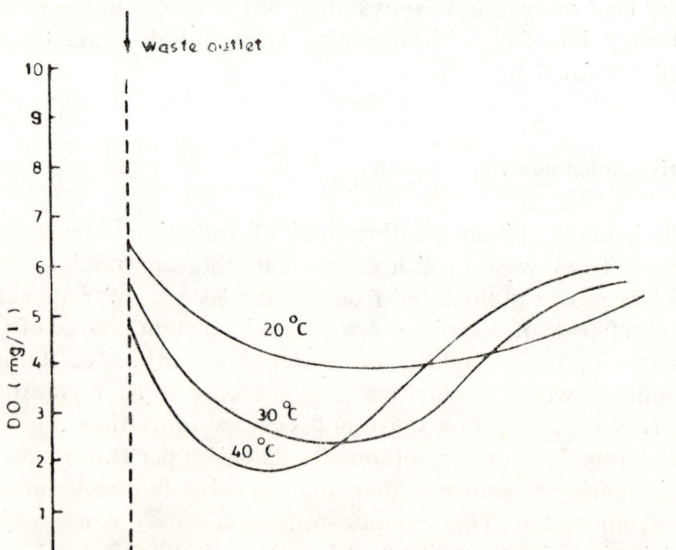

Fig. 7.10 Relation between temperature and oxygen profile

In addition to low DO levels the metabolic activity of micro-organisms increases with temperature at a rate which is approximately doubled for every 10°C increase in temperature. Thus, an increase in temperature produces simultaneously a decrease in the availability of oxygen and an increase in the rate at which it is consumed. If an organic load is imposed under such conditions of increased temperature, a sharp increase in oxygen sag, or deficit, in water will result (Fig. 7.10).

The life cycles and natural processes of many aquatic life forms are closely and delicately geared to water temperature. Some cold water fish like trout may die if the water temperature is above 25°C and their eggs will not hatch at temperatures above 14.5°C. Most fresh water fauna populations decline with rising temperatures and above 40°C only a few species can exist. An increase in temperature also increases the toxicity of some chemical pollutants. Poisons are, therefore, more toxic in summer months.

Oil

Oil is an important commodity involved in some way or other in virtually every activity of contemporary life. Therefore, there is obvious concern about its polluting effects. Oil and oil wastes enter rivers and other water bodies from several sources like industrial effluents, oil refineries and storage tanks, automobile waste oil, and petrochemical plants. All these make a significant contribution to the pollution of the aquatic environment. Similarly, normal tanker operations such as ballasting and tank washings, discharge from offshore wells, and spillage from oil tanker accidents cause marine pollution and shore contamination.

Since oil is virtually insoluble in water, it floats and spreads rapidly into a thin layer. The lighter, low molecular-weight elements, which are most toxic to organisms, soon evaporate and others are degraded biologically but only at a slow rate. At sea, oil slicks are responsible for the deaths of many birds. The oil penetrates the bird feathers thereby affecting their insulation and buoyancy. Thus, the birds become colder and more susceptible to diseases, and experience difficulty in floating and flying.

The birds ingest oil when they dive to feed, and also when they preen their contaminated plumage. This ingestion of oil produces toxic effects. Very little is known about the toxic effects of oil on marine and fresh water organisms. Organisms which either float near or on the surface and those which come to the surface for air are most likely to be affected.

An oil slick on the surface of water can prevent oxygen transfer from the atmosphere and lead to very low DO levels in the water due to microbial oxidative attack on the hydrocarbon molecules. Films of oil on the surface may be driven in-shore through the action of wind and tides, and may accumulate along the shoreline where they pose aesthetic problems.

7.4 WATER POLLUTION LAWS AND STANDARDS

In India, specific laws have been passed by both State and Central governments to control water pollution. All these laws are based on "standards", a set of parameters used to define "water quality". These standards are usually based on the water quality in the streams, and specify the quality of effluents discharged. The Indian Standard Institution (ISI) has the principal role in specifying the norms for various effluents so that the ambient water quality standards are maintained.

A Water Pollution Act was passed by the parliament in 1974. The Act makes provision for the constitution of Central and State Boards. The Central Board advises the Central government on matters pertaining to the prevention and control of water pollution and is bound by the directives of the Central government in discharging its duties. The State Board, on the other hand, is bound by the directives of both the Central and State governments. Maharashtra is perhaps the only state in India where a law for the control of water pollution was enacted as early as 1969. Usually the water pollution control comes under the State list, and most of the States follow the guidelines set by the Central Water Pollution Control Board. The Maharashtra Water Pollution Board has two types of pollution monitoring. In the first type, surprise checks are conducted once in three months to see whether permissible limits are exceeded; then the offending party is served with a warning notice to explain as to why action should not be taken against it. The case is then taken to the Board for discussion and in the case of unsatisfactory explanation to court for prosecution. In the second type of monitoring, various streams and rivers are checked once in a month at various points to see if water quality standards are maintained.

There are, however, several enforcement problems. Since the Central Act is rather silent on funds for the Boards, the State Boards were unable to enforce the Act effectively. Hence, the Water (Prevention and Control of Pollution) Cess Act was passed in 1977 to meet the expenses of the Central and State Boards. The Water Board has no power to take direct action against the erring party. The court procedures are time-consuming and the delays often prevent quick and preventive action thereby defeating the sole purpose of the Act. Because of the problems inherent in the implementation of the Act, amendments were proposed for strengthening the working of the State Boards.

Inspite of the legislative measures, the pollution of our waterways continues unabated. This is due to the lack of civic sense among people and due to the lack of necessary infrastructure for enforcing implementation of the laws efficiently.

REFERENCES

1. *Indian agriculture in brief* (Directorate of Economics and Statistics, Government of India, New Delhi), 10th Edn, 1970, p. 18.
2. B.S. Nag and G.N. Kathpalia, "Water resources of India", in *Proc. of the second world congress on water resources.* Vol. 2 (CBIP, New Delhi), 1973.
3. A. Lahiri and R. Nagarajan, "Water resources in India and irrigation development", in *India in Perspective,* Vol. II (Arnold—Heinemann. New Delhi), 1977, p. 161.
4. G.N. Kathpalia and N.S. Varadan, Paper presented in the symposia on *Integrated development of surface and subsurface water resources* (CBIP, New Delhi), 1971.
5. D.K. Todd, *The water encyclopedia* (Water Inf. Cent., Port Washington, N.Y.), 1970.
6. P.H. McGauhey, *Engineering management of water quality* (McGraw-Hill, New York), 1968.
7. Metcalf and Eddy, Inc., *Wastewater engineering* (McGraw-Hill, New York), 1972.
8. R.N. Siddiqi, "Characteristics of domestic and municipal sewage in India", *Journal of the Inst. of Engrs, India* (*Public Health Div.*), 1974-75, p. 85.
9. *Report of the committee on urban wastes. Govt. of India* (Min. of Works and Housing, New Delhi), 1975.
10. H.M. Dix, *Environmental pollution* (John Wiley & Sons, New York), 1981, p. 174.
11. T.J. Mulligan and R.D. Fox, "Treatment of industrial wastewaters", *Chem. Engg.* Vol. 83 (1976), Oct. 18.
12. A.L.H. Gameson and G.A. Truesdale, "Some oxygen studies in stream", *J. Inst. Water Engrs.*, Vol. 13 (1959), p. 175.
13. H.D. Young, *Satistical treatment of experimental data* (McGraw-Hill, New York), 1962.
14. E.W. Moore, H.A. Thomas and W.B. Snow, "Simplified method for the analysis of BOD data", *Sewage and Industrial Wastes* Vol. 22(10) (1950), p. 1343.
15. E.C. Tsivoglou, *Oxygen relationships in streams* (Robert A. Taft Sanitary Engg. Center, Technical Report), W-58-2, 1958.
16. J.P. Sheehy, "Rapid methods for solving monomolecular equations", *J. WPCF,* Vol. 32(6), 1960.
17. H.A. Thomas (Jr), "Graphical determination of BOD curve constants", *Water and Sewage Works,* Vol. 97 (1950), p. 123.
18. G.M. Fair, J.C. Geyer and D.A. Okun, *Elements of water supply and wastewater disposal* (John Wiley & Sons, New York), 1971.
19. G.S. Schroepfer, *Advances in water pollution control,* Vol. 2 (Pergamon, Oxford), 1964.

20. J.D. O'Conner and W.E. Dobbins, "Mechanism of reaeration in natural streams", Trans, Am. Soc. Civil Engrs., Vol. 123 (1958), p. 655.
21. W.B. Langbein and W.H. Durum, *The aeration capacity of streams*, Circular 542 (U.S. Geological Survey, Washington, D.C.), 1967.
22. J.E. Zajic, *Water pollution—disposal and reuse*, Vol. 1 (Marcel Dekker, New York), 1971, p. 12.
23. A. Wint, "The disposal of toxic wastes", in *Industrial effluent treatment*, Vol. 1, edited by J.K. Walters & A. Wint (Elsevier Applied Science Publishers Ltd., London), 1981.
24. W.W. Eckenfelder and D.J. O'Connor, *Biological waste treatment* (Pergamon Press, New York), 1961, Chap. 3.

SUGGESTIONS FOR FURTHER READING

1. V. Dagli (ed.), *Natural resources in the Indian economy* (Vora & Co. Pvt. Ltd., Bombay), 1971.
2. G.M. Woodwell, "Toxic substances and ecological cycles", *Scientific American*, Vol. 216(3) (1967), pp. 24-31.
3. S. Jackson and V.M. Brown, "Effect of toxic wastes on treatment processes and water courses", *J. Water Pollution Control*, Vol. 69 (1970), p. 292.
4. C.N. Sawyer, "Basic concepts in eutrophication", *J. Water Pollution Control*, Vol. 38 (1966), p. 737.
5. M. Eisenbud, *Environmental radioactivity*, 2nd Edn (Academic Press, New York), 1973.
6. J.R. Clark, "Thermal pollution and aquatic life", *Scientific American* Vol. 220(3) (1969), pp. 18-27.
7. R.W. Holcomb, "Oil in the ecosystems", *Science*, Vol. 166 (1969), pp. 204-206.
8. H.E. Stockinger, "Toxicology and drinking water contaminants", *J. Amer. Water Works Assoc.*, Vol. 69 (1977), pp. 399-402.
9. A.P. Dufour, "Disease outbreaks caused by drinking water", *J. Water Pollution Control Federation, Literature Review Issue*, Vol. 54 (1982), pp. 980-983.

PROBLEMS

1. Differentiate between water quality characteristics of surface water and ground water resources.

2. Calculate the solubility of oxygen in mg/l in equilibrium with water at 20°C and a total pressure of 1 atmosphere. Use Henry's law. (9.07 mg/l)
3. The variation of dissolved oxygen concentration with temperature may be represented by[24],

$$C_S = 14.6 - 0.3943T + 0.007714\,T^2 - 0.0000\,646\,T^3$$

where C_S is the saturation level of oxygen in water at 1 atmosphere in mg/1, and T is the temperature in °C. Calculate the value of C_S as a function of T and compare with Fig. 7.1.
4. What proportion of total BOD of a wastewater would be utilized in five days with k_1' values of 0.05, 0.1, 0.15, 0.20 and 0.25? (0.438; 0.684; 0.882; 0.9; 0.944)
5. The ultimate BOD for some waste is 300 mg/l. For values of k_1' equal to 0.1, 0.15, 0.2, and 0.3, plot BOD utilized curve as a function of time. What is the 5 day BOD in each case?
6. Assuming values of k_1', plot the ratio of the 5 day BOD to the ultimate BOD versus k_1', and explain the behaviour of the curve.
7. The BOD results given below are observed on a sample of wastewater.

t. days	0	1	2	4	6	8	10
BOD, mg/l	0	6.5	11	18	22	24	26

(a) Plot the BOD curve.
(b) Calculate the parameters k_1', and L_u.

8. Sketch an oxygen sag curve for a stream having two sources of sewage pollution. Assume that the downstream source is located slightly beyond the point at which the sag curve from the first source is at its minimum.
9. A stream with a depth of 2 m flows with a velocity of 0.25 m/s. The dissolved oxygen concentration in the stream at 15°C is 4.2 mg/l at a particular point.
   (a) What is the initial oxygen deficit? (5.9 mg/l)
   (b) What is the rate of reaeration? (20.65 mg/l per day)
   (c) What is the amount of oxygen added in 15 minutes?
10. A large stream has a rate of reaeration $k_2 = 0.55$ and a rate of deoxygenation $k_1 = 0.23$ per day. The DO deficit of the mixture of stream water and wastewater at the point of reference, D_0, is 4.0 mg/l and the ultimate BOD of the waste, L_u, is 75 mg/l. Calculate (1) the DO deficit at a point one day distant from the point of reference; and (2) the critical deficit and the critical time. (14 mg/l; 17.7 mg/l; 2.48 days).

11. Write a short (two-page) summary of the history of man's knowledge of water-borne diseases.
12. Cooling towers are used in many industrial plants as a means for increasing water recycle within the plant. Outline and briefly discuss the advantages and disadvantages of this approach.
13. A power plant uses 1000 metric tonnes of coal as fuel in the boiler. The steam temperature at the turbine is 677°C and the temperature of cooling water supplied to condenser is 15°C from a nearby river. The heating value of coal is 7000 kcal/kg. The efficiency of the common energy converters is as follows: chemical to thermal: 88%; thermal to mechanical: 42%; mechanical to electrical: 98%.

How much waste heat is produced? Out of this waste heat how much is lost to the atmosphere through the boiler stack, and how much thermal pollution load is imposed on the water environment? Draw the energy flow diagram. (4.465×10^9 kcal; 0.84×10^9 kcal; 3.573×10^9 kcal).

8. Wastewater Sampling and Analysis

Wastewater sampling and analysis form an important part of any water pollution control programme because the identification of the pollutants and pinpointing their concentrations are vital to the selection of a proper treatment process. The sampling method used should be one that gives a representative picture of the quality of the waste stream; otherwise, the time and effort involved in the analysis will be completely wasted.

8.1 SAMPLING

Generally, two types of samples can be taken: (1) the grab sample, and (2) the composite sample. The grab sample shows only the prevailing conditions at the time of sampling and cannot represent average conditions. Hence, it should not be used as a basis for treatment. However, grab samples are useful in determining the effects of extreme conditions of the waste during the time composite samples are being collected or when the wastewater flow is intermittent.

Composite samples provide more meaningful data when the composition of the waste varies with either time or flow. A composite sample can be obtained by collecting individual samples at frequent intervals and mixing them together. Composite samples can be collected either manually or with automatic samplers on the basis of flow or time.

When wastewater flow and composition are relatively uniform, grab samples of a fixed volume can be manually taken at given time intervals and composite sample obtained. If the flow rate varies, the volume of the grab sample collected is proportional to the flow. Usually, the flow rate of wastewater at sampling point is determined with a flow meter each time a portion of the composite sample is collected. Manual sampling is particularly advantageous when visible changes such as colour variation, suspended solids, floating oil, etc., occur in the waste stream. When such changes do take place, operating personnel can collect grab samples and analyze the effect of these changes on the waste stream.

Automatic devices can be used to obtain composite samples, particularly when there is a need to collect wastewater samples at regular, frequent intervals throughout the day. Some of these devices continuously collect one large composite sample, while others collect small samples in individual containers. In automatic devices the sample pickup line is generally placed below the surface of the water to obtain as representative a sample as possible. Any contaminant which might form a stratified layer in the waste stream is completely missed by such devices.

Depending on the type of analysis to be performed, a minimum volume of 1 to 2 litres is collected in grab sampling, while the total composite volume is between 2 and 4 litres. However, it is better to obtain a sample that is too large rather than one that is too small as the analyst may wish to run additional tests or check the reliability of his earlier results. The time allowed between collection and analysis of a sample depends on the nature of the sample, the type of analysis to be made and the conditions of storage. It is of prime importance that special preservation methods are used for samples that are not to be analyzed immediately. This ensures that the characteristics to be analyzed are not changed between collection and analysis. Table 8.1 gives special handling requirements for several wastewater characteristics.

Table 8.1 Preservation of wastewater samples (refs. 1, 2)

Parameter	Preservation method	Maximum holding period
BOD	Refrigeration at 4°C	6 hours
COD	2 ml H_2SO_4/l	7 days
Colour	Refrigeration at 4°C	24 hours
Cyanide	*p*H of the sample raised to 10 or higher with NaOH	24 hours
Fluoride	None required	7 days
Metals	No specific preservation, sample should be acidified	6 months
Nitrogen (Kjeldahl, ammonia, nitrite)	Add 40 mg, $HgCl_2$/l; refrigeration at 4°C	7 days except for Kjeldahl nitrogen which is unstable
Phenol	1.0 g $CuSO_4$/l+H_3PO_4 to lower pH to less than 4; refrigeration at 4°C	24 hours
Phosphorus	40 mg $HgCl_2$/l; refrigeration at 4°C	7 days
Sulphide	2 ml zinc acetate/l	7 days
Odour	Refrigeration at 4°C	7 days
Turbidity	None available	7 days
Coliform bacteria	Sterilized bottle, no specific preservative, refrigeration at 4°C	36 hours

The sample from industrial wastewaters should be representative of the industry. For example, in case there are great variations in flow and/or characteristics, the sampling frequency should be so adjusted as to obtain a representative sample. Often, industries find it desirable to continuously monitor water quality of rivers and lakes which are used as sources for water supply or for disposal of waste effluents. In the latter case, monitoring is done so that the quality and quantity of the discharge can be proportioned according to the assimilative capacity of the water body and for maintaining its quality standards. Several municipalities also use continuous monitoring systems for their treatment facilities. Many continuous monitors are so sophisticated that not only the analyses can be performed automatically and continuously, but also the analytical data can be processed and translated by a computer for further use and application.

8.2 METHODS OF ANALYSIS

Various methods are available for analyzing wastewaters. These methods largely involve standardized procedures that are often complicated and time-consuming. Sometimes simpler rapid methods can be substituted with little, if any, loss in either precision or accuracy. Test procedures often followed in practice are illustrated here briefly for specific constituents. Detailed individual procedures are available in standard references[3, 5].

8.3 DETERMINATION OF ORGANIC MATTER

Dissolved Oxygen (DO)

Surface waters of good quality should be saturated with dissolved oxygen. A fall in DO level is one of the first indications that a water body is polluted by organic matter. The DO level in water depends on physical, chemical and biochemical activities prevailing in the water body and, thus, it is one of the important parameters for assessing the purity of the water body. It is usually determined by Winkler's method[4], which is based on the reaction of dissolved oxygen with manganese ions to form a precipitate of manganese dioxide.

$$Mn^{2+} + O_2 \rightarrow MnO_2 \downarrow \tag{8.1}$$

The manganese dioxide is then treated with iodide ions when iodine is liberated in an amount chemically equivalent to the original dissolved oxygen.

$$MnO_2 + 2I^- + 4H^+ \rightarrow Mn^{2+} + I_2 + 2H_2O \tag{8.2}$$

The liberated iodine is determined, usually by titrating it with sodium thiosulphate.

$$2\ S_2O_3^{2-} + I_2 \rightarrow S_4O_6^{2-} + 2\ I^- \tag{8.3}$$

The presence of nitrites or iron in +2 oxidation state in the solution can interfere with the original DO determination. To eliminate these interferences, several modifications of the basic method have been proposed[5]. These include the use of azide, permanganate and the alum to remove interferences due to nitrite, ferrous iron and suspended solids, respectively.

The interference problems can also be overcome by using DO analyzers with membrane electrodes. The plastic membrane has the selective ability of allowing oxygen to diffuse through but preventing interfering ions, such as nitrites. The dissolved oxygen, after diffusing through the membrane, reacts with the metal electrode and causes a cell current which is directly proportional to the oxygen concentration in the sample. The analyzer is calibrated by measuring the DO of a sample of known oxygen content, which is determined by the Winkler method.

Biochemical Oxygen Demand (BOD)

The most widely used and accepted measure of biodegradable organic content of wastewater is the 5-day, 20°C BOD value. The brief analytical procedure is outlined below:

1. Two standard 300-ml BOD bottles are filled completely with the wastewater of which the BOD is to be measured and the bottles are sealed.
2. Oxygen content of one bottle is determined immediately.
3. The other bottle is incubated at 20°C for 5 days in total darkness, after which its oxygen content is measured.
4. The difference between the two DO values is the amount of oxygen that is consumed by microorganisms during the 5 days and is reported as the BOD_5 (5-day BOD) value of the sample.

In practice, however, BOD measurements are more complicated than the simple procedure given above. The wastewaters may have a high oxygen demand, high enough to deplete all the dissolved oxygen in the sample before the end of 5 days, thus making the test indeterminate. So, often the sample is diluted with high purity water to prevent total depletion of DO. The diluting water is made up of deionized water to which appropriate nutrients, phosphate buffer, trace elements, and seed organisms (usually settled dometic sewage) are added. A blank is run on the diluting water so that the oxygen demand of the seem material can be subtracted from the results. The BOD_5 is calculated from the following equation:

$$\text{BOD}_5 \text{ (in mg/l)} = \text{D}^* \, [(\underbrace{\text{DO}_{t=0} - \text{DO}_{t=5}}_{\text{sample}}) - (\underbrace{\text{DO}_{t=0} - \text{DO}_{t=5}}_{\text{blank}})] \quad (8.4)$$

where D^* = dilution factor.

When the standard 5-day/20°C conditions are used, approximately two-thirds of the carbonaceous material is broken down; an incubation of about 20 days is needed for nearly complete breakdown. Nitrogenous nutrients can create problems in the BOD test. Because of the slowness of the nitrification process, the oxygen demand of the nitrifying bacteria is assumed to be negligible in the standard 5-day incubation period at 20°C. The actual environmental conditions of temperature, biological population, oxygen concentration, etc., are impossible to reproduce in the test; hence, care must be exercised in extrapolating the test results to the actual stream oxygen demands. In addition, many industrial wastewaters contain toxic materials which interfere with the growth of microorganisms thus making the BOD test unreliable or even inapplicable without the addition of suitable inoculum. Similarly, the presence of algae in the wastewater affects the BOD test by leading to higher BOD values even when the test is performed in darkness.

The BOD analysis, despite its limitations, has been in use for many years for monitoring the quality of stream pollution involving oxygen sag, but it is now being replaced by the chemical oxygen demand (BOD) test for research and plant control purposes.

Chemical Oxygen Demand (COD)

In the COD test, the oxidizing bacteria of the BOD test are replaced by a strong oxidizing agent under acidic conditions. A sample of the wastewater containing organic material is mixed with an excess of potassium dichromate and sulphuric acid and the mixture is heated under total reflux conditions for a period of 2 hours. During digestion, the chemically oxidizable organic material reduces a stoichiometrically equivalent amount of dichromate, the remaining dichromate is titrated with standard ferrous ammonium sulphate solution. The amount of potassium dichromate reduced gives a measure of the amount of oxidizable organic material. Dichromate has advantages over other oxidants in oxidizing power and applicability to a wide variety of samples.

The COD test does not distinguish between organic materials that are biodegradable and those that are not, and, hence, gives a measure of the total oxidizable organic material in the sample. Due to this, the COD test results are higher than those of BOD tests carried out on the same sample. If inorganic substances such as chlorides and nitrites are present in the wastewater, they interfore with the COD test since they are also oxidized by dichromate and create an inorganic COD that leads to an error in the measurement. Chloride interference can be eliminated by adding mercuric sulphate to the sample prior to the addition of other reagents, and nitrite interference can be overcome by adding sulphamic acid to the dichromate solution.

The COD test is much more useful than the BOD test for estimating the oxygen requirements of certain industrial wastewaters. It is valuable for wastes where BOD test is not applicable due to the presence of toxic substances, low rate of oxidation, or other similar factors. Ratios of BOD to COD can be employed to get an indication of the degree of biotreatability of the waste. Ratios of 0.8 or higher indicate wastes that are highly amenable to biological treatment, while lower ratios indicate that the wastes are not favourable to biological treatment.

Total Organic Carbon (TOC)

Total organic carbon test is based on the oxidation of the carbon of the organic matter to carbon dioxide, which is measured by a non-dispersive infrared analyzer. Alternatively, the carbon dioxide can be reduced to methane, which is then measured by a flame ionization detector.

In this test, a few micro-litres (5 to 10 μl) of the aqueous sample are injected into a combustion tube containing a catalyst and heated to 900°C in a constant flow of air. Water is vaporized and the carbonaceous matter is oxidized to CO_2 and steam. Outside the combustion tube, the steam is condensed and removed. CO_2 is swept into a non-dispersive infrared analyzer,

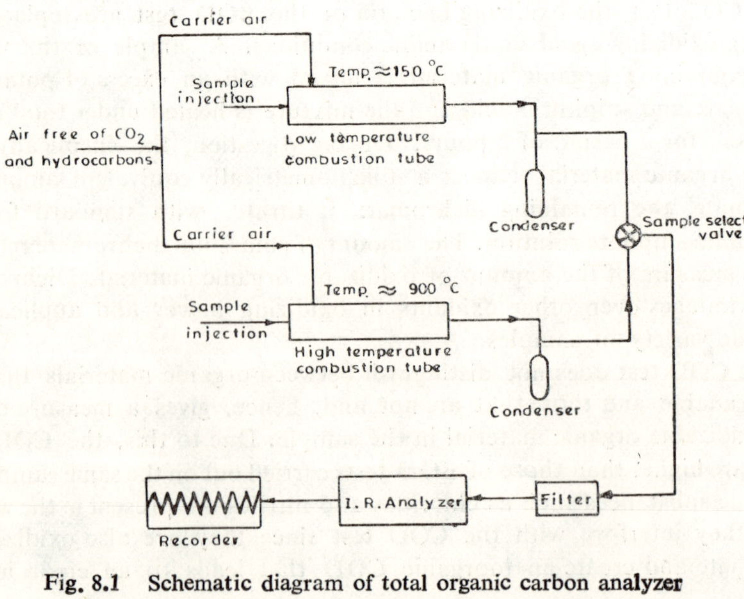

Fig. 8.1 Schematic diagram of total organic carbon analyzer

which measures the amount of CO_2. The concentration of CO_2 is directly proportional to the concentration of total carbon present in the original sample and it includes both organic and inorganic carbon. Inorganic carbon can be measured separately using an acid catalyst at 150°C, which is below the temperature at which organic matter is oxidized. Organic carbon content can then be obtained by subtracting the inorganic carbon from the total result. A flow diagram for a dual combustion tube total carbon analyzer is shown in Fig. 8.1.

The TOC test can be performed in a relatively short period of time (few minutes) compared to BOD and COD measurements and, hence, offers a valuable supplement to BOD and COD estimations. Because of this advantage, an empirical correlation between TOC and COD or BOD, that is specific to a particular plant operation, can be established and from this it is possible to obtain a tentative estimate of the plant performance quickly.

8.4 DETERMINATION OF INORGANIC SUBSTANCES

Nitrogen

The important chemical species containing nitrogen in wastewater systems are ammonia, organic nitrogen, nitrite and nitrate. Ammonia exists in aqueous solution as one of the intermediate compounds formed due to microbiological activity. Nitrogen, which is tied up in high energy compounds such as amino acids and amines, is known as organic nitrogen. Usually, organic nitrogen is a potential source of ammonia because deamination reactions that occur during the metabolism of organic compounds release ammonium ions. These two forms of nitrogen are often combined in one measure, known as Kjeldahl nitrogen.

Aerobic decomposition eventually leads to nitrogen in the nitrite (NO_2^-) and finally nitrate (NO_3^-) forms. For domestic wastewaters in U.S.A., the normal proportion of nitrogen-containing compounds is relatively constant: NH_4^+ 50-60%; organic nitrogen 40-59%; and nitrites and nitrates 0-5%. Although the contribution of nitrites and nitrates in domestic wastewaters is negligible, there may be a measurable concentration of these ions following biological treatment due to the occurrence of nitrification. All the above forms of nitrogen can be measured analytically by colorimetric techniques.

Kjeldahl nitrogen is determined by digesting the sample in sulphuric acid when the organic nitrogen is converted to ammonia. The total ammonia content is measured after neutralizing the excess sulphuric acid. The standard technique employed for the determination of ammonia is the Nessler method[6]. In this technique, the neutralized sample is first distilled to separate the ammonia from interfering substances. The ammonia in the distillate is reacted with Nessler's reagent (potassium mercuric iodide, K_2HgI_4), which produces a yellowish-brown colloidal dispersion. The colour intensity is directly proportional to the amount of NH_3 present. The colour can be compared visually to standards or can be estimated photometrically.

Another method of measuring the ammonia content in the sample is the phenol-hypochlorite method[7]. In this method, ammonia is treated with phenol and hypochlorite in the presence of a catalyst to produce the blue coloured solution of indophenol. The intensity of colour is proportional to the ammonia concentration.

Determination of nitrogen present as nitrite is based on the diazotization of sulphanilic acid with nitrite followed by its combination with N-(1-napthyl)-ethylenediamine dihydrochloride to form a purple azo-dye[8]. The intensity of the purple colour is directly proportional to the nitrite concentration and can be read on a photometer. The method is very sensitive but is time consuming. A simple and rapid method[9] consists of the addition of 1,5-dimethyl-2-phenyl-3-pyrazolone to the nitrite solution which forms a nitroso derivative, the extinction coefficient of which can be measured spectrophotometrically.

Nitrogen present as nitrate can be analyzed by reducing it to nitrite or ammonia, which is then determined by employing one of the sensitive methods previously described. One such method in popular use is the cadmium reduction method where the nitrates are quantitatively reduced by cadmium to nitrite. The nitrite thus produced is measured by the diazotization method. Since this method gives the sum of nitrate and nitrite nitrogen, a separate analysis of the nitrite is required so that this value can be subtracted from the results of the cadmium procedure. Nitrates can also be reduced to nitrites with zinc under controlled conditions and the nitrites determined colorimetrically. Other techniques of the determination of nitrogen are the Brucine, phenoldisulphonic acid, and the chromotropic acid methods[5].

Phosphorus

Phosphorus exists in wastewaters as orthophosphate (PO_4^{3-}, HPO_4^{2-}, $H_2PO_4^-$), polyphosphate, and organic phosphate. Orthophosphates are assimilated by bacteria during their growth process. However, polyphosphates must undergo enzymatic hydrolysis to the ortho form before they can be assimilated. The organic phosphate is present in molecules such as RNA, DNA and nucleotides. This is also an important constituent of industrial wastes.

The orthophosphate is determined using colorimetric methods. These methods involve treating the sample with ammonium molybdate under acidic conditions to form molybdophosphoric acid, which is reduced using vanadium, or stannous chloride or ascorbic acid when a coloured complex is formed. The intensity of the colour is proportional to the phosphate concentration in the solution. The vanadium method is quite useful for determining phosphate concentrations in the range of 0.2-18 mg. The stannous chloride method is highly sensitive and concentrations as low as 3 μg/l can be detected. The ascorbic acid method on the other hand is slightly less sensitive and the minimum detectable concentration is approximately 20 μg/l (ref. 10).

Polyphosphates are hydrolyzed into orthophosphates in an acidic medium and the excess acid is neutralized before the addition of ammonium molybdate solution. The amount of polyphosphate is obtained by subtracting the amount of orthophosphate originally present in the sample from the result.

The measurement of organic phosphorus is done by first oxidizing the organic matter so that phosphorus is released as phosphate ion. The total phosphorus can be measured by the procedure previously outlined for orthophosphate and the amount of the organic phosphorus present is obtained by subtracting the inorganic content from the result.

Trace Elements

The standard methods used for determination of toxic heavy metals like cadmium, chromium, lead and zinc are the atomic absorption spectrophotometry and colorimetry. The atomic absorption system consists of a light source which emits light of a particular wavelength, an atomizer burner unit for introducing the sample into the flame, a prism to separate and isolate the emission lines, and a detector. A schematic diagram of an atomic absorption spectrophotometer is given in Fig. 8.2. In this technique, light from a source lamp of an exact wavelength necessary to excite the atoms of a particular

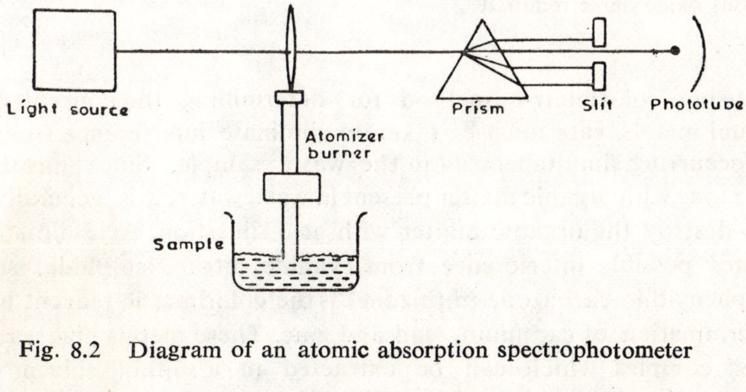

Fig. 8.2 Diagram of an atomic absorption spectrophotometer

trace element traverses the sample which is atomized in a flame, and passes through a prism where the desired wavelength is isolated. The light intensity passing through the flame is reduced as light is absorbed by each atom of the element; the amount absorbed is detected by the photomultiplier tube. This decrease in intensity with the sample is a measure of the concentration of the element. The technique is quite rapid, highly selective, and automated for monitoring effluent streams. The detection limits for some selected toxic metals are given in Table 8.2.

Table 8.2 Atomic absorption detection limits (ref. 11)

Element	Detection limit (ppm)	Analytical wavelength (nm)	Suggested resolution (Å)
Arsenic*	0.1	1937	7
Boron**	6.0	2497	7
Barium**	0.05	5536	4
Beryllium**	0.002	2349	20
Cadmium*	0.001	2288	7
Chromium	0.005	3579	2
Mercury	0.5	2537	20
Lead	0.03	2833	7
Silver	0.005	3281	7

* Argon-hydrogen flame required
**Nitrous oxide flame required

In using colorimetric method for determining the concentration of individual metals, care must be taken to eliminate interference from other metals occurring simultaneously in the waste sample. Since metals form complex ions with organic matter present in wastewater, it is generally necessary to destroy the organic matter with acid digestion. Acid digestion also eliminates possible interference from cyanide, nitrite, sulphide, sulphite, etc. Diphenylthio-carbazone (dithizone) is the colorimetric reagent used for the determination of cadmium, lead and zinc. These metals give a red/pink coloured complex which can be extracted in a suitable solvent such as carbon tetrachloride or chloroform. The colour intensity is then determined by a colorimeter or spectrophotometer. Chromium determination involves oxidation of the metal to the hexavalent state and its reaction with diphenylcarbazide in acidic medium when a red-violet product is obtained whose intensity can be measured photometrically.

Alkalinity

Alkalinity in wastewaters is due to the presence of bicarbonates, carbonates and hydroxides of metal ions such as calcium, magnesium, sodium and potassium. Of these, bicarbonates are most common, since they are formed by the action of carbon dioxide on the basic materials such as the salts of calcium and potassium in the soil. In addition, salts of weak acids and ammonia present in wastewaters also contribute to the total alkalinity.

Alkalinity is measured by titrating the sample with a standard acid such as $N/50$ H_2SO_4; the results are expressed in terms of $CaCO_3$. If the *p*H of the original sample is higher than 8.3, the titration is carried out using phenolphthalein indicator until it becomes colourless. This occurs at a *p*H value of 8.3. This end point is known as phenolphthalein alkalinity or bicarbonate equivalence point, which is observed in the case of hydroxide, or carbonate, or both.

Usually, bicarbonates are present in the original sample along with hydroxides and carbonates. In such cases the sample is titrated with methyl orange indicator and the titration is carried out beyond the phenolphthalein end point. When a *p*H of about 4.5 is reached, there occurs a colour change from yellow to red indicating carbonic acid equivalence point. This result is used to calculate the total alkalinity, which may be due to any of the three alkaline ions, bicarbonates, carbonates or hydroxides.

Determination of alkalinity is of importance in water softening, chemical treatment of wastewater, corrosion control, and in removal of ammonia by air stripping.

8.5 PHYSICAL CHARACTERISTICS

Suspended Solids

The determination of suspended matter is extremely important in the analysis of wastewaters. The suspended matter contained in domestic wastewaters and in many industrial wastes is largely organic and is, therefore, responsible for a significant portion of the oxygen demand. Hence, it is one of the major parameters used to evaluate the quality of domestic and industrial wastes.

Analytically, the term total solids in wastewater refers to the matter that remains as residue after evaporation and drying at 103 to 105°C. The total solids can be classified as filterable solids and suspended solids. The filterable solids are those that are able to pass through a specified filter and consist of colloidal and dissolved solids. The colloidal solids range in size from 10^{-3} to 10^{-6} mm, while dissolved solids are smaller than 10^{-6} mm and exist as molecules and ions in solution. The suspended solids are retained on the filter and their minimum diameter is about 1 μm. Of the suspended solids, the settleable solids are those which settle in a cone-shaped container (Imhoff cone) in a definite time period.

The suspended solids (also filterable solids) may be further classified on the basis of their volatility at 600°C. The organic matter that is oxidized at this temperature is known as "volatile suspended solids" (VSS), and the inorganic fraction that remains as ash is termed "fixed suspended solids" or "non-volatile suspended solids" (NVSS). Fig. 8.3 gives an approximate classification of solids found in medium strength wastewaters.

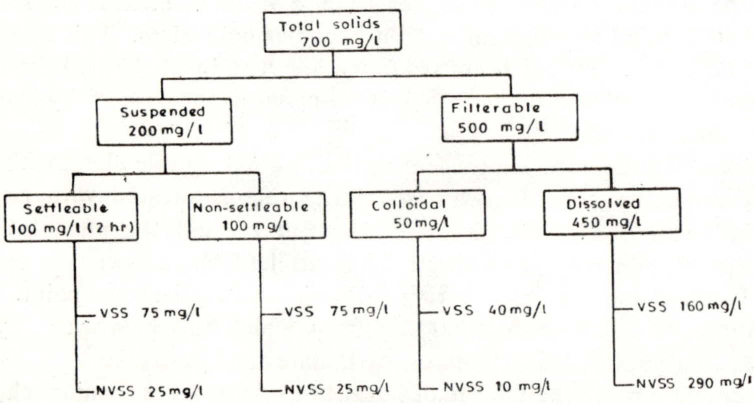

Fig. 8.3 Solids content of medium strength domestic wastewater (ref. 12, reproduced with permission of McGraw Hill. N.Y. 1972.)

The suspended solids concentration is an easily measured parameter. In almost all the cases gravimetric methods are used such as filtration through a Gooch crucible, but the determination is subject to considerable error if proper precautions are not taken. An alternative procedure for measurement of suspended solids makes use of a continuous flow nephelometer. The measurement of light transmitted or light scattered by suspended particles can be correlated with the actual suspended solids concentration.

Colour and Odour

Colour and odour are important parameters from the standpoint of aesthetics. Colour may be of organic or mineral origin. Organic sources include algae, tannins, humic compounds, etc.; inorganic sources are iron and manganese compounds, chemicals and dyes from various industries. Objectionable odours in water are caused by algae and decaying vegetation. Some inorganic substances such as mercaptans, amines and sulphides also cause bad odour.

Colour is measured by comparison with known standards. Coloured water made with potassium chloroplatinate when tinted with cobalt chloride closely resembles colour of many natural waters which are yellow-brownish in appearance. A stock solution containing 500 mg/l of potassium chloroplatinate, with cobalt chloride added as a tinting material, is usually prepared and this is assigned colour of 500 units. Then a series of working standards is prepared from it by dilution. The intensity of colour in water is measured against these standards by direct visual comparison.

Suspended matter in water can interfere with the measurement of colour. In such cases the sample is centrifuged to separate the solids and analysis is made on the clear portion. Removal of the suspended matter from the sample by filtration is not recommended because of the possibility that the filter may adsorb some colour and thereby cause error in the measurement.

The odour causing substances are usually present in very small quantities and are often very complex. It is, therefore, often impossible and impractical to isolate and identify positively the odour causing materials. Evaluation of odour is, thus, dependent on the olfactory senses of the tester and on his ability to distinguish between different levels and kinds of odours. The testing is based entirely on arbitrary comparison since no absolute unit or base exists for odour.

The standard test method is the threshold dilution method in which the odour bearing sample is diluted successively with odour-free water, until a dilution is obtained which has a barely perceptible odour. The result is expressed in terms of the "threshold odour number (TO)", which is given by,

$$\text{TO} = \frac{\text{ml of sample} + \text{ml odour-free water}}{\text{ml of sample}} \tag{8.5}$$

It has been recommended that odour tests be run at 40 and 60°C. Since TO is a function of temperature, the test temperature should be reported in all cases.

8.6 BACTERIOLOGICAL MEASUREMENTS

From public health standpoint the bacteriological quality of water is as important as the chemical quality. A variety of procedures have been devised to measure the bacteriological quality, but the most popular method is based on determining the content of "indicator organisms" in it. The principal indicator organisms used are the coliform group of bacteria. This group is normally nonpathogenic, present in polluted waters when pathogens are, or might be present, unable to multiply under conditions when pathogens do not multiply, and readily identified and enumerated by simple laboratory techniques.

The coliform group of bacteria has the ability to ferment lactose or lauryl tryptose broth and produce gas. This offers a simple test of the presence of coliform organisms. However, some other organisms also ferment the broth under certain conditions and, therefore, additional growth reactions must be carried out to confirm the presence of coliform group. The test is conducted by inoculating the broth by multiple portions of a series of decimal dilutions of the water sample. The result is usually expressed as the number of coliform organisms per unit volume of the sample.

Two methods are generally used for obtaining the number density of coliform organisms in the test sample. One is the membrane-filter technique and the other is the most probable number (MPN) technique. In the membrane-filter technique, a known volume of water sample is passed through the filter that has a very small pore size. The coliform bacteria are captured by the filter and the filter is then exposed to nutrients which promote the growth of coliform while inhibiting that of other organisms. After 24 or 48 hours of incubation, the number of coliform colonies is counted and their density is determined. The coliform colonies appear pinkish in colour and their count is made with the aid of an optical device. The coliform density is reported in terms of (total) coliforms per 100 ml.

The most probable number of coliform organisms is a statistical estimate of their density and is based on the examination of a number of portions of different sizes of the water sample for the presence of coliforms. From a knowledge of the number of portions of different sizes giving positive or negative results for coliforms, the MPN can be determined from the following equation[13],

$$Y = \frac{1}{a}[(1-e^{-N_1\lambda})^p \, (e^{-N_1\lambda})^q][(1-e^{-N_2\lambda})^r \, (e^{-N_2\lambda})^s]$$

$$[(1 - e^{-N_3\lambda})^t \, (e^{-N_3\lambda})^u] \qquad (8.6)$$

where N_1, N_2, N_3 = sizes of portions examined, ml

p, r, t = number of portions of respective sizes giving positive results.

q, s, u = number of portions of respective sizes giving negative results.

λ = concentration of coliforms/ml

Y = probability of occurrence of a particular result.

a = constant for any particular set of conditions and, therefore, may be omitted for computing λ.

The MPN is that value of λ which gives a maximum value of Y. The calculation of MPN is illustrated in the following example.

Example 1

A sample of wastewater is tested for coliform organisms by the MPN method with the following series of observations. Determine the coliform density per 100 ml.

Size of portion (ml)	No. positive	No. negative
10	4	1
1.0	4	1
0.1	2	3

Solution

From Eq. (8.6), $N_1 = 10$, $N_2 = 1$, $N_3 = 0.1$, $p = 4$, $r = 4$, $t = 2$, $q = 1$, $s = 1$ and $u = 3$. The probability of occurrence, Y, of a particular result can be obtained by selecting the values of λ. The value of λ that will make Y maximum gives the MPN. Let $\lambda = 0.46$, 0.47, and 0.48 respectively. Substituting the values for N_1, N_2, N_3, p, r, t, q, s, and u in Eq. (8.6), we get

$$aY = (1 - e^{-10\lambda})^4 (e^{-10\lambda}) (1 - e^{-\lambda})^4 (e^{-\lambda}) (1 - e^{-0.1\lambda})^2 (e^{-0.1\lambda})^3$$

Substituting the selected values of λ, we can calculate aY.

λ	aY
0.46	1.954×10^{-7}
0.47	1.980×10^{-7}
0.48	1.973×10^{-7}

From the above result, the most probable number is 0.47 per ml at which aY is maximum. In other words the MPN per 100 ml is 47.

Although the coliform bacteria meet some of the criteria for indicator organisms, they do have certain limitations. Coliforms often grow in treatment facilities or water bodies and the tests can reveal higher populations even though pathogens are far less likely to multiply under the same environmental conditions. In addition, some viruses and several other organisms may survive for longer periods than coliforms, and reduction of coliforms to low concentrations or even to zero may not always indicate that the water is safe. Nevertheless, in spite of these and other limitations, the coliform test has provided a valuable and reliable technique for evaluating water purity.

8.7 WATER QUALITY STANDARDS

The analyses required of water samples depend on the intended use of the water. For example, if its intended use is drinking, water should meet certain quality criteria with respect to the appearance (turbidity, colour), potability (taste, odour), health (bacteria, nitrates, chlorides, etc) and toxicity (metals, organics). These and similar criteria are established by health or other regulating agencies to ensure that the water quality in a resource is suitable for the proposed use.

Based on the criteria, quality standards are set, which reflect the current state of knowledge of various water constituents. These standards are continuously revised as more and more is learnt about the effects of water constituents on proposed uses. Hence, these standards should not be used as absolute limits, but only as guidelines that can be used for preliminary judgements. Table 8.3 summarizes several quality criteria and their standards for drinking water as suggested by the following agencies:

(1) Indian Council of Medical Research (ICMR)
(2) World Health Organization (WHO)
(3) United States Public Health Service (USPHS)

Standards have been prepared for raw waters to be used as a source for various industrial applications. The specific purpose for which the water is used usually controls the requisite water quality. Water used in food and beverage industries will need to meet standards similar to those for drinking water, while water for cooling purposes can contain much higher concentrations of impurities. Table 8.4 illustrates the water quality characteristics of raw surface waters that have been used as source for various industrial operations. The values listed are the maximum concentrations of different constituents in the raw surface waters used in the specified industry, and not the maximum concentrations that could be tolerated.

Table 8.3 **Standards for drinking water**
A=Recommended maximum concentration[a] (mg/l except where shown otherwise)
B=Maximum permissible concentration[b] (mg/l except where shown otherwise)

	ICMR		WHO		USPHS	
	A	B	A	B	A	B
Physical:						
Turbidity (units)	5	25	5	25	5	—
Colour (units)	5	25	5	50	15	—
Odour	Nothing disagreeable		Unobjectionable		TO=3	—
Chemical:						
*p*H, units	7–8.5	(6.5 or 9.2)	7–8.5	(6.5 or 9.2)	—	—
Total solids	—	—	500	1500	500	—
Calcium	75	200	75	200	—	—
Magnesium	50	150	50	150	—	—
Iron	0.3	1.0	0.3	1.0	0.3	—
Manganese	0.1	0.5	0.1	0.5	0.05	—
Copper	1.0	3.0	1.0	1.5	1.0	—
Sulphate	200	400	200	400	250	—
Phenols	0.001	0.002	0.001	0.002	0.001	—
Fluorides	1.0	2.0	0.5	1.0–1.5	0.6–1.7	—
Nitrates	20	50	—	50–100	45	—
Toxic:						
Arsenic	—	0.2	—	0.2	0.01	0.05
Barium	—	—	—	—	—	1.0
Cadmium	—	—	—	—	—	0.01
Chromium	—	0.05	—	0.05	—	0.05
Cyanide	—	0.01	—	0.01	0.01	0.2
Lead	—	0.1	—	0.1	—	0.05
Selenium	—	0.05	—	0.05	—	0.01
Silver	—	—	—	—	—	0.05
Bacteriological:	—	1 coliform per 100 ml	—	1 coliform per 100 ml	—	1 coliform per 100 ml

[a]Constituents should not be present in excess of listed concentrations where other more suitable supplies are or can be made available.
[b]Constituents in excess of the concentrations listed shall constitute grounds for rejection of the supply.

Table 8.4 Maximum concentrations of constituents in raw water supplies for various industries, mg/*l* (refs. 14, 15)

Constituent	Boiler water (0-1500) psig	Cooling water (once through)	Textile industry	Pulp and paper	Chemical industry	Petroleum industry
Silica (SiO_2)	150	50	—	50	—	85
Aluminium	3	3	—	—	—	—
Iron	80	14	0.3	2.6	5	15
Manganese	10	2.5	1.0	—	2	—
Calcium	—	500	—	—	25	220
Magnesium	—	—	—	—	100	85
Ammonia	—	—	—	—	—	40
Bicarbonate (HCO_3)	600	600	—	—	600	480
Sulphate	1400	680	—	—	850	900
Chloride	19000	600	—	200	500	1600
Nitrate	—	30	—	—	—	8
Phosphate	—	4	—	—	—	—
Dissolved solids	35000	1000	15	1080	2500	3500
Suspended solids	15000	1500	1000	—	10000	5000
Hardness	5000	850	120	475	1000	900
Alkalinity	500	500	—	—	500	500
*p*H (units)	—	5.0–8.9	6.0–8.0	4.6—9.4	5.5–8.0	6.0–9.0
Colour (units)	1200	—	—	360	500	25
COD	100	—	—	—	—	—
Temperature (°F)	120	100	—	95	—	—

REFERENCES

1. L.D. Banefield and C.W. Randall, *Biological Process Design for Wastewater Treatment* (Prentice-Hall, Inc., Englewood Cliffs, N.J.), 1970.

2. J.G. Rabosky and D.L. Koraido, "Gaging and sampling industrial wastewaters". *Chem. Engg.*, Vol. 74, 1973.
3. B.G. Liptek (ed.), *Environmental Engineer's Handbook: Water Pollution,* Vol. I (Chilton Book Company, Randor, Pa.), 1974, Chapter 4.
4. L.W. Winkler, "The determination of dissolved oxygen in water", *Bericht der Deutschen chemischen Gasellshaff.* Vol. 21 (1888), p. 2843.
5. *Standard Methods for the Examination of Water and Wastewater,* 13th Edn (American Public Health Assoc., Yearbook Publication, Chicago), 1971.
6. A.A. Sutcliffe and G.A. Jones, "Research into some methods of analysis of sewage", *J. of Water Pollution Control,* Vol. 67 (1968), p. 209.
7. J.A. Russell, "The colorimetric estimation of small amounts of ammonia by the phenol-hypochlorite reaction", *J. of Biological Chemistry,* Vol. 156 (1944), p. 457.
8. C.N. Sawyer and P.L. McCarty, *Chemistry for Environmental Engineering,* 3rd Edn (McGraw-Hill, New York), 1978, p. 450.
9. K.G. Weiss and D.F. Boltz, "The ultra violet spectrophotometric determination of nitrite by the antipyrine method", *Analytica Chimica Acta,* Vol. 55 (1971), p. 77.
10. J.E. Harwood, R.A. Van Steenderen and A.L. Kuhn, "A rapid method for orthophosphate analysis at high concentrations in water", *Water Research,* Vol. 3 (1969), p. 417.
11. H.L. Kahn, *Advan. Chem., Ser.,* Vol. 73 (1968), p. 183.
12. Metcalf and Eddy Inc., *Wastewater Engineering* (McGraw-Hill, New York), 1972.
13. J.W. Clark, W. Viessman and M.J. Hammer, *Water Supply and Pollution Control,* 3rd Edn (IEP—A Dun-Donnelley, New York), 1977, p. 264.
14. J.C. Lamb, *Water Quality and its Control* (John Wiley & Sons, New York). 1985, p. 301.
15. J.W. Clark, W. Viessman and M.J. Hammer, *Water Supply and Pollution Control,* 3rd Edn (IEP—A Dun-Donnelley, New York), 1977, p. 250.

SUGGESTIONS FOR FURTHER READING

1. K.H. Mancy and W.J. Weber (Jr.), *Analysis of Industrial Wastewaters* (Wiley-Interscience, New York), 1971.
2. K.V. Melbourne, K.G. Robertson and A.B. Oaten, "Developments in instruments for water pollution control", *J. of Water Pollution Control,* Vol. 71 (1972), p. 278.

3. R.L. Cooper and K.C. Wheatstone, "The determination of phenols in aqueous effluents", *Water Research,* Vol. 7 (1973), p. 1375.
4. P.J. Hewitt and H.B. Austin, "Determination of cyanide in industrial effluents", *J. of Water Pollution Control,* Vol. 71 (1972), p. 381.
5. E.F. McFarren, R.J. Lishka and J.H. Parkar, "Criterion for judging acceptability of analytical methods", *Analytical Chemistry,* Vol, 42 (1970), p. 358.
6. C.H. Tate and R.R. Trussel, "Developing drinking water standards", *J. of American Water Works Assoc.,* Vol. 69 (1977), pp. 486-498.
7. R.M. Santaniello, "Water quality criteria and standards for industrial effluents", Part 2, Chapter 4, in *Industrial Pollution Control Handbook,* edited by H.F. Lund (McGraw Hill, New York), 1971.

9. Wastewater Treatment

9.1 BASIC PROCESSES OF WATER TREATMENT

The purpose of wastewater treatment is to remove the contaminants from water so that the treated water can meet the acceptable quality standards. The quality standards usually depend upon whether water will be reused or discharged into a receiving stream. Available wastewater treatment processes can be broadly classified as physical, chemical or biological. These processes, which consist of a series of unit operations, are applied in different combinations and sequences depending upon the prevailing situations of influent concentration, composition and condition and specifications of the effluent.

Physical processes are based on exploitation of the physical properties of the contaminants and are generally the simplest forms of treatment. These principally comprise screening, sedimentation, flotation and filtration. Chemical processes utilize the chemical properties of the impurities or of the added reagents. Commonly used chemical processes are precipitation, coagulation, and disinfection. Other physical and chemical processes such as air stripping, carbon adsorption, oxidation and reduction, ion exchange and membrane processes like reverse osmosis and electrodialysis are also important in certain particular cases. Biological processes utilize biochemical reactions; typical examples are biological filtration and the activated sludge process.

The wastewater treatment processes are generally grouped according to the water quality they are expected to produce. These processes are usually grouped as the primary treatment, the secondary treatment, and the tertiary or the advanced waste treatment. Primary treatment removes identifiable suspended solids and floating matter. In the secondary treatment, also known as the biological treatment, organic matter that is soluble or in the colloidal form is removed. Advanced waste treatment may involve physical, chemical or biological processes or their various combinations depending on the impurities to be removed. These processes are employed to remove residual soluble non-biodegradable organic compounds, including surfactants, inorganic nutrients and salts, trace contaminants of various types, and dissolved inorganic salts. The advanced waste treatment processes are expensive, and are used only when water produced is required to be of higher quality than that produced by conventional secondary treatment so that the treated water can be reclaimed and put to some form of direct re-use.

9.2 PRIMARY TREATMENT

Pretreatment

The primary treatment comprises a pretreatment step and a sedimentation step. Wastewater is pretreated to remove large floating and suspended solids which could interfere with the normal operation of subsequent treatment processes. The pretreatment operation may also include flow measurement and sometimes prechlorination to prevent any odours that may emanate during subsequent processes.

Pretreatment consists of screening and grit removal. Screens of various sizes and shapes are used, depending on the nature of solids to be removed, and cleaning is done either manually or mechanically. Fixed bar screens (Fig. 9.1) are the most common type of screens used in domestic wastewater

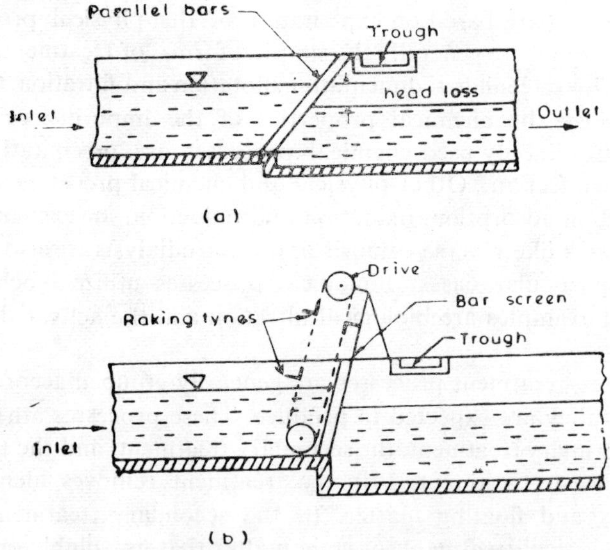

Fig. 9.1 (a) Manually-raked bar screen
(b) Mechanically-raked bar screen

treatment facilities. Bar screens are made up of parallel metal bars and have apertures in the range 20–60 mm for coarse screens and 10–20 mm for fine and medium screens. Where a fine screen is employed, it is usually preceded by a coarse screen to remove large solids in order to avoid clogging problems.

The screening process produces objectionable screenings which must be

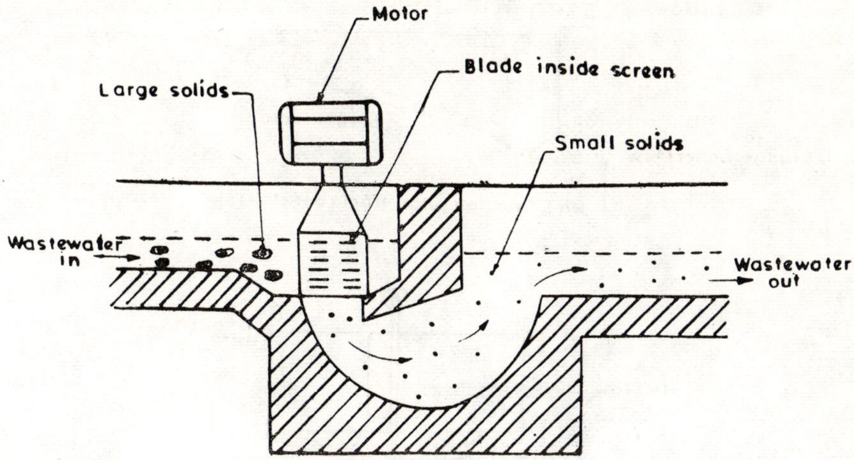

Fig. 9.2 Typical comminutor

disposed of in a satisfactory manner. Methods of disposal include burial, incineration, grinding and digestion. To avoid the disposal problems, some treatment plants use a device known as a comminutor (Fig. 9.2) instead of the screens. The comminutor grinds large solids which can then be satisfactorily handled in the sedimentation tank. After screening, the wastewater enters a grit chamber for the removal of inorganic grit, consisting of sand, gravel, cinders and pebbles. Grit chambers are provided to protect pumps from abrasion and to reduce the formation of heavy deposits in pipes and channels. Grit chambers are generally designed as long chambers and the velocity of waste flow is reduced to 30 cm/sec, which allows settling of the grit material. The grit can be removed by scrappers. When comminutors are used in place of bar screens, they are generally placed after the grit chambers.

Sedimentation

In this step, the settleable solids are removed by gravitational settling under quiescent conditions. The sludge formed at the bottom of the tank is removed as underflow either by vacuum suction or by raking it to a discharge point at the bottom of the tank for withdrawal. The clear liquid produced is known as the overflow and it should contain no readily settleable matter. If the sedimentation tank is poorly designed the overflow may contain solid particles or the underflow may be more dilute than desired.

The sedimentation operation in waste treatment applications may be carried out in rectangular horizontal flow, circular radial flow, or vertical

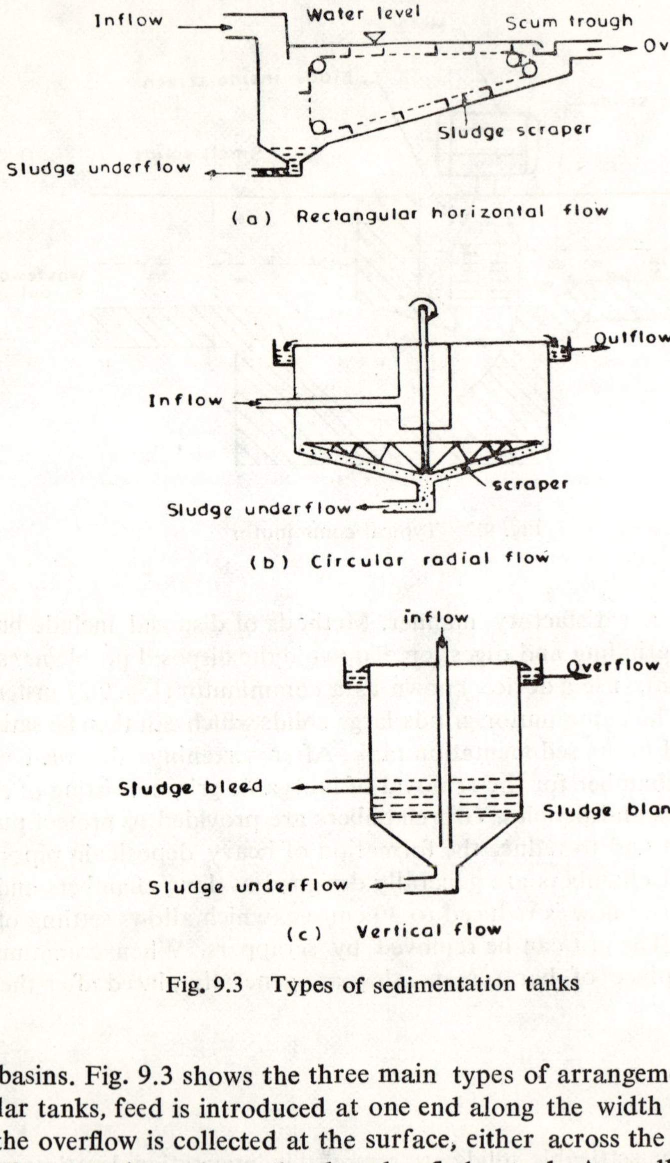

Fig. 9.3 Types of sedimentation tanks

flow basins. Fig. 9.3 shows the three main types of arrangements. In rectangular tanks, feed is introduced at one end along the width of the tank and the overflow is collected at the surface, either across the other end or at different points along the length of the tank. An endless conveyor scrapes the floating material into a screen though while it also pushes the settled solids into a sludge hopper.

In the circular radial flow tanks the feed is introduced through a centre well and the clarified effluent is collected at weirs along the periphery of the tank. Sludge removal is effected by means of a rotary sludge scraper which forces the settled sludge down a slopping bottom into a central hopper, from which it is withdrawn. Scum is removed by a surface skimming board which is attached to the rotary mechanism and positioned in such a manner that the scum can be collected into a trough situated at the surface.

Vertical flow tanks are often used in small treatment plants where the feed is applied at a point or points along the bottom, and clarified effluent is collected at the top. A sludge blanket is maintained in the lower part of the tank through which the suspension rises. It is important to control the sludge withdrawal and bleed carefully to avoid losing the blanket, which acts as a filter for small particles.

Basic Theory and Design

Two types of particle behaviour can be distinguished in sedimentation. These are the discrete particle and the flocculent particle behaviour. In discrete settling behaviour particles do not interfere with one another, and their sizes, shapes and masses remain constant during settling, whereas in flocculent behaviour the particles agglomerate during settling and thus do not have constant characteristics. The settling of particles in a suspension depends upon their concentration and their flocculating properties. Generally, three types of settling are recognized:

1. Settling of dilute suspensions of discrete particles
2. Settling of dilute suspensions of flocculent particles
3. Zone settling which includes hindered settling and compressive settling

Discrete Settling

The basic theory of sedimentation assumes discrete settling. Sedimentation of a particle from such suspensions is unhindered by the presence of other settling particles and is a function only of the properties of the fluid and the particle in question. When such a particle is placed in a liquid of lower density, it will accelerate untill a limiting terminal velocity is reached. The terminal settling velocity (v_t) for spherical particles can be calculated using Eqs. (5.12) to (5.17) by replacing the density and viscosity of the gas by those of the liquid. The settling of sand particles in grit chambers is a typical example of this type of settling.

The usual procedure in designing sedimentation tanks operating on a continuous flow basis with discrete suspension of particles is to select a critical particle with a terminal velocity v_t and to design the tank so that all particles with settling velocity $\geqslant v_t$ will be removed. Suppose Fig. 9.4 represents an ideal rectangular sedimentation tank. The critical particle will be one that enters at the top of the settling zone, at point A, and settles with a velocity just sufficient to reach the sludge zone at point B. The particle falls through a depth H in the retention time t_0:

$$t_0 = \frac{H}{v_t} \tag{9.1}$$

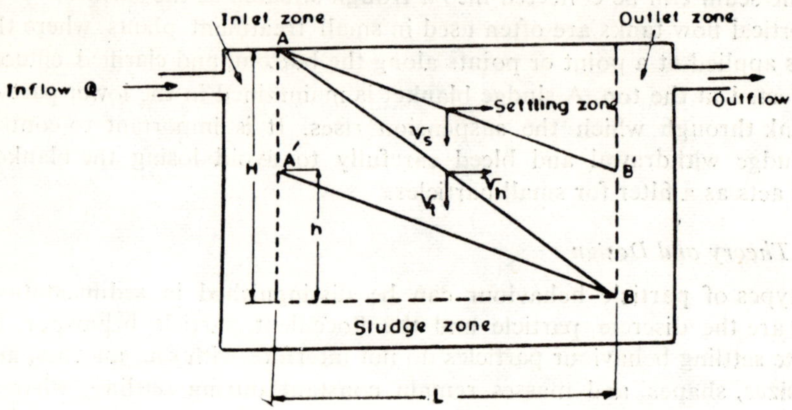

Fig. 9.4 Discrete settling in an ideal sedimentation tank

In the same time the particle travels a distance L with a horizontal velocity v_h,

Therefore, $$t_0 = \frac{L}{v_h} \tag{9.2}$$

Equating the two expressions and considering that $v_h = Q/WH$, where Q is the inflow rate, and W is the width of the tank, we get the terminal settling velocity, v_t

$$v_t = \frac{Q}{LW} = \frac{Q}{A} \tag{9.3}$$

where A is the surface area of the settling zone.

Equation (9.3) gives the terminal settling velocity of the slowest-settling particles that could be removed completely in an ideal sedimentation basin. The parameter Q/A, which is called the overflow rate or surface loading rate, is an important design parameter and forms the basis on which the performance of the sedimentation tank is evaluated. It follows from the equation that for discrete sedimentation behaviour, the rate of removal for a given discharge is independent of the depth but is directly related to the surface area. It can also be shown that for radial circular flow and vertical flow situations, the same expression holds true[1].

In wastewater treatment the sedimentation tank is fed with suspension of particles of varying sizes. Suppose that a particle of terminal velocity v_s is introduced at the top at a point A, where $v_s < v_t$. This particle will not be removed during the time provided for settling since it does not reach the bottom of the settling zone (path AB'). However, if it is introduced at a distance $\leqslant h$ from the bottom, where $h = v_s t_0$, it will be removed (path $A'B$). Hence, the fraction of the particles having a settling velocity v_s that can be removed is given by,

$$x_t = \frac{v_s}{v_t} \tag{9.4}$$

For a given inflow rate Q, where

$$Q = v_t A, \tag{9.5}$$

only those particles that have settling velocities $v_s \geqslant v_t$ will be completely removed. Such particles constitute the fraction $1 - x_t$ of the total originally present in the suspension. The fraction removed of the particles having velocities $v_s < v_t$ will be

$$\int_0^{x_t} \left(\frac{v_s}{v_t}\right) dx.$$

The overall fraction (R) removed from the suspension, will be

$$R = (1 - x_t) + \frac{1}{v_t} \int_0^{x_t} v_s \, dx\,. \tag{9.6}$$

The second term in Eq. (9.6) can be determined by graphical integration of a settling analysis curve, such as the shaded portion of Fig. 9.5.

In an ideal sedimentation tank with a radial flow pattern, discrete particles with settling velocities $v_s < v_t$ will also be removed in proportion to x_t. By contrast, in an ideal upflow tank, no particles with settling velocities $v_s < v_t$ will be removed.

Example 1

A settling analysis performed on a dilute suspension of discrete particles yields the following data which are recorded for samples collected at the depth of 1 metre in a batch sedimentation column.

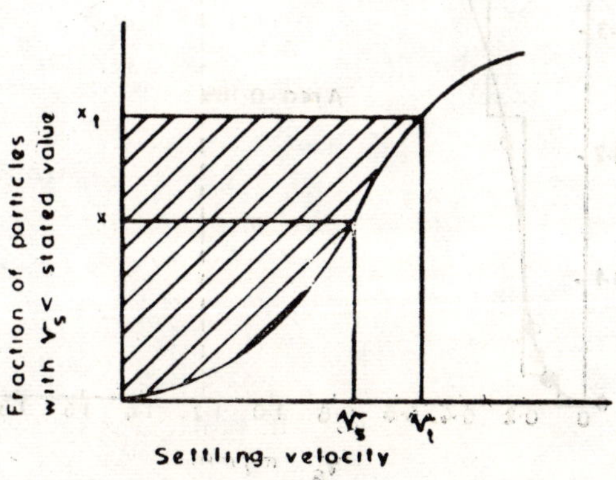

Fig. 9.5 Settling analysis curve for a discrete suspension

Settling time, min.	0.5	1.0	2.0	4.0	6.0	8.0
Weight fraction remaining	0.53	0.48	0.39	0.21	0.06	0.02

Determine the per cent removal of suspended solids in an ideal horizontal flow sedimentation tank operating at (1.2 m^3/min)/m^2.

Solution

1. The settling velocities of particles settling 1 m during the stated settling time are obtained and these are plotted in Fig. 9.6 as a function of the fraction of particles remaining.
2. The settling velocity of the critical particle is equal to the outflow rate

$$v_t = \frac{Q}{A} = 1.2 \text{ m/min}$$

3. From Fig. 9.6, $x_t = 0.5$. This means half of the particles in the suspension have settling velocities less than 1.2 m/min.

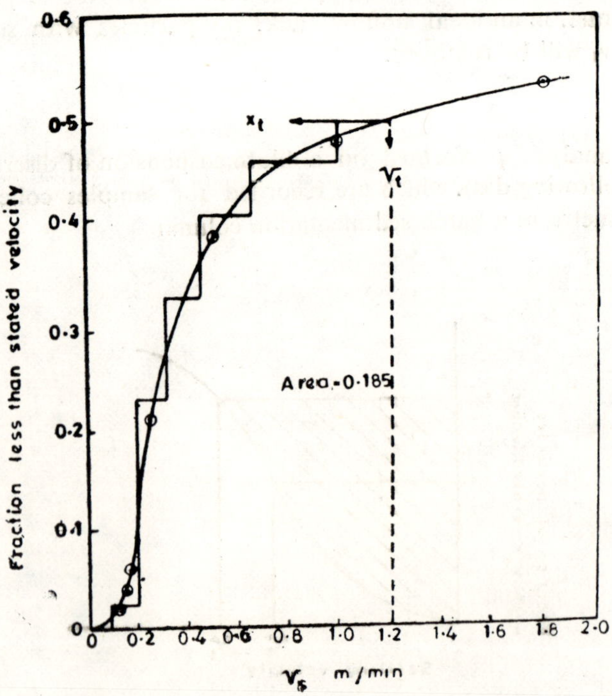

Fig. 9.6 Distribution of settling velocities

4. $\int_0^{x_t} v_s \, dx$, given by area in Fig. 9.6 is 0.185.

5. The overall fraction removed is, then,

$$R = (1 - 0.5) + \frac{1}{1.2}(0.185) = 0.5 + 0.155$$

$$= 0.655 \text{ or } 65.5\%$$

Flocculent Settling

The process of flocculent settling takes place when the settling velocity of the particles increases due to coalescence with other particles. This type of phenomenon is clearly observed in the primary clarifiers. The opportunity for coalescence increases with the increase in bed depth, and as a result the particle removal efficiency depends on both the overflow and the bed depth. The path followed by flocculent particles, unlike that of discrete particles, is a curve as shown in Fig. 9.7. No satisfactory formulation is available for evaluating the flocculation effect on sedimentation. A settling column analysis is usually performed by placing the test suspension in a column that has the same depth as that of the sedimentation tank in question.

The column is provided with sampling ports at various depths, and samples are withdrawn from different depths at different time intervals and analyzed for suspended solids. The percentage removal of suspended solids at each time interval and depth can be obtained. Typical test results of a settling column are shown in Fig. 9.8. The numerical values of the per

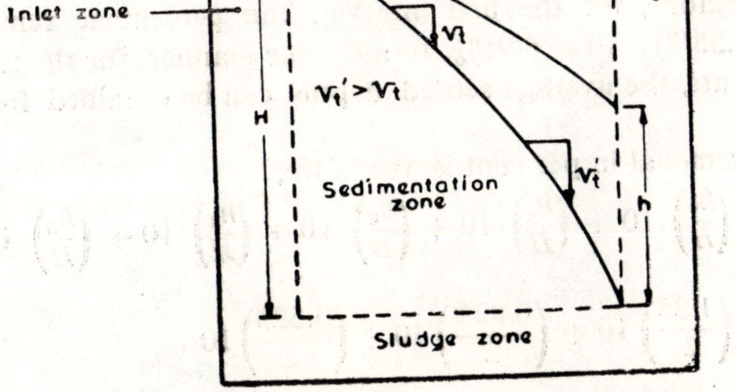

Fig. 9.7 Settling of flocculent particles

cent solids removed are shown at diffierent points identifying the time and depth of the sample. Curves of equal per cent removal can then be sketched as indicated in the diagram. The overall removal of solids (say) for a sedimentation tank 2 m deep and detention time of 31 min can be determined using Fig. 9.8 as follows:

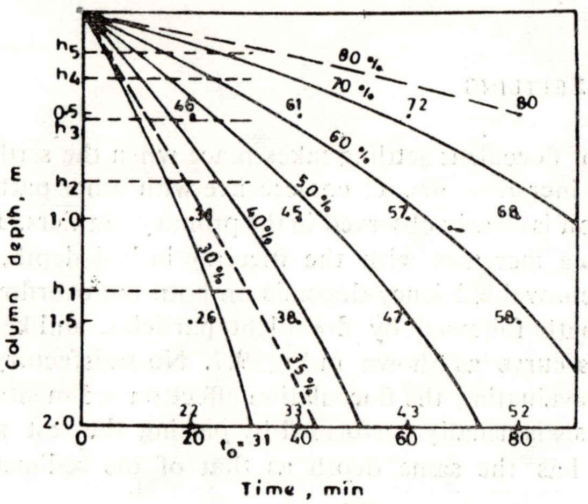

Fig. 9.8 Settling profile for flocculent particles

From the figure, it is apparent that 30% of the particles will have settling velocities greater than 2/31 m/min. This means 30% of the suspended solids are completely removed. Considering 10% increments, particles in the range of 30-40% will be removed in proportion to v_{1s}/v_t or in the proportion of average settled depth (h_1) to the total depth (H). The average settled depth (h_1) is estimated by drawing a curve corresponding to 35% constant removal in Fig. 9.8. The depth h_1 corresponding to $t_0 = 31$ min is 1.35 m. Therefore, for the first interval, the percentage removal is $(h_1/H)\times 10 = (1.35/2)\times 10 = 6.75\%$. In a similar manner, for the succeeding 10% increments, the average settled depths can be obtained from the figure.

The overall removal in per cent is, therefore,

$$R = 30 + \left(\frac{h_1}{H}\right)10 + \left(\frac{h_2}{H}\right)10 + \left(\frac{h_3}{H}\right)10 + \left(\frac{h_4}{H}\right)10 + \left(\frac{h_5}{H}\right)10 \tag{9.7}$$

$$= 30 + \left(\frac{1.35}{2}\right)10 + \left(\frac{0.825}{2}\right)10 + \left(\frac{0.525}{2}\right)10$$

$$+ \left(\frac{0.325}{2}\right)10 + \left(\frac{0.2}{2}\right)10$$

$$= 30 + 6.75 + 4.12 + 2.62 + 1.62 + 1.0 = 46.11\%$$

The total removal of particles $R = 46\%$ (approximately).

Zone Settling

This type of settling takes place in the secondary clarifier of the activated sludge process where the suspension is dense and the particles are so closely spaced that the velocity fields of the fluid displaced by the adjacent particles during settling overlap. There is an upward displacement of the fluid with a reduction in the settling velocities of the particles, and this effect is known as hindered settling.

The most commonly encountered form of hindered settling occurs at very high particle concentrations where the suspension tends to settle *en masse* with distinct interface between the settling solids and the clear liquid above it. This is termed as zone settling. Four zones are distinctly observable in a typical settling process. Fig. 9.9 represents the relative positions of different zones as a function of time in a thick suspension. Initially the solids

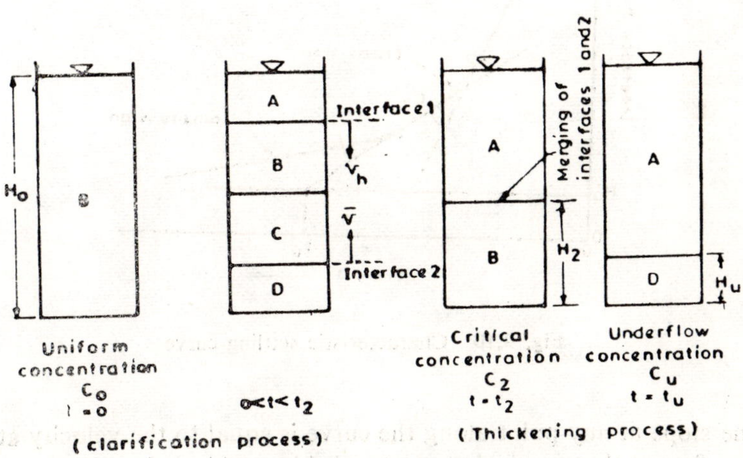

Fig. 9.9 Zone settling at different time intervals

concentration is uniform (B) but as time passes a clear liquid (A) forms at the top and an interface develops between A and B. Hindered settling conditions prevail in zone B and the particles settle as a blanket with a constant velocity v_h.

With the formation of zone A and the interface 1, sludge (D) begins to build up simultaneously at the bottom and is subject to compressive settling as the solids are supported mechanically by those beneath them. A transition zone C is formed in between zones B and D, in which the settling velocity decreases due to an increase in the concentration of solids as the suspension changes gradually in concentration from that of the zone B to that of the zone D. The interface 2 between C and D rises with a constant velocity $\bar{v}$. Ultimately, at time t_2, the interfaces 1 and 2 merge and only the zones representing the clarified liquid and the sludge remain. At this time, the settled sludge exhibits a uniform concentration C_2, which is termed the critical concentration. The sludge begins to thicken and eventually reaches an ultimate concentration C_u termed as the underflow concentration. Fig. 9.10 shows the typical characteristic settling curve for concentrated suspensions where the interface varies with time.

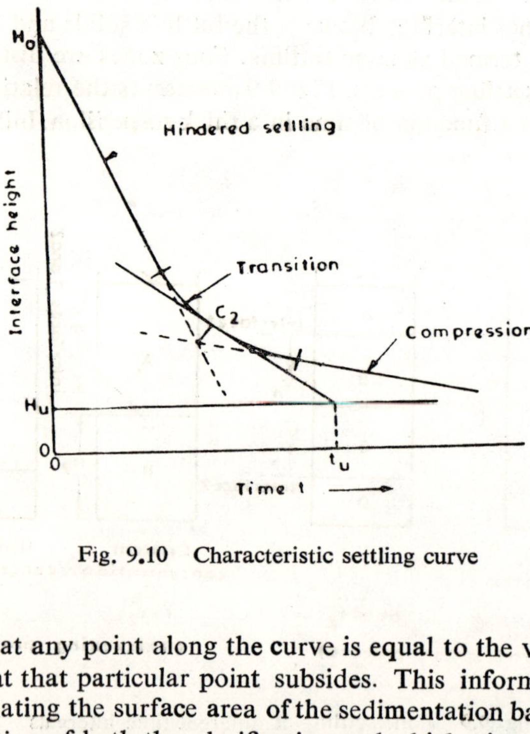

Fig. 9.10 Characteristic settling curve

The slope at any point along the curve is equal to the velocity at which the interface at that particular point subsides. This information could be used for estimating the surface area of the sedimentation basin. The surface area is a function of both the clarification and thickening capacities of the system. The design surface area is usually specified by comparing the area based on clarification with that computed on the basis of thickening. The larger of the two is accepted as the controlling requirement.

The area required for clarification is given by:

$$A_c = \frac{Q}{v_h} \tag{9.8}$$

where A_c = clarification area, Q = overflow rate, and v_h = subsidence velocity in the zone of hindered settling. v_h can be obtained from the hindered settling portion of the curve (Fig. 9.10) by computing its slope.

The area required for thickening can be obtained by the graphical procedure developed by Talmadge and Fitch[2]. According to this procedure, the area for thickening is given by:

$$A_t = \frac{Q_t t_u}{H_0} \tag{9.9}$$

where A_t = surface area required for thickening, Q_t = inflow rate of suspension entering tank, H_0 = initial height of interface in settling column, and t_u = time required to attain desired underflow concentration C_u. The critical concentration C_2 is obtained by bisecting the angle formed by the tangents to the hindered settling and compression portions of the settling curve. The value of t_u can be determined by first constructing a horizontal line at the depth H_u, which is the final height of the sludge zone. H_u can be calculated from a simple material balance of the total solids in the suspension:

$$H_0C_0 = H_uC_u \tag{9.10}$$

or

$$H_u = \frac{H_0C_0}{C_u} \tag{9.11}$$

Next, a tangent to the settling curve is constructed at point C_2; its intersection with the horizontal line from H_u gives the value of t_u as shown in Fig. 9.10. The following example illustrates the procedure.

Example 2

A settling test conducted in a 40 cm high cylindrical jar gave the curve shown in Fig. 9.11. The initial concentration of solids was 3000 mg/l. Determine the thickner area if an underflow concentration of 20,000 mg/l is desired. The sludge is to be settled in a continuous flow unit operated at a rate of 0.03 m³/s.

Solution

1. Compute H_u from Eq. (9.11)

$$H_u = \frac{H_0C_0}{C_u} = \frac{(40)\,(3000)}{20{,}000} = 6 \text{ cm}$$

2. Construct a horizontal line through H_u = 6 cm. Locate point C_2 by bisecting the angle formed by the tangents to the hindered and compression portions of the curve on Fig. 9.11. Draw a tangent to the

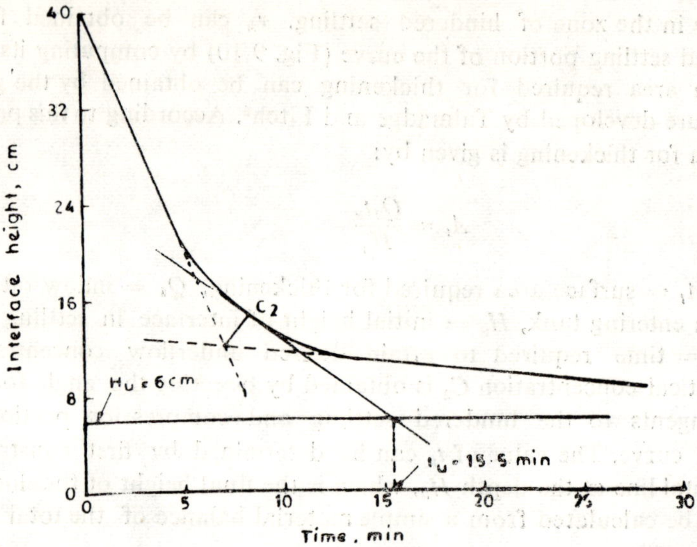

Fig. 9.11 Settling curve for example 2

settling curve at C_2. The intersection of the tangent at C_2 and the horizontal line through $H_u = 6$ cm gives t_u. Thus, $t_u = 15.5$ min.

3. The area for thickening is

$$A_t = \frac{Q_i t_u}{H_0} = \frac{0.03\ (15.5)\ (60)}{(40/100)} = 69.75\ \text{m}^2$$

4. From the hindered settling portion of the curve, compute the subsidence velocity v_h:

$$v_h = \frac{(40-20)/100}{5(60)} = 0.000666\ \text{m/s}$$

5. The overflow rate is proportional to the volume above the sludge zone

$$Q = \frac{0.03\ (40-6)}{40} = 0.0255\ \text{m}^3/\text{s}$$

Therefore,

$$A_c = \frac{Q}{v_h} = \frac{0.0255}{0.000666} = 38.29\ \text{m}^2$$

6. Take $A = A_t = 69.75$ m² as the controlling requirement.

Flotation

Flotation may be used in place of sedimentation, primarily for treating industrial wastewaters containing finely divided suspended solids and oily matter. Flotation technique is used in paper industry to recover fine fibres from the screened effluent and in the oil industry for the clarification of oil-bearing waste. It is also used for treating effluents from tannery, metal finishing, cold-rolling, and pharmaceutical industries. An increasingly important application is the thickening of the sludge obtained from activated sludge process.

Particles of density very close to that of water are very difficult to settle in normal sedimentation tanks and take a long time for separation. In such cases, the separation can be speeded up by aerating the effluent whereby air bubbles are attached to the suspended matter. This has the effect of increasing the buoyancy of the particles; as a result, the particles float to the surface where they can be readily removed. To aid in the flotation process, chemical coagulants such as aluminium and ferric salts or polymer coagulant-aids are often used. These chemicals increase the flocculent structure of the floated particles so that they can easily entrap the air bubbles.

Two methods of flotation are currently available: (1) dispersed-air flotation, and (2) dissolved-air flotation. In dispersed-air flotation, air is introduced directly into the liquid through a revolving impeller or through diffusers. The air bubbles generated in dispersed-air flotation systems are normally about 1 mm in diameter and they usually cause turbulence which breaks up fragile floc particles. Due to this, dispersed-air flotation is not a favoured technique in the treatment of municipal wastewaters, although it finds a limited application in treating industrial wastes containing oil, grease and fine powders.

In dissolved-air flotation, air is intimately brought into contact with the wastewater at a pressure of several atmospheres when air is dissolved. The

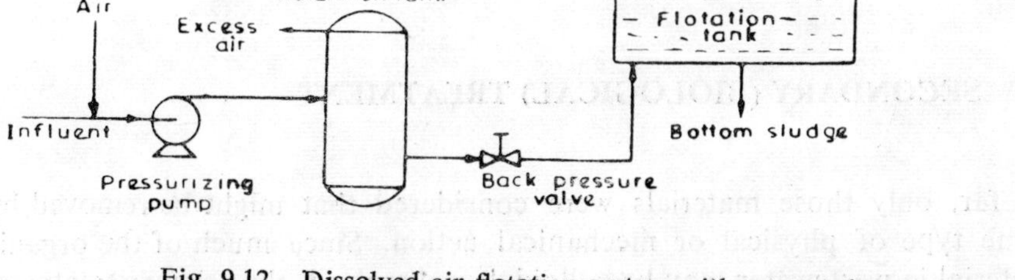

Fig. 9.12 Dissolved-air flotation system without recycle

pressure on the liquid is reduced to atmospheric level through a back-pressure valve, thereby releasing micron-sized bubbles. Suspended solids and oil are carried to the surface of the flotation tank by these minute air bubbles. A typical flotation system is illustrated in Fig. 9.12. Here, the entire flow is pressurized and held in the retention tank so that the air gets dissolved in the liquid. This is also known as the total pressurization or once-through system.

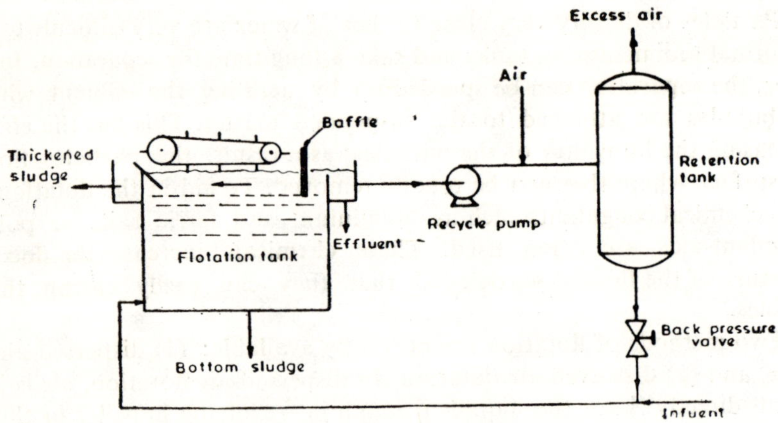

Fig. 9.13 Dissolved-air flotation system with recycle

The intense mixing of air and wastewater in the pressurization system often degrades flocculent suspensions or oil emulsions following chemical treatment. A portion of the clear effluent is recycled for pressurization (Fig. 9.13) to prevent such degradation. Compressed air is introduced into the discharge of the recycle pump and intimate contact is achieved in the retention tank. The recycled flow is then returned through a backpressure valve (where the pressurized air is released) and mixed with the influent for flotation. The amount of pressurization flow is based on the air to solids ratio required for treatment. Ratios ranging from 0.01 to 0.05 have been found to be effective and a design value of 0.02 is appropriate for many applications. The residence time in the flotation tank is about half an hour.

9.3 SECONDARY (BIOLOGICAL) TREATMENT

So far, only those materials were considered that might be removed by some type of physical or mechanical action. Since much of the organic material in wastewater may be colloidal or dissolved, the primary treatment processes are largely ineffective in removing it. This organic material still represents a high demand for oxygen which must be reduced further so that the effluent may be rendered suitable for discharge into the water bodies.

Biological or secondary treatment, as it is commonly referred to, is very similar in concept to the natural biodegradation of organic matter by aerobic bacteria. In biological treatment, oxygen supplied to the bacteria is consumed under controlled conditions so that most of the BOD is removed in the treatment plant rather than in the watercourse. Thus, the principal requirements of a biological waste treatment process are an adequate amount of bacteria that feed on the organic material present in wastewater, oxygen, and some means of achieving contact between the bacteria and the organics.

Two of the most commonly used systems for biological waste treatment are the activated sludge system and biological-film system. In the activated sludge system the wastewater is brought into contact with a diverse group of microorganisms in the form of a flocculent suspension in an aerated tank. Whereas in the biological-film system, also known as trickling filters, the wastewater is brought into contact with a mixed microbial population in the form of a film of slime attached to the surface of a solid support medium. In both cases the organic matter is metabolized to more stable inorganic forms. The most popular means of treating domestic sewage has been the biological-film system because of its ease of operation. However, the activated sludge process is gaining popularity and is preferable to the biological-film system because of its reliability and suitability for handling large volumes of wastewater, and because of the high degree of treatment achieved.

Activated Sludge Process

The process flow diagram for a typical activated sludge plant is given in Fig. 9.14. The essential features of the process are: an aeration stage, solids-liquid separation following aeration, and a sludge recycle system. Wastewater after primary treatment enters an aeration tank where the organic matter is brought into intimate contact with the sludge from the secondary clarifier. This sludge is heavily laden with microorganisms which are in an active state of growth. Air is introduced into the tank, either in the form of bubbles through diffusers or by surface aerators. The microorganisms utilize the oxygen in the air and convert the organic matter into stabilized, low-energy compounds such as NO_3, SO_4, CO_2, and synthesize new bacterial cells. The effluent from the aeration tank containing the flocculent microbial mass, known as the sludge, is separated in a settling tank, sometimes called a secondary settler or a clarifier. In the settling tank the separated sludge exists without contact with the organic matter and becomes activated. A portion of the activated sludge is recycled to the aeration tank as a seed; the rest is wasted. If all the activated sludge is recycled, then the bacterial mass would keep increasing to the stage where the system gets clogged with solids. It is, therefore, necessary to "waste" some of the microorganisms, and this wasted sludge is the one which is processed and disposed.

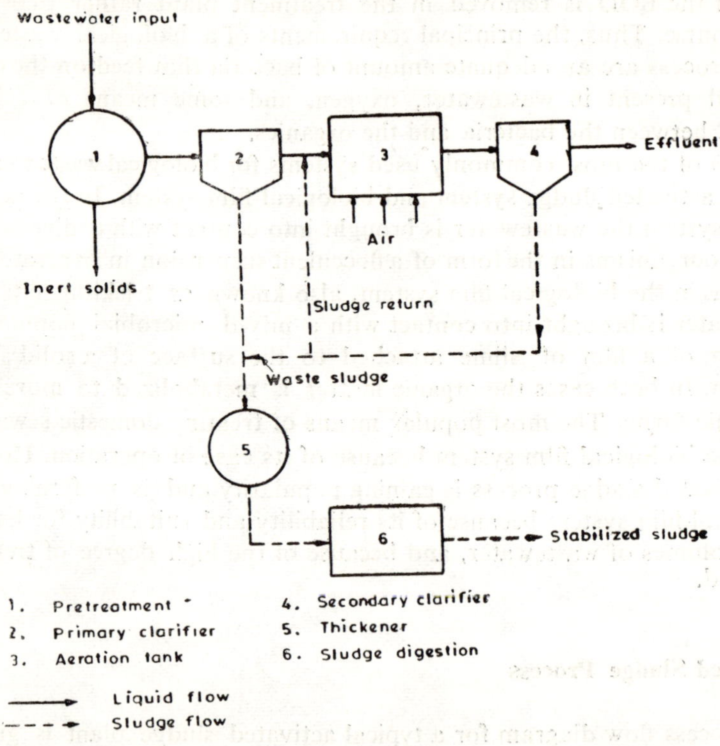

Fig. 9.14 Flow diagram of an activated sludge treatment plant

Basic Theory and Design

In the activated sludge system the major design parameter is the loading, or the amount of organic matter (food) added relative to the microorganisms (activated sludge) available. This ratio is also known as the food-to-microorganisms ratio (F/M). Unfortunately, measurement of either F or M accurately is difficult and, hence, the ratio is usually expressed as the amount of BOD utilized per unit mass of active biological solids. The combination of the liquid and microorganisms in the aeration tank is known as "mixed liquor", and the suspended solids are called "mixed liquor suspended solids" (MLSS).

In a biological treatment process, the relationship between the rate of growth of microorganisms and the rate of substrate utilization can be expressed by[3]:

$$\frac{dX}{dt} = Y\frac{dS}{dt} - K_d X \qquad (9.12)$$

where X = concentration of biological solids (microorganisms); dX/dt = net growth rate of microorganisms; Y = growth coefficient, mass of microorganisms produced per mass of substrate utilized; S = concentration of organic food substrate utilized by microorganisms; and K_d=microbial decay coefficient, expressed as time^{-1}.

Dividing both sides of Eq. (9.12) by X, we get

$$\frac{dX/dt}{X} = Y\frac{dS/dt}{X} - K_d \qquad (9.13)$$

In the above equation, $(dX/dt)/X$ is the specific growth rate, often represented by μ. The inverse of μ is referred to as the solids retention time or the mean cell residence time, θ_c

$$\theta_c = \frac{X}{(dX/dt)} \qquad (9.14)$$

The term $(dS/dt)/X$ is the substrate utilization rate per unit amount of biomass and is termed as the specific substrate utilization rate. It can be approximated by the following expression[4]:

$$\frac{(dS/dt)}{X} = \frac{kS}{K_s + S} \qquad (9.15)$$

where k = maximum specific substrate utilization rate, time^{-1};

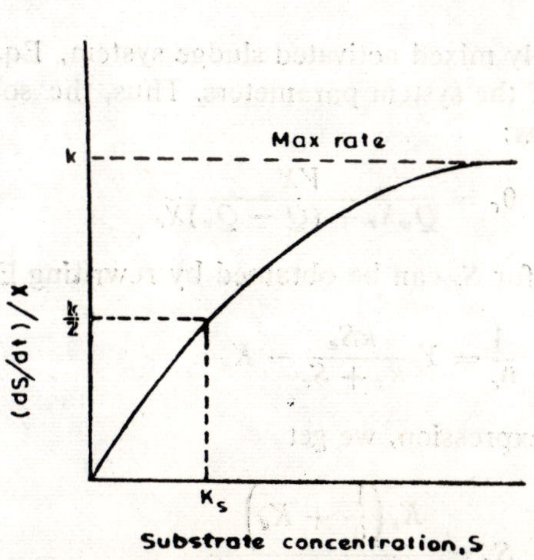

Fig. 9.15 Rate of substrate utilization per unit mass of microorganisms versus substrate concentration

and K_s = saturation constant which has a value equal to the substrate concentration when $(dS/dt)/X = \frac{1}{2}k$.

This is illustrated in Fig. 9.15.

Substituting Eq. (9.14) and (9.15) into Eq. (9.13), we get

$$\frac{1}{\theta_c} = Y \frac{kS}{K_s + S} - K_d \tag{9.16}$$

In the activated sludge system shown in Fig. 9.16, it is assumed that the contents in the aeration tank are completely mixed and that there are no microbial solids in the raw wastewater influent. It is further assumed that the influent substrate concentration, S_0, remains constant and that the system operates under steady state conditions. The solids are wasted from the sludge recycle line, although they may also be wasted from the aeration tank.

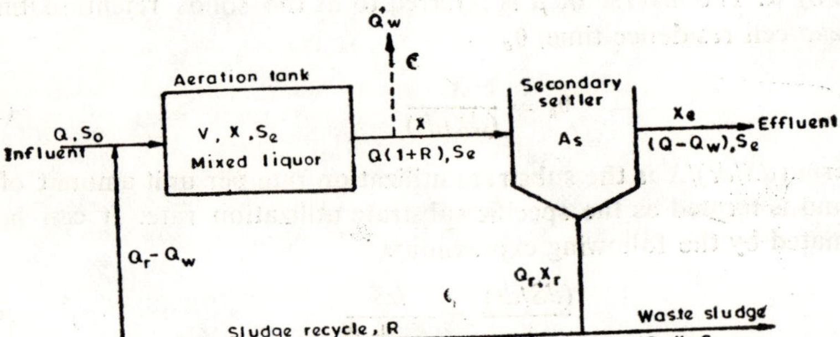

Fig. 9.16 Activated sludge plant material balance

For a completely mixed activated sludge system, Eq. (9.14) can be rewritten in terms of the system parameters. Thus, the solids retention time can be expressed as:

$$\theta_c = \frac{VX}{Q_w X_r + (Q - Q_w) X_e} \tag{9.17}$$

An expression for S_e can be obtained by rewriting Eq. (9.16) as:

$$\frac{1}{\theta_c} = Y \frac{kS_e}{K_s + S_e} - K_d \tag{9.18}$$

Rearranging this expression, we get

$$S_e = \frac{K_s\left(\frac{1}{\theta_c} + K_d\right)}{Yk - \left(\frac{1}{\theta_c} + K_d\right)} \tag{9.19}$$

Multiplying the numerator and the denominator by θ_c, we get

$$S_e = \frac{K_s(1+\theta_c K_d)}{\theta_c Yk - (1+\theta_c K_d)} \tag{9.20}$$

or

$$S_e = \frac{K_s(1+\theta_c K_d)}{\theta_c(Yk - K_d) - 1} \tag{9.21}$$

The relationship between X and S_e in the aeration tank can be obtained by first considering a substrate material balance around the aeration tank. This step gives the amount of substrate utilized per unit time and per unit volume of the aeration tank:

$$\frac{dS}{dt} = \frac{Q(S_0 - S_e)}{V} \tag{9.22}$$

Then, substituting Eq. (9.22) in Eq. (9.13) and utilizing Eq. (9.14), we get

$$\frac{1}{\theta_c} = Y\frac{Q}{V}\frac{(S_0 - S_e)}{X} - K_d \tag{9.23}$$

Rearranging the above equation for X, we get

$$X = \frac{YQ(S_0 - S_e)}{V\left(\frac{1}{\theta_c} + K_d\right)} \tag{9.24}$$

or

$$X = Y\left(\frac{Q}{V}\right)\frac{\theta_c(S_0 - S_e)}{(1+\theta_c K_d)} \tag{9.25}$$

The rate V/Q is the period of aeration or hydraulic retention time, represented by the symbol θ. Therefore,

$$\theta = \frac{V}{Q} \tag{9.26}$$

Substituting for θ in Eq. (9.25), we get

$$X = Y\frac{\theta_c(S_0 - S_e)}{\theta(1+\theta_c K_d)} \tag{9.27}$$

Table 9.1 Biokinetic parameters for domestic wastewater

Parameter	Unit	Range of values
Y	mg VSS/mg COD	0.35–0.45
K_d	day^{-1}	0.05–0.10
k	mg COD/mg VSS-day	6–8
K_s	mg/1 COD	25–100

Before these equations can be used in the design, values of the biokinetic parameters Y, k, K_s, and K_d must be available. Experimental procedures for evaluating the constants are given in standard references[3,5]. The values proposed by Lawrence[6] for the treatment of domestic wastewater by activated sludge systems at 20°C are given in Table 9.1.

Process Control

In recycle systems, since θ_c is directly related to the effluent quality (S_e), control of θ_c will directly lead to the control of the process efficiency. This can be achieved by regulating the amount of sludge wasted from the system as shown by Eq. (9.17). Since X_e is usually small, Eq. (9.17) can be rewritten as

$$\theta_c \simeq \frac{VX}{Q_w X_r} \tag{9.28}$$

Equation (9.28) shows that if θ_c is used as a control parameter, both the mixed liquor and return sludge concentration should be known. Equation (9.28) is applicable to the system where the wasting of the sludge is from the recycle line. The sludge can also be wasted from the aeration tank (as shown by a dashed line in Fig. 9.16. Then only Q_w' and V need to be known because Eq. (9.17) would become

$$\theta_c = \frac{VX}{Q_w' X + (Q - Q_w')X_e} \tag{9.29}$$

and assuming X_e to be very small, we can write Eq. (9.29) as

$$\theta_c \simeq \frac{V}{Q_w'} \tag{9.30}$$

Equation (9.30) is independent of the concentration of suspended solids and, hence, wasting of the sludge from the aeration tank offers an opportunity for better plant control.

The other process control parameter is the F/M ratio since this is also directly related to the effluent quality and, hence, will directly control the process efficiency. However, plant control on the basis of F/M ratio is tedious and impractical as it involves considerable laboratory work because of the necessity of knowing the BOD and the MLSS in the system. Therefore, the θ_c method is simpler. Both theoretical and practical considerations suggest that it is one of the best methods for controlling an activated sludge system.

Example 3

A completely mixed activated sludge process is to be used to treat a wastewater flow of 1000 m³/hr having a BOD of 250 mg/l. Design criteria are as follows: $Y = 0.4$, $k = 8$ day^{-1}, $K_d = 0.1$ day^{-1}, $K_s = 75$ mg/l, $\theta_c = 5$ days, and $X = 2000$ mg/l MLSS.

Calculate the substrate exit concentration and the volume of the aeration tank.

Solution

Equation (9.20) gives

$$S_e = \frac{75\,(1 + 5\times0.1)}{5(0.4)\,(8) - (1 + 5\times0.1)} = 7.75 \text{ mg/l}$$

From Eq. (9.27), the period of aeration can be calculated:

$$\theta = Y\,\frac{\theta_c\,(S_0 - S_e)}{X\,(1 + \theta_c K_d)} = \frac{0.4\,(5)\,(250 - 7.75)}{(2000)\,(1 + 5\times0.1)}$$

$$= 0.1615 \text{ days} = 3.87 \text{ hr.}$$

Using Eq. (9.26), we can obtain the volume of the aeration tank,

$$V = Q\theta = (1000)\,(3.87) = 3870 \text{ m}^3$$

Loading Criteria

The food-to-microorganisms (F/M) ratio expresses the substrate loading per unit time per unit mass of biomass in the aeration tank. For a completely mixed system, the F/M ratio can be expressed as

$$\frac{F}{M} = \frac{QS_0}{VX} \tag{9.31}$$

Since F/M ratio is involved in determining the biomass activity, it must be related to the cell residence time θ_c. Substituting Eq. (9.31) in Eq. (9.23), we get

$$\frac{1}{\theta_c} = Y\left(\frac{F}{M} - \frac{QS_e}{VX}\right) - K_d \tag{9.32}$$

When the effluent concentration is small compared with the influent concentration, then the term QS_e/VX can be neglected. Equation (9.32) can be approximated as

$$\frac{1}{\theta_c} = Y\left(\frac{F}{M}\right) - K_d \tag{9.33}$$

According to Eq. (9.31), the F/M ratio and θ_c are inversely related. At very high organic loadings (i.e., large F/M ratios and short θ_c values) the substrate removal efficiency is low (60 to 80%). Such a situation is commonly encountered in the so-called "high-rate" processes where the aeration periods are short and the sludge production is high. On the other hand, in the extended aeration, where the aeration period is quite high, the organic loading is very low (i.e. small F/M ratio and long θ_c) and consequently the process efficiency is moderately high (90 to 95%). Here the sludge production is minimal. The process is sometimes called complete oxidation, although there will always be some inert biological residue, and inorganic solids emanating from the system.

Table 9.2 gives typical design and operating values for the conventional and the above mentioned two variations of the activated sludge process.

Table 9.2 Activated sludge design criteria (refs. 3, 7)

Parameter	Conventional	High rate	Extended aeration
BOD removal per cent	85–95	60–80	90–95
Aeration period, θ, hr	4–8	2–4	18–30
Organic loading, F/M, kg BOD_5 per day per kg MLVSS	0.2–0.5	1.5–5.0	0.05–0.15
Mean cell residence time, θ_c, day	5–15	0.2–0.5	20–30
Sludge recycle ratio, R	0.25–0.50	0.25–0.50	Upto 100
Excess sludge production, kg sludge solids per kg BOD_5 destroyed	0.5	0.5	<0.5
Oxygen requirement:			
Minimum allowable dissolved oxygen, mg/l	0.5–1.0	0.5–1.0	0.5–1.0
Air flow rate (m^3/day)/kg BOD_5/day	90–100	30–45	125
Secondary clarifier design:			
Sludge detention time, hr	1.5–2.0	2.0–2.5	2.5–4.0
Unit area for sludge thickening, m^2/(kg per day)	0.013	0.013	0.013
Overflow rate, (m^3/day)/m^2	25–33	25	25

These values may serve only as guidelines for industrial waste treatment.

Air Supply and Process Modifications

In the initial stages of contact between microorganisms and the organic matter in the wastewater, the process requires a great deal of oxygen. In the conventional systems aeration and mixing are achieved at a fairly uniform rate over the length of the aeration tank. As a result, extremely low oxygen concentrations prevail at the inlet end of the tank (Fig. 9.17)

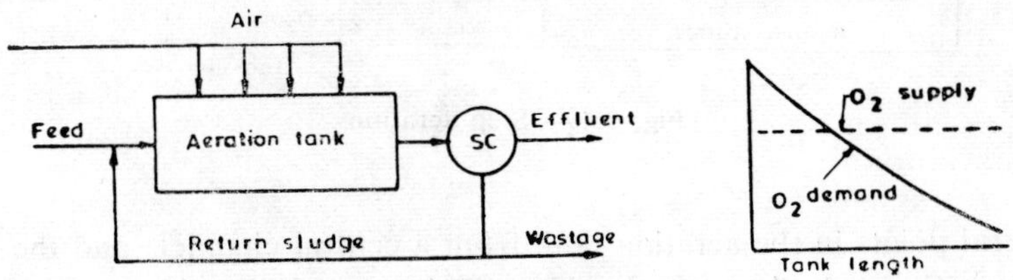

Fig. 9.17 Conventional system

and this condition can be detrimental to the microbial population. Near the outlet end, the oxygen supply may be more than that required for the process.

Several variations of the conventional system have been suggested to meet the rate of oxygen consumption at minimal costs. In one modification of the basic scheme, aeration is intensified at the inlet end and reduced progressively along the length of the aeration tank (Fig. 9.18). This system is known as the tapered aeration and is a major improvement in reducing the cost of pumping air, as air is used more effectively.

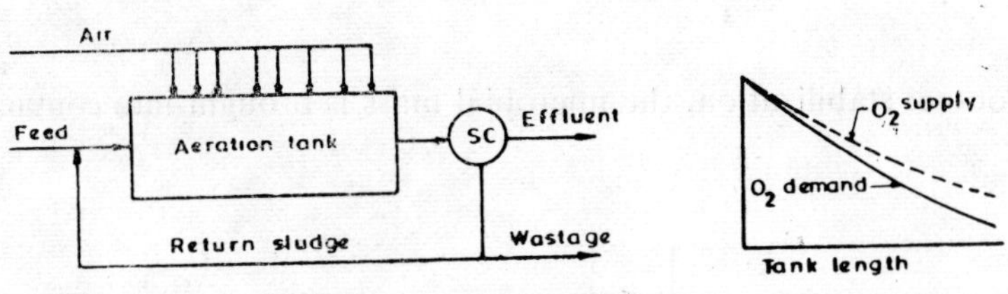

Fig. 9.18 Tapered aeration

Step aeration, shown in Fig. 9.19, is another method of equalizing oxygen supply and demand. In this system, instead of varying the rate of oxygen supply, fresh feed is introduced at several points along the aeration tank. In this manner, a more even distribution of oxygen demand is achieved throughout the aeration stage. Baffles are provided to divide the aeration tank into several channels. Each channel constitutes one step of the process, and the steps are linked together in series.

The third modification of the conventional process is the complete mix process. Here fresh feed and recycled sludge are combined and introduced

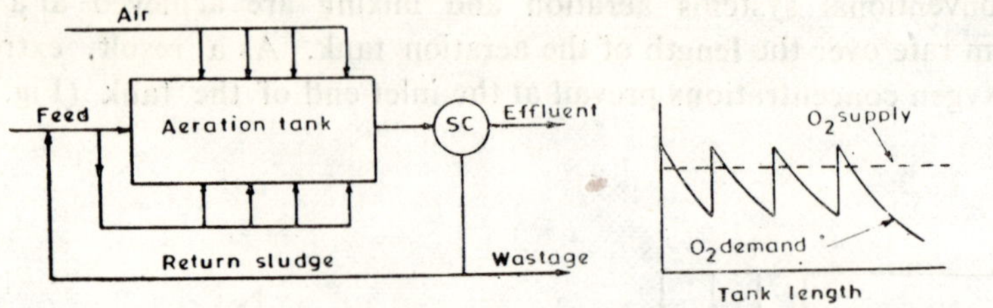

Fig. 9.19 Step aeration

at several points in the aeration tank from a central channel, and the effluent leaves the tank from both sides. This ensures a uniform supply and demand of oxygen along the tank is indicated in Fig. 9.20. It should be recalled that the model equations developed earlier assume complete mixing.

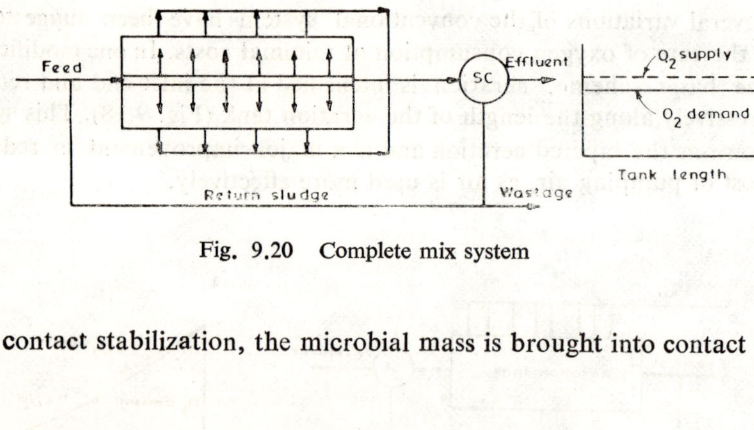

Fig. 9.20 Complete mix system

In contact stabilization, the microbial mass is brought into contact with

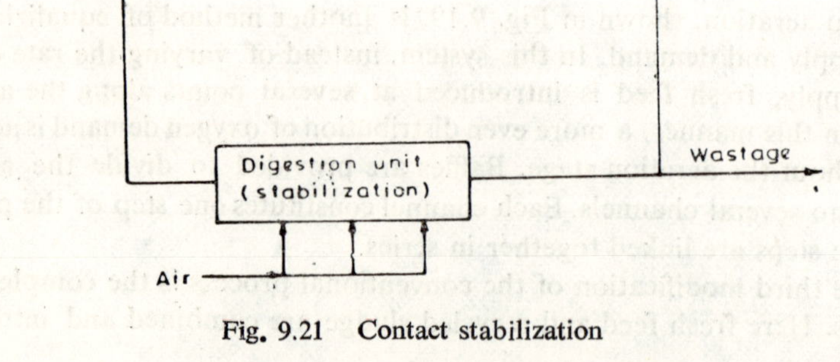

Fig. 9.21 Contact stabilization

the wastewater for short periods of time ($\frac{1}{2}-1$ hr) in the biosorption unit (Fig. 9.21). The resulting sludge is settled and finally transferred to an aerobic digestion unit for stabilization (2–3 hr retention). In the digestion unit the microorganisms are allowed to consume the organic matter that was removed from the wastewater in the biosorption unit. Since the stabilization is carried out on the return sludge stream which has a much higher solids concentration than that of the mixed liquor in the biosorption unit, the volume in the aeration tank is reduced. The stabilized sludge is then recycled to the aeration tank.

Air may be replaced by pure oxygen in activated sludge systems. In this process (Fig. 9.22) pure oxygen is supplied into well mixed and converted chambers and is recirculated. This way about 90% utilization of

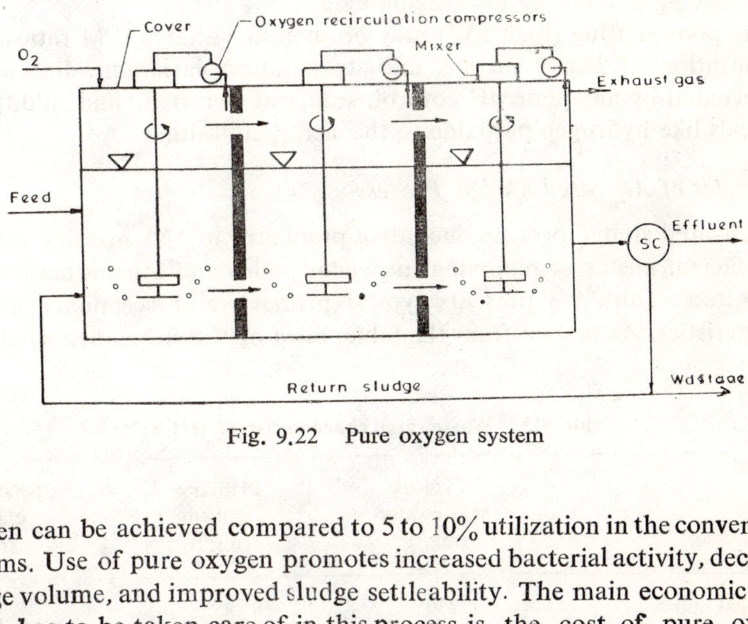

Fig. 9.22 Pure oxygen system

oxygen can be achieved compared to 5 to 10% utilization in the conventional systems. Use of pure oxygen promotes increased bacterial activity, decreased sludge volume, and improved sludge settleability. The main economic factor which has to be taken care of in this process is the cost of pure oxygen. A good estimate of the oxygen requirement for an activated sludge system can be obtained from Eq. (9.34):

$$\begin{bmatrix} \text{Mass } O_2 \\ \text{required} \\ \text{per day} \end{bmatrix} = \begin{bmatrix} \text{Mass ultimate} \\ \text{BOD removed} \\ \text{per day} \end{bmatrix} - 1.42 \begin{bmatrix} \text{Total active mass of} \\ \text{organisms wasted} \\ \text{per day} \end{bmatrix} \tag{9.34}$$

Sludge Volume Index (SVI)

The success of the activated sludge system depends on many factors, among which the separation of the solid-liquid phase in the secondary settler is one of the more important ones. The settling characteristics of the activated sludge are dependent upon the physiological condition of the microorganisms present. This condition is often characterized by a biomass composed almost totally of filamentous organisms which do not settle easily. The sludge is then called a "bulking" sludge, having poor settling characteristics.

A conventional measure of the settling ability of the sludge is the sludge volume index (SVI), defined as the volume in millilitres occupied by one gram of sludge after it has settled in one litre cylinder for 30 minutes. SVI has units of ml/g and it can be calculated as follows:

$$\text{SVI} = \frac{\text{Sludge volume after settling for 30 min., ml/l}}{\text{MLSS concentration, mg/l}} \times 1000 \qquad (9.35)$$

SVI varies from 40 to 100 for a good sludge, but may exceed 200 for a poor sludge having a tendency towards bulking.

The poor settling (high SVI) may be due to variable F/M ratios, high concentration of heavy metals, and temperature variations. Bulking can be prevented by adequate *p*H control, sufficient aeration, and addition of chemicals like hydrogen peroxide to the aeration basin.

Efficiencies of Activated Sludge Process

The activated sludge process, including pretreatment and primary settling, is an efficient means of removing suspended solids and organic matter from wastewaters. Table 9.3 presents typical primary and secondary effluent characteristics. As is seen from the table, most of the suspended solids and

Table 9.3 Wastewater characteristics (ref 8)*

	Typical wastewater (mg/l)	Primary effluent (mg/l)	Secondary effluent (mg/l)
Suspended solids	230	85	20
BOD	200	135	20
Ammonia nitrogen	18	16	15
Total nitrogen	30	25	20
Phosphorus	13	11	10
Ultimate oxygen demand	383	276	99

*Based on average treatment efficiencies. These values may vary from plant to plant, and within any one plant with time.

the BOD materials are removed after secondary treatment. However, the activated sludge process does a poor job in removing both nitrogen and phosphorus from wastewater. The total nitrogen content is reduced by approximately one-third. The influent ammonia is oxidized by nitrifying bacteria and this reduction is partially balanced by the production of ammonia by the sludge mass. The net result is that there is a slight reduction in the ammonia concentration.

Primary sedimentation in conventional treatment settles only a small percentage of the phosphorus present in wastewater, since its major part is dissolved. In the activated sludge process the microbial floc takes the soluble phosphate as a nutrient but the removal efficiency is small (less than 30%), so that most of the phosphorus remains uneffected by the secondary treatment. The ultimate oxygen demand (UOD) is reduced by about 75% in the conventional secondary treatment. Most of this reduction is due to the oxidation of the organic carbonaceous material and the remainder is due to removal of ammonia. The extended aeration process may be expected to raise the removal efficiency of UOD to about 90%.

Trickling Filters

As mentioned earlier, the second commonly used biological waste treatment process is the trickling filter method. It has good adaptability to handle peak shock loads and the ability to function satisfactorily after a short period of time. However, like all biological units, trickling filters are affected by temperature; therefore, cold weather slows down biological activity in the filter. Milk processing, paper mill and pharmaceutical wastes are among those treated by trickling-filters. Conventional trickling filters normally consist of a rock bed, 1 to 3 metres in depth, with enough openings between rocks to allow air to circulate easily. The influent is sprinkled over the bed packing (Fig. 9.23) which is coated with a biological slime. As the liquid trickles over the packing, oxygen and the dissolved organic

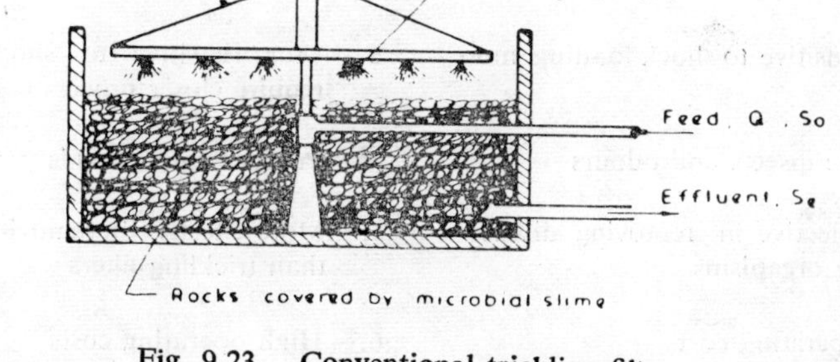

Fig. 9.23. Conventional trickling filter

matter diffuse into the film to be metabolized by the microorganisms in the slime layer. End products such as CO_2, NO_3, etc., diffuse back, out of the film and appear in the filter effluent.

As the microorganisms utilize the organic matter, the thickness of the slime film increases to a point where it can no longer be supported on the solid media and gets detached from the surface. This process is known as sloughing. A settling tank following the trickling filter removes the detached bacterial film and some suspended matter.

Conventional rock filters have been designed based on hydraulic and BOD loading, and are classified as "standard-rate" or "high-rate". Standard rate filters have low hydraulic and organic loading rates [0.3 – 1.0 m^3 (liquid)/m^3 (bed) – day and 0.07 – 0.10 kg BOD/m^3 (bed) – day respectively] while these are significantly higher for high-rate filters (hydraulic $>$ 3 and organic $>$ about 0.6). Standard rate filters are suitable for treating weak wastewaters whereas high rate systems are used for partial or 'roughing' treatment of wastes before the waste is sent for further treatment or discharged into a sewer.

More recently, plastic filters have been used for trickling filters. The plastic packing is light and can be stacked many times higher than the conventional rock bed. It also has a much higher degree of treatment capability because of the increase in the available surface area and is mostly used in high-rate processes.

Table 9.4 gives the major differences between conventional trickling filters and activated sludge systems, and also lists their advantages and disadvantages.

Table 9.4 Comparison of trickling filters with activated sludge systems

Trickling filters	Activated sludge systems
1. Bacterial growth is fixed on the media	1. Bacterial growth is suspended as a dispersed floc
2. All solids from the settler are wasted	2. Solids from the settler are partially recycled
3. Less sensitive to shock loading, more stable	3. More sensitive to shock loadings, require closer process control
4. Produce insects and odours	4. Produce spray clouds
5. Less effective in removing disease-causing organisms	5. More effective in removing pathogens than trickling filters
6. Low operating costs	6. High operating costs

Sludge Treatment and Disposal

Handling and disposal of sludge from biological wastewater treatment plants is an important problem and represents about half the cost of most sewage treatment plants. The concentration of solids in the primary sewage sludge is about 5%; the activated sludge contains less than 1% solids; and the sludge from trickling filters has about 2% solids. This means that the sludge is composed almost entirely of water and volume reduction is the key to economic disposal. In addition to reducing its high water content, the sludge must be stabilized so that its biological activity and tendency towards putrefaction are reduced drastically.

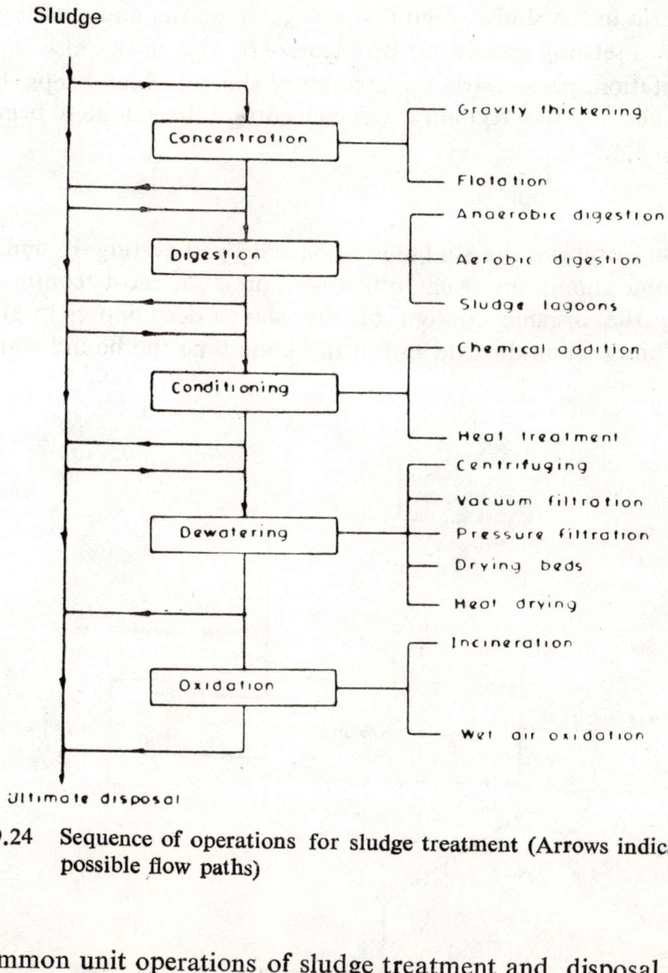

Fig. 9.24 Sequence of operations for sludge treatment (Arrows indicate possible flow paths)

The common unit operations of sludge treatment and disposal involve concentration or thickening, digestion, conditioning, dewatering, oxidation and safe disposal. Fig. 9.24 represents the general flow sheet of sludge treatment techniques in approximately the order in which they could be applied in a treatment scheme.

Concentration

The purpose of concentration or thickening is to remove water from the sludge and reduce its volume as much as possible so that the sludge can be handled more efficiently. The common methods for thickening are gravity settling and flotation. In gravity thickeners the sludge is subjected to gentle agitation by means of a slow stirrer which enhances settling. In this manner the combined sludge from primary and secondary settlers can be thickened so as to contain 5-9% solids. Often the thickening of activated sludge is complicated by anaerobic action, particularly under warm conditions when the bacteria in the sludge decompose organic matter and release gases. This also creates settling and odour problems. The sludge can also be thickened by air flotation, particularly the secondary sludge, which keeps the system aerobic. The flotation technique can concentrate the sludge to bring its solids content to 4%.

Digestion

After concentration, the sludge is stabilized by digesting it under aerobic or anaerobic conditions. Anaerobic digestion is the most common method in which the organic content of thc sludge decomposes to give mainly methane and carbon dioxide and at the same time the bound water is relea-

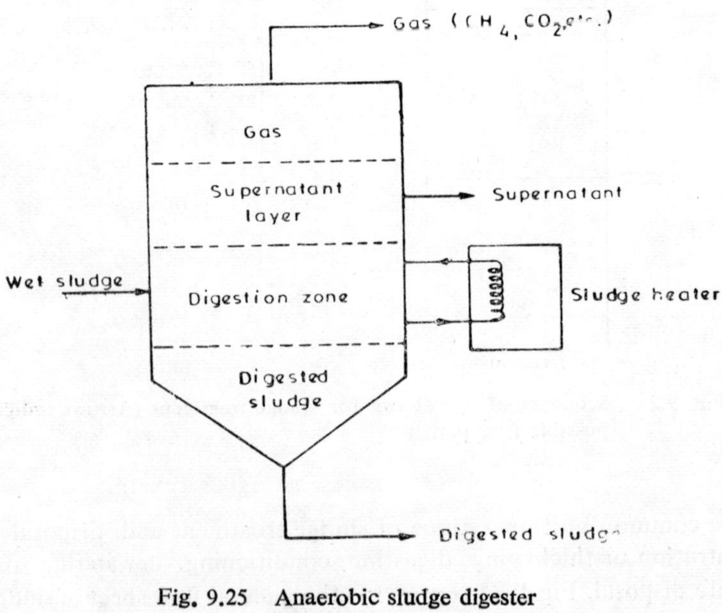

Fig. 9.25 Anaerobic sludge digester

sed from the sludge. Properly digested sludge is black with a faint smell of tar, and is stable. In a typical sludge digester, shown in Fig. 9.25, raw sludge is fed into the active digestion zone and gas lifts the sludge particles and other materials which form a supernatant layer on the top of the digestion zone. The gas is collected at the top and the digested sludge is withdrawn from the bottom. The normal detention period in standard digesters varies from 30 to 70 days depending upon the temperature conditions.

Aerobic digestion is also used, and it may be considered similar to extended aeration. The sludge is aerated in a tank for about 20 days at ambient temperatures. During the process the bacterial cells are destroyed and a substantial portion of the sludge is oxidized resulting in the reduction of the solid content by about 30%. Sometimes shallow lagoons are employed as digesters. Large land areas are required for sludge lagoons and odour problems may occur frequently.

Conditioning

The sludge after stabilization may be conditioned to improve its dewatering characteristics. This is done by adding chemicals like iron salts, alum, lime and polyelectrolytes. These chemicals bind the sludge particles together and encourage the release of absorbed water. Physical conditioning methods such as heat treatment are becoming popular. The sludge is heated under pressure and after a period of time the gel structure of the sludge breaks down so that the water is released. Heat treatment has the advantage of sterilizing the sludge; at the same time the sludge is partially oxidized and completely stabilized.

Dewatering

The thickened sludge is dewatered for efficient handling and disposal. Dewatering is accomplished by mechanical methods, the most common being centrifugation and filtration, which includes pressure filtration and vacuum filtration. In centrifugation, conditioned sludge is added to a rotating bowl that separates the sludge into a cake and a dilute stream. The solid cake is transported within the bowl and is removed by a screw conveyor at one end of the bowl; the liquid is removed at the opposite end (Fig. 9.26). Centrifugation is a compact method which requires careful control of process variables.

Filtration, using plate-and-frame pressure filters or rotary drum vacuum filters, is widely used for dewatering digested sludge. In pressure filtration, the sludge is pumped slowly with increasing pressure into filter plates supporting a cloth, which retains the solids. It is a batch process, and after the dewatering period the plates are separated and the sludge cake removed. Pressure filtration can produce a cake with a solids content of 25–50%. In contrast to pressure filtration, vacuum filtration is a continuous process

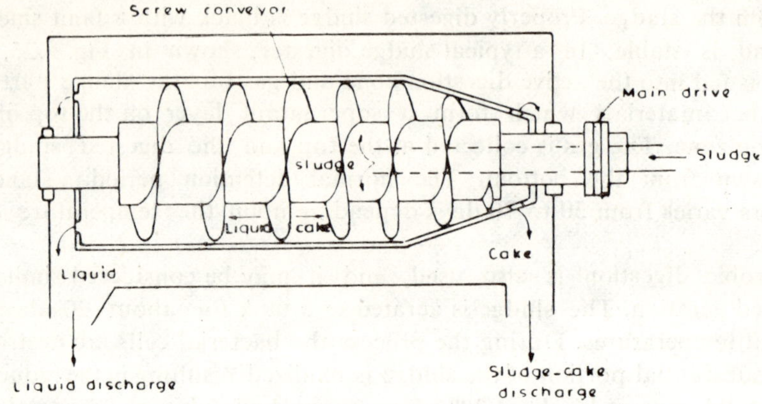

Fig. 9.26 Solid-bowl centrifuge

where a rotating drum, which is covered with filter cloth, is partially submerged in the sludge. On applying a vacuum of 80–90 kPa inside the drum, the liquid is sucked into the drum. The cake is deposited on the outside of the drum and is removed by a scraping mechanism. Vacuum filtration yields a dewatered sludge with a solids content of about 25%.

Drying beds are also commonly used for dewatering. The bed consists of a filtering medium on which the sludge is applied to a depth of upto 250 mm, depending on the solids content. Dewatering takes place by a combination of drainage and evaporation. Removal of dried sludge is carried out mechanically. Another technique, heat drying, may be utilized in applications where the sludge is to be incinerated or when a saleable commodity can be produced. A major problem associated with this process is the control of gases and ash particles which may be generated in drying.

Oxidation

Before the final disposal, some sludges may be oxidized to reduce the organic content, with the consequent destruction of bacteria and a significant reduction in their volumes. Incineration and wet oxidation are the two common methods employed for sludge oxidation. Incineration is usually performed in a multiple hearth furnace, although fluidized beds or flash dryers may also be used. In the multiple hearth furnace the sludge passes downward through a series of hearths. Vaporization of moisture occurs in the upper hearths, followed by incineration in the lower ones. The combined efficiency of evaporation and incineration in multiple hearth furnaces is about 55%. The fluidized bed system consists of a bed of sand fluidized by air. When the sludge is introduced, the sludge particles are dried almost instantly as they are dispersed, and are oxidized.

Wet oxidation is a process in which the sludge is ground, mixed with air in stoichiometric proportions, and then subjected to high temperature and pressure in a reactor (Fig. 9.27). The sludge is oxidized under these conditions to produce a mixture of gas, liquid and ash which exits the

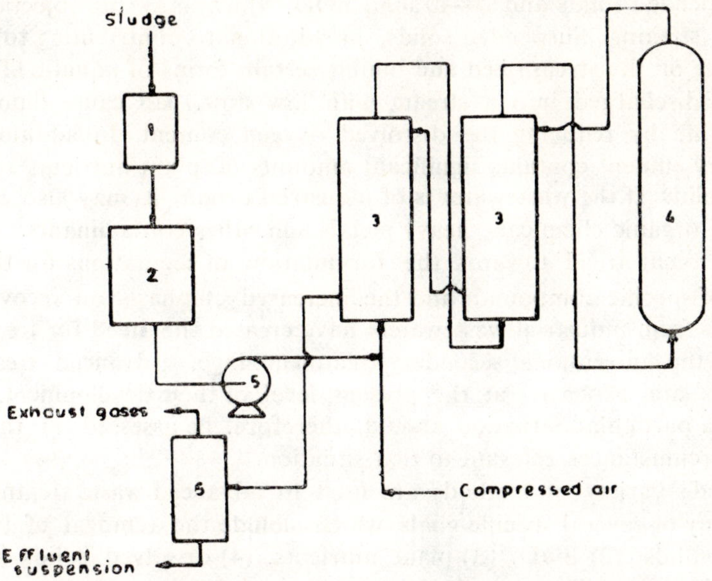

1. Grinder
2. Sludge storage tank
3. Heat exchangers
4. Reactor
5. Sludge pump
6. Separator

Fig. 9.27 Wet oxidation process flow diagram

reactor. Under the conditions of high temperature (260°C or more) and high pressure (8000–12000 kPa) there is no vaporization of the sludge liquid. Initially, external heat is applied to start the oxidation of the organic matter in the sludge. The rate of oxidation increases until it reaches an equilibrium value. The products from the top of the reactor are passed through heat exchangers to recover heat which in turn is utilized by the incoming feed.

Ultimate Sludge Disposal

Several methods are employed for the ultimate disposal of sludge. Wet digested sludge may be sprayed onto cropland where it functions as a fertilizer or fertilizer base. It may be lagooned; however, when the lagoons fill they must be abandoned. Dried sludge may be used as a landfill or a soil conditioner. Wet or partially dewatered sludge or ash from incineration may be transported from the shore to dumping grounds at sea.

9.4 ADVANCED WASTEWATER TREATMENT

The effluent from a typical secondary treatment plant still contains 20–40 mg/l suspended solids and 20–40 mg/l BOD, which may be objectionable in some streams. Suspended solids, in addition to contributing to BOD, may settle on the stream bed and inhibit certain forms of aquatic life. The BOD, if discharged into a stream with low flow, can cause damage to aquatic life by reducing the dissolved oxygen content. In addition, the secondary effluent contains significant amounts of plant nutrients and dissolved solids. If the wastewater is of industrial origin, it may also contain traces of organic chemicals, heavy metals and other contaminants.

The recent trend towards the formulation of regulations for the discharge of specific compounds and the increased emphasis on recovery of valuables from industrial wastewaters have created the need for treatment beyond the conventional secondary treatment stage. Advanced treatment processes are expensive at the present level of their development. Their need in a particular situation should, therefore, be assessed in the light of the circumstances relevant to that situation.

A wide variety of methods are used in advanced waste treatment to satisfy any of several specific goals, which include the removal of (1) suspended solids, (2) BOD, (3) plant nutrients, (4) dissolved solids and (5) toxic substances. These methods may be introduced at any stage of the total treatment process as in the case of industrial wastewaters or may be used for complete removal of pollutants after the secondary treatment. In this section only a few representative methods are discussed to illustrate the problems and the potential solutions associated with advanced waste treatment.

Removal of Suspended Solids

Removal of suspended solids in the advanced treatment implies the removal of those materials that have been carried over from a secondary settler. Of the several methods proposed, the two methods most commonly utilized in this application are microstraining, and chemical coagulation followed by settling and mixed-media filtration.

Microstraining

Microstraining utilizes a rotating drum-type filter to screen suspended solids. The filtering media consist of a finely woven stainless steel fabric with a mesh size of 23 to 35 microns. The fabric is mounted on the periphery of the drum and water is allowed to pass from inside to the outside as shown in Fig. 9.28. Back-washing is accomplished by high pressure water jets placed at the highest point of the drum. The solids which are retained on the

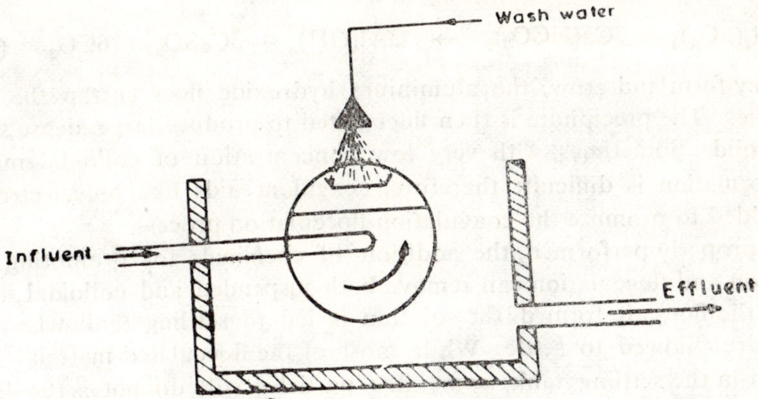

Fig. 9.28 Microstrainer

fabric are washed into a trough, which recycles the solids to the sedimentation tank.

Microstrainers can handle high flow rates of influent but cannot handle fluctuations in solid content and shock loads. Furthermore, the removal of suspended solids depends on their filterability characteristics; colloidal suspensions usually pass through the unit unhindered.

Coagulation and Filtration

Coagulation is the method in which certain chemicals are rapidly dispersed in wastewater to change the characteristics of the suspended particles so that they coalesce and form flocs which sink rapidly. Coagulation is employed to improve or make possible the removal of negatively charged colloidal suspensions which do not normally settle out and cannot be removed by conventional physical treatment. Coagulation is done by the addition of positive ions, e.g., Al^{3+}, which reduce the electrokinetic repulsion between the particles.

In industrial wastewater treatment, coagulation is frequently used for oily emulsions, and finely divided and nonsettleable solids such as pigments, paper fibre, meat and tannery effluents, and mineral beneficiation slimes. The most widely used coagulants for wastewater treatment are aluminium and iron salts such as aluminium sulphate (alum), ferric sulphate, and ferric chloride. At high (alkaline) *p*H prevalent in the water these salts produce insoluble aluminium hydroxide or ferric hydroxide flocs. Typical reactions are:

$$Al_2(SO_4)_3 + 6H_2O \rightarrow 2Al(OH)_3 + 3H_2SO_4 \quad (9.36)$$

$$3H_2SO_4 + 3Ca(HCO_3)_2 \rightarrow 3CaSO_4 + 6H_2CO_3 \quad (9.37)$$

$$6H_2CO_3 \rightarrow 6CO_2 + 6H_2O \quad (9.38)$$

The overall reaction may be represented as:

$$Al_2(SO_4)_3 + 3Ca(HCO_3)_2 \rightarrow 2Al(OH)_3 + 3CaSO_4 + 6CO_2 \quad (9.38)$$

As they form and grow, the aluminium hydroxide flocs entrap the solid particles. The precipitate is then flocculated to produce large, dense settleable solids. Sometimes, with very low concentration of colloidal matter, floc formation is difficult; therefore, coagulant aids like polyelectrolytes are added to promote the coagulation-flocculation process.

If properly performed, the addition of chemicals for promoting coagulation and flocculation can remove both suspended and colloidal solids. After the flocs are formed, the solution is led to settling tank where the flocs are allowed to settle. While most of the flocculated material is removed in the settling tank, some floc particles still do not settle in the settling basin. These flocs are removed by the filtration process, which is usually carried out using beds of porous media such as sand or coal. The current trend is to use a mixed-media filter, graded coarse to fine, in the direction of water flow.

In comparison to a single layer filter, the mixed-media filter has the advantage of effectively utilizing the filter bed depth. The mixed-media filter consists of fine garnet in the bottom layer, silica sand in the middle layer and coarser coal in the top layer. This type of filter bed, having decreasing particle size from top to bottom, allows both surface straining and "in-depth" filtration, without causing premature surface plugging or solids breakthrough.

Removal of Dissolved Solids

The dissolved solids are of both organic and inorganic types. The most common method used for the removal of soluble organics from wastewater is adsorption on activated carbon. Solvent extraction is also used to recover certain organic chemicals like phenol and amines from industrial wastewaters. A number of methods have been investigated for the removal of inorganic constituents from wastewater. Three methods which are finding wide application in advanced waste treatment are ion exchange, electrodialysis and reverse osmosis.

Adsorption on Activated Carbon

Adsorption is a surface phenomenon wherein the dissolved substances in wastewater are attracted to and adhere to the surface of the adsorbent. Activated carbon, with its extensive internal microporous structure, is the most commonly used adsorbent either in powdered or granular form. Adsorption on activated carbon is probably the most economical and technically attractive method available for removing soluble organics such as phenols, chlorinated hydrocarbons, surfactants, and colour and odour producing substances from wastewater. The basic concepts related to the adsorption process and the design of adsorption columns are treated in Chapter 5.

Adsorption systems based on granular activated carbon can either be of fixed-bed type (flow down) or expanded-bed type (flow up). The adsorbers can be arranged in several different process configurations, as shown in Fig. 9.29. In series operation, the wastewater enters the first column and leaves the second column whereas in parallel operation, each column receives the same feed and the effluents from each column are combined to form a composite product. When the carbon bed is saturated with

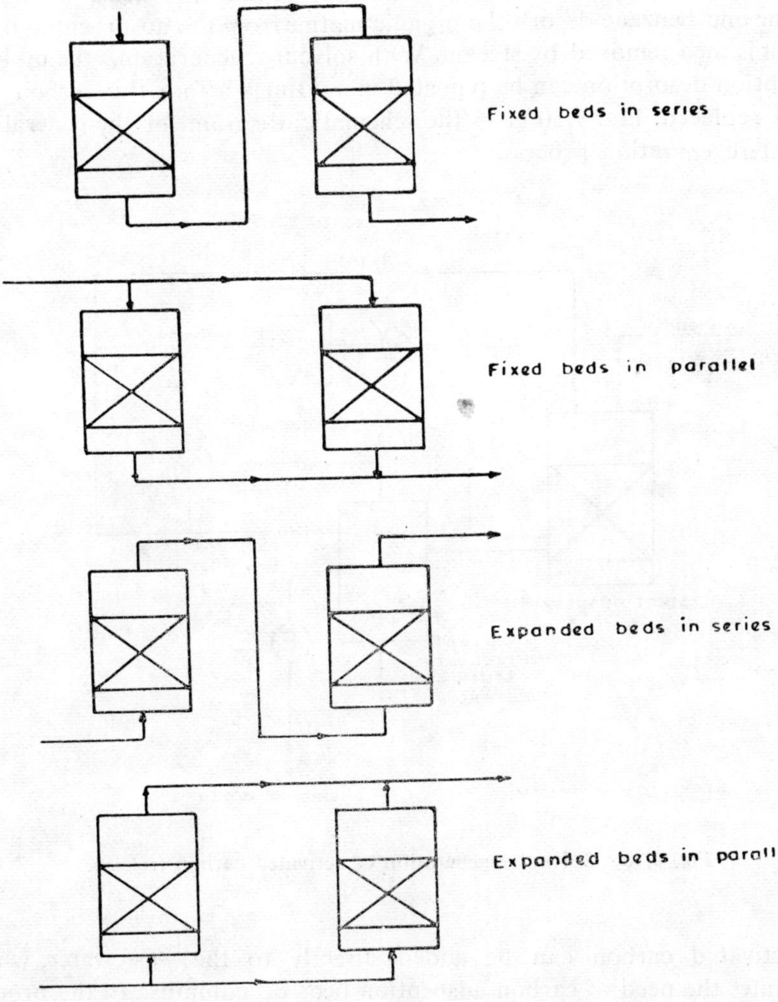

Fig. 9.29 Adsorber configurations for granular carbon waste treatment

chemicals, it is regenerated. In the series operation, the first column is removed from service when the bed is exhausted and a fresh column is placed at the end of the series. The second column then becomes the lead column. In parallel operation the beds become saturated simultaneously and are replaced by a fresh batch.

After the bed becomes saturated, it is necessary to regenerate it for reuse. In thermal regeneration the activated carbon is dewatered and then passed through a multiple-hearth furnace where the adsorbed organics are oxidized at temperatures of 815 to 925°C. The regenerated carbon is then water-quenched and stored for return to the treatment system. Instead of thermal regeneration, where carbon losses ranging from 5 to 10% per cycle occur because of oxidation and physical attrition, solvent regeneration of adsorbents has received more attention recently[9]. Solvents such as methanol, acetone and benzene desorb the organic matter from the adsorbent and the solvent is then removed by stream. With solvent regeneration, the cycle of adsorption-desorption can be repeated many times before the carbon has to be replaced. Fig. 9.30 gives the schematic diagram of the generalized solvent regeneration process.

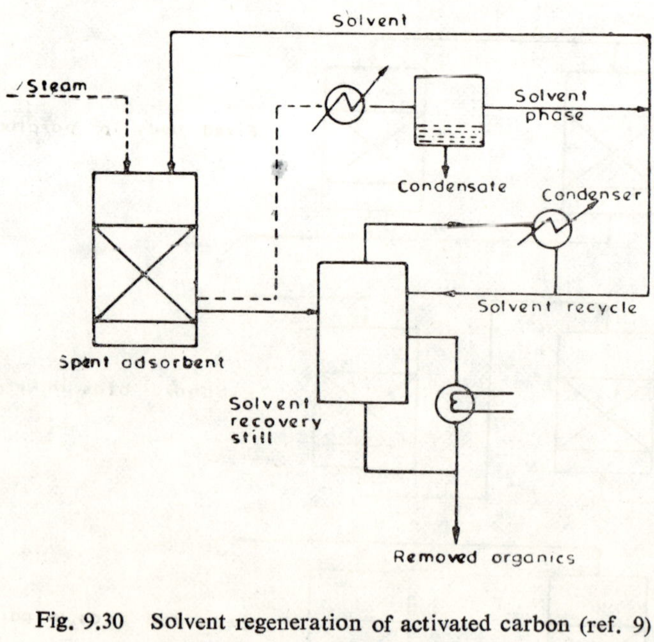

Fig. 9.30 Solvent regeneration of activated carbon (ref. 9)

Activated carbon can be added directly to the wastewater, which eliminates the need of carbon adsorption beds or columns. In the process developed by Du Pont[10], powdered activated carbon is added directly to the aeration tanks of the activated sludge system. In this way, the simpler organic molecules which are difficult to adsorb on activated carbon are easily removed by the biological system whereas the high molecular weight compounds are removed by the activated carbon. The spent carbon can be regenerated and the activated sludge incinerated simultaneously in a multiple-hearth furnace. A flow diagram of DuPont's process is shown in Fig. 9.31.

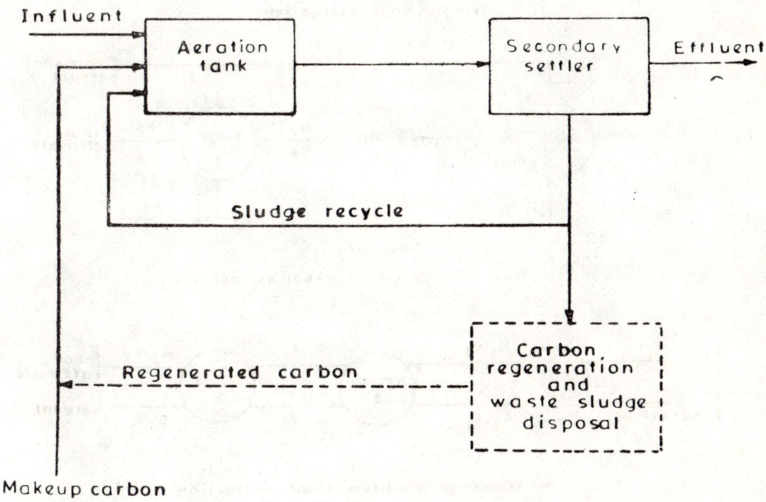

Fig. 9.31 DuPont's powdered activated carbon process (ref. 10)

Solvent Extraction

Solvent extraction has long been practised to recover phenolic materials from wastewaters of refineries and coke plants. For solvent extraction, solvents such as benzene are used from which phenol is recovered by distillation after the addition of alkali. In liquid extraction, the wastewater is intimately brought in contact with a solvent having low aqueous solubility and high affinity for the solute. After contact of the wastewater with the solvent, two immiscible layers are formed and an equilibrium is reached between the two layers. The solvent-rich product is called the extract and the residual liquid from which the solute has been removed is called the raffinate.

For example, in the extraction of phenol from wastewater using benzene as the solvent, the extract is predominantly a mixture of phenol and benzene, and the raffinate is predominantly water, containing a little bit of phenol and solvent. The contact between the feed and the solvent could be accomplished in several configurations: single-stage contact, multistage cross current extraction, and multistage counter-current extraction. (Fig. 9.32)

In Fig. 9.32, F is the feed rate in kg/hr, x_F is the weight fraction of solute in the feed, R_1, R_2, and R_3 are the flow rates of the raffinate in kg/hr, x_1, x_2 and x_3 are the weight fractions of the solute in the raffinate, E_1, E

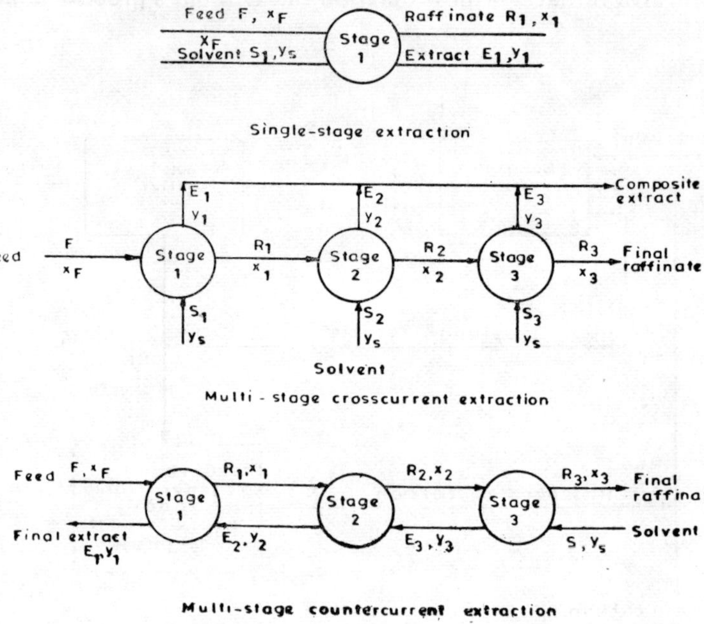

Fig. 9.32 Extractor configurations

and E_3 are the flow rates of the extract in kg/hr, y_1, y_2 and y_3 are the weight fractions of the solute in the extract, S is the solvent flow rate in kg/hr, and y_s is the weight fraction of the solute in the solvent. Since the extract and raffinate solutions are in equilibrium with each other, the concentration of the solute in the two streams can be represented by the distribution equation,

$$y = Kx \tag{9.39}$$

where K is the distribution coefficient. Values of K are given in Table 9.5 for selected aqueous systems.

Now for the single-stage extractor, a solute material balance gives

$$Fx_F + S_1 y_s = R_1 x_1 + E_1 y_1 \tag{9.40}$$

Since the raffinate and the extract are in equilibrium,

$$y_1 = Kx_1 \tag{9.41}$$

Table 9.5 Values of distribution coefficient for selected aqueous systems (ref. 7)†

Components*		Distribution coefficient (K) at 25°C
Solute B	Solvent A	
Phenol	Carbon tetrachloride	0.36
	Chloroform	2.8
	Benzene	2.3
	Xylene	1.4
Acetone	Carbon tetrachloride	0.44
	Chloroform	5.5
Acetic acid	Carbon tetrachloride	0.059
	Chloroform	0.028**
	Ethyl ether	0.45
	Benzene	0.023
Benzoic acid	Chloroform	2.4
	Benzene	1.8
Diethylamine	Chloroform	2.2
	Benzene	0.63
	Toluene	0.63
	Xylene	0.20

† Reprinted with permission of John Wiley & Sons, N.Y., 1977.
* Solute B distributed between nearly immiscible A and C; component C is water.
**Value at 20°C.

Substituting Eq. (9.41) in Eq. (9.40) and solving for x_1, we get

$$x_1 = \frac{Fx_F + S_1y_S}{R_1 + KE_1} \tag{9.42}$$

If a pure solvent is used, then $y_S = 0$ and Eq. (9.42) becomes

$$x_1 = \frac{Fx_F}{R_1 + KE_1} = \frac{x_F}{\left(\frac{R_1}{F} + \frac{E_1}{F}K\right)} \tag{9.43}$$

where x_1 is the solute (contaminant) concentration in the raffinate leaving the system. Extending this to a multistage crosscurrent extraction containing (say) 3 stages, we get

$$x_2 = \frac{x_1}{\left(\frac{R_2}{R_1} + \frac{E_2}{R_1} K\right)} = \frac{x_F}{\left(\frac{R_1}{F} + \frac{E_1}{F} K\right)\left(\frac{R_2}{R_1} + \frac{E_2}{R_1} K\right)} \tag{9.44}$$

Similarly,

$$x_3 = \frac{x_2}{\left(\frac{R_3}{R_2} + \frac{E_3}{R_2} K\right)} = \frac{x_F}{\left(\frac{R_1}{F} + \frac{E_1}{F} K\right)\left(\frac{R_2}{R_1} + \frac{E_2}{R_1} K\right)\left(\frac{R_3}{R_2} + \frac{E_3}{R_2} K\right)} \tag{9.45}$$

where x_3 is the solute concentration in the raffinate leaving the system. For a multistage countercurrent process containing 3 stages, the material balance equations are:

Stage 1:

$$x_1 = \frac{x_F + \frac{E_2}{F} y_2}{\left(\frac{R_1}{F} + \frac{E_1}{F} K\right)} \tag{9.46}$$

$$y_2 = K x_2 \tag{9.47}$$

Stage 2:

$$x_2 = \frac{x_1 + \frac{E_3}{R_1} y_3}{\left(\frac{R_2}{R_1} + \frac{E_2}{R_1} K\right)} \tag{9.48}$$

$$y_3 = K x_3 \tag{9.49}$$

Stage 3:

$$x_3 = \frac{x_2}{\left(\frac{R_3}{R_2} + \frac{E_3}{R_2} K\right)} \quad \text{for a pure solvent.} \tag{9.50}$$

From these equations x_3, the solute concentration in the raffinate leaving the system, can be obtained by trial-and-error procedure. The extract streams E_1, E_2 and E_3, and the raffinate streams R_1, R_2 and R_3 have to be evaluated by a graphical procedure, and these methods are well documented in standard texts like that by Treybal[11].

The increased emphasis on solvent extraction in wastewater treatment is exemplified by the variety of its applications. These include use of volatile solvents under pressure for the treatment of wastewaters obtained from the manufacture of acetic acid, caprolactum, acrylonitrile, and chlorinated hydrocarbons. New chemical extractants have been developed, which have the ability to separate efficiently both organic and inorganic chemicals from wastewaters. Amines have been used as extractants in a carrier solvent for the recovery of metal cyanides from plating waste streams[12].

Ion Exchange

Ion exchange has been used extensively to remove hardness, and iron and manganese salts from drinking water supplies. It has also been used to selectively remove specific impurities, and to recover valuable trace metals like chromium, copper, lead, cadmium, and nickel from industrial waste discharges. The process takes advantage of the ability of certain natural and synthetic materials to exchange one of their ions with another contained in the water passing through them.

In the water softening process, the hardness producing elements such as calcium and magnesium are replaced by sodium ions. A cation resin operating on the sodium cycle is normally used. The typical exchange reaction is:

$$R^-Na_2 + \left.\begin{matrix} Mg \\ Ca \end{matrix}\right\} \begin{matrix} Cl_2 \\ SO_4 \\ (HCO_3)_2 \end{matrix} \rightarrow R^-\left\{\begin{matrix} Mg \\ Ca \end{matrix}\right. + \left\{\begin{matrix} 2NaCl \\ Na_2SO_4 \\ 2NaHCO_3 \end{matrix}\right. \qquad (9.51)$$

where R represents the solid phase of the exchange resin. The product water thus has high sodium content, which is not likely to be troublesome unless the original water is very hard. When the exchanger is saturated, it has to be regenerated to allow reuse of the expensive resin. Regeneration can be achieved using sodium chloride solution which removes the calcium and magnesium ions in the form of their soluble chlorides and restores the resin to its original condition:

$$R^- \left\{\begin{matrix} Mg \\ Ca \end{matrix}\right. + 2NaCl \rightarrow R^- Na_2 + \left.\begin{matrix} Mg \\ Ca \end{matrix}\right\} Cl_2 \qquad (9.52)$$

Demineralization of wastewater is achieved in two steps (the cation and the anion exchange steps). The positive ions such as Ca, Na or Mg are removed in the cation exchanger using a hydrogen cycle. A typical exchange reaction is

$$R^- H_2 + \left.\begin{matrix} Ca \\ Mg \\ 2\,Na \end{matrix}\right\} \begin{matrix} Cl_2 \\ SO_4 \\ CO_3 \end{matrix} \rightarrow R^- \left\{\begin{matrix} Ca \\ Mg \\ 2\,Na \end{matrix}\right. + \left\{\begin{matrix} 2\,HCl \\ H_2SO_4 \\ H_2CO_3 \end{matrix}\right. \qquad (9.53)$$

In the anion exchange step the negative ions such as Cl, SO_4 or CO_3 are removed and demineralized water is produced:

$$R^+ OH + \left.\begin{matrix} HCl \\ H_2SO_4 \\ H_2CO_3 \end{matrix}\right\} \rightarrow R^+ \left\{\begin{matrix} Cl \\ SO_4 \\ CO_3 \end{matrix}\right. + H_2O \qquad (9.54)$$

The regeneration is done by treating the cation exchangers with a strong acid and the anion exchangers with caustic soda. The steps are:

$$R^- \begin{cases} Ca \\ Mg \\ 2\,Na \end{cases} + 2\,HCl \rightarrow R^-\,H_2 + \begin{cases} CaCl_2 \\ MgCl_2 \\ 2\,NaCl \end{cases} \tag{9.55}$$

$$R^+ \begin{cases} Cl \\ SO_4 \\ CO_3 \end{cases} + NaOH \rightarrow R^+\,OH + Na \begin{cases} Cl \\ SO_4 \\ CO_3 \end{cases} \tag{9.56}$$

Suspended solids in wastewater can clog the exchangers and cause operational problems. Hence, the feed water to the exchanger should be free of suspended matter. Another serious problem is the resin binding caused by residual organic matter in the feed water. Although the ion exchange process is very effective and produces high quality effluents, the cost of this treatment is fairly high at present. It has been successfully applied to the recovery of chromate from wastewater in pigment manufacturing. Resins are available for particular applications, and one natural zeolite, clinoptilolite, seems to adsorb both phosphate and ammonium ions and may prove to be of great value in wastewater treatment[13].

Reverse Osmosis

In the reverse osmosis process, demineralized water is produced by forcing water through semipermeable membranes at high pressure. In ordinary osmosis, if a vessel is divided by a semipermeable membrane (one that is permeable to water but not to dissolved material), and one compartment is filled with water and the other with a concentrated salt solution, water diffuses through the membrane towards the compartment containing salt solution until the difference in water levels on the two sides of the membrane creates a sufficient pressure to counteract the original water flow (Fig. 9.33a). The difference in levels represents the osmotic pressure of the solution. The process can be reversed by applying sufficient pressure to the concentrated solution to overcome the osmotic pressure and force the net flow of water through the membrane towards the dilute phase. The solute concentration (impurity) builds up on one side of the membrane while relatively pure water passes through the membrane. In order to obtain adequate solvent (water) flux through the membrane, pressures of the order of 4000 to 7000 kN/m^2 are required. Fig. 9.33b presents the principle of operation of the reverse osmosis unit.

The solvent flux through the membrane is proportional to the pressure gradient and is given by

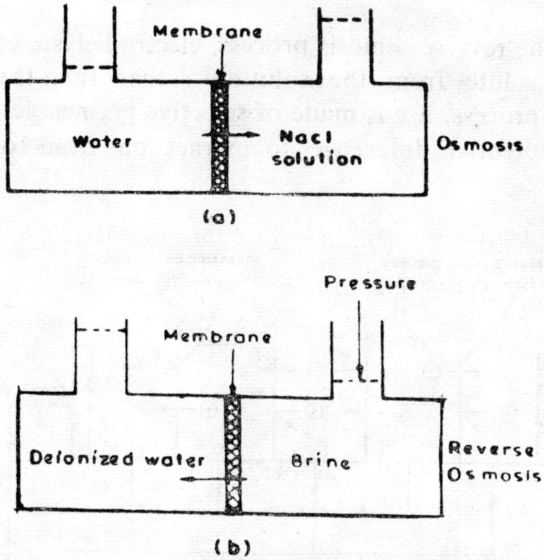

Fig. 9.33 Osmosis and reverse osmosis

$$F_m = W\,(\Delta P - \Delta P_0) \tag{9.57}$$

where F_m = solvent flux through the membrane, ΔP = imposed pressure differential across the membrane, ΔP_0 = osmotic pressure differential across the membrane and W = membrane coefficient.

In most cases, some amount of solute passes through the membrane by molecular diffusion, and its flux is given by

$$F_i = K_i\,\Delta C_i \tag{9.58}$$

where F_i = solute (impurity) flux through membrane, ΔC_i = concentration difference between solutions across the membrane, and K_i = overall mass transfer coefficient that includes effects of membrane thickness

Equations (9.57) and (9.58) show that the water flux is dependent upon the applied pressure, whereas the solute flux is not directly pressure-dependent. As the pressure is increased, both the water flux and the degree of salt rejection tend to improve because of the constant solute flux.

The critical element in the reverse osmosis process is the permeable membrane. Some of the best membranes are cast from a mix of cellulose acetate, acetone, formamide, and magnesium perchlorate. In addition to its use for desalting, reverse osmosis has been satisfactorily applied to the separation of toxic ions from plating wastes, concentration of radioactive wastes, and removal of organics from vegetable and animal wastes. Some fouling of membranes can be brought about by suspended and colloidal particles and, hence, pretreatment is necessary to remove these particles.

Electrodialysis

In contrast to the reverse osmosis process, electrodialysis employs the removal of the solute from the solution rather than the removal of the solvent. In this process, use is made of selective permeable membranes and an electrical potential difference to extract ions from solution. Fig. 9.34

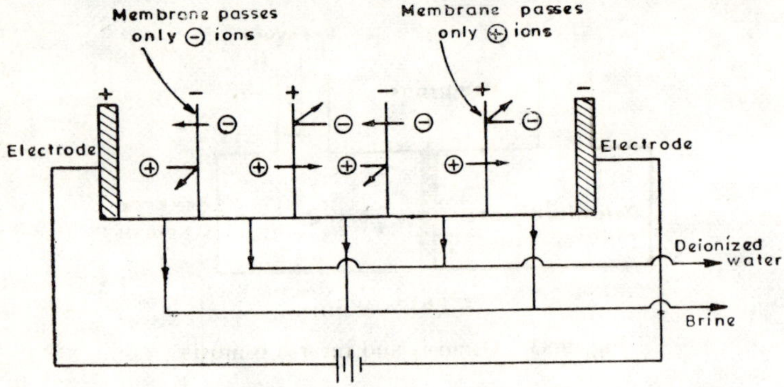

Fig. 9.34 Electrodialysis cell

shows a simple electrodialysis cell in which wastewater may be deionized. As shown in the figure two types of membranes (anionic and cationic) are arranged alternatively to form many compartments between the electrodes placed at the two ends. When voltage is applied across the cell containing mineralized water, the anions migrate to the positive electrode and the cations migrate to the negative electrode. This causes solution in alternate compartments to become more concentrated while that in the remaining becomes more dilute. The electric power requirement is proportional to the number of ions removed from the water.

In the electrodialysis process, organic molecules are not removed and they can collect on and clog the membranes. Another disadvantage of this method is that it still leaves concentrated wastewater (brine) to be disposed of by some appropriate scheme. The process does not require any chemical additives and has low energy requirements and as such it can be an economically feasible means of demineralization.

Nitrogen Removal

All forms of nitrogen in wastewater effluents are potentially harmful. Plants can utilize the inorganic forms as nutrients; ammonia can be utilized by certain bacteria as an energy source with the result that oxygen is depleted in the aquatic systems. Similarly, the organic nitrogen puts a heavy demand of oxygen on the receiving water bodies. The organic nitrogen can be metabolized in the secondary treatment plants by nitrifying bacteria, but this step does not remove nitrogen from wastewater. Nitrification merely converts the organic nitrogen to nitrate, thereby reducing oxygen demand in the effluent.

Two popular methods of nitrogen removal from wastewater are (1) ammonia stripping and (2) biological nitrification and denitrification.

Ammonia Stripping

At neutral *p*H, nitrogen is present in wastewater in the soluble ionic, ammonium form (NH_4^+). The technique of stripping can be exploited only if the ammonium ion is changed to ammonia gas. The reaction that governs this conversion is,

$$NH_4^+ + OH^- \rightleftharpoons NH_4OH \rightleftharpoons NH_3\uparrow + H_2O \qquad (9.59)$$

Increase in the *p*H (the OH^- concentration) shifts the reaction to the right to create more free ammonia. Fig. 9.35 shows the amount of ammonia present in the free state as a function of *p*H and temperature. As can be

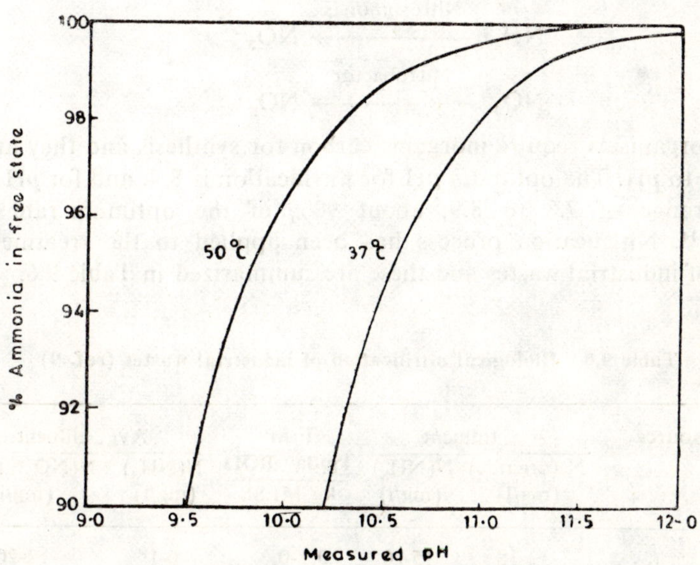

Fig. 9.35 Conversion of NH_4^+ to free ammonia as a function of *p*H and temperature (ref. 7, reproduced with permission of John Wiley & Sons, Inc., N.Y., 1977)

seen from the figure, adjustment of the *p*H of wastewater to a value above 11, will lead to significant removal of ammonia. The free ammonia is air-stripped in a stripping tower.

The ammonia stripping process is particularly well suited to follow a lime coagulation process in which the *p*H of wastewater is adjusted to about 9.5. However, further addition of lime is required to elevate the *p*H to above 11. A major problem in this process is the formation of calcium carbonate scale. This can be controlled by regular maintenance, which consists of washing and scrapping the tower to remove scale. The ammonia removal efficiencies at the *p*H values of 11.5 and 10.5 are about 86% and 80% respectively. However, removal efficiencies drop considerably in cold weather.

Nitrification-Denitrification

This is essentially a modification of the activated sludge process and is specifically geared to the removal of nitrogen. In nitrification, the waste is thoroughly treated in the aeration chamber when NH_4^+ is converted by bacteria to nitrate ions. This usually involves longer detention times than those for the conventional activated sludge process, and F/M ratios are maintained around 0.3. During nitrification, bacteria such as nitrosomonas and nitrobacter catabolize ammonia to nitrite and nitrates according to the following two equations:

$$NH_4^+ \xrightarrow{\text{Nitrosomonas}} NO_2^- \tag{9.60}$$

$$NO_2^- \xrightarrow{\text{Nitrobactor}} NO_3^- \tag{9.61}$$

The organisms require inorganic carbon for synthesis and they are very sensitive to *p*H. The optimum *p*H for nitrification is 8.4, and for *p*H values in the range of 7.8 to 8.9, about 90% of the optimum rate can be achieved[14]. Nitrification process has been applied to the treatment of a variety of industrial wastes and these are summarized in Table 9.6.

Table 9.6 Biological nitrification of industrial wastes (ref. 9)

Waste source	Influent		F/M	Avg. effluent	
	N (organic) (mg/l)	$N(NH_3)$ (mg/l)	kg/day BOD / kg MLSS	$N(NH_3)$ (mg/l)	$N(NO_3^-+NO_2^-)$ (mg/l)
Domestic wastes	12-18	15-20	0.1-0.3	0-1	12-20
Organic acids	3-8	25	0.12	low	30
Pharmaceutical and organic chemicals	80	25	0.05-0.1	2	34
Synthetic fibres and polymers	80	low	0.15	5	18
Nylon	200	50	0.15-0.2	2	230

The NO_2^- and NO_3^- forms are removed from the system by the process of denitrification. It is an anaerobic process in which heterotrophic bacteria found in activated sludges reduce NO_2^- and NO_3^- to nitrogen gas. These organisms require an organic carbon source for cell growth, and because the effluent from the nitrification tank is very low in organic carbon, some organic compound must be added. Methanol is usually recommended for this purpose. The organisms metabolize methanol according to the following reaction:

$$NO_3^- + \tfrac{5}{6}\,CH_3OH \rightarrow \tfrac{1}{2}\,N_2 + \tfrac{5}{6}\,CO_2 + \tfrac{7}{6}\,H_2O + OH^- \qquad (9.62)$$

The equation indicates that 5/6 mole of methanol is required to reduce completely one mole of nitrate to molecular nitrogen or 1.9 mg/l of CH_3OH for 1 mg/l of nitrate on the weight basis. If properly operated, the nitrification-denitrification process has been shown to convert or remove nitrogen in an extremely efficient manner. A three-stage nitrification-denitrification process is shown in Fig. 9.36.

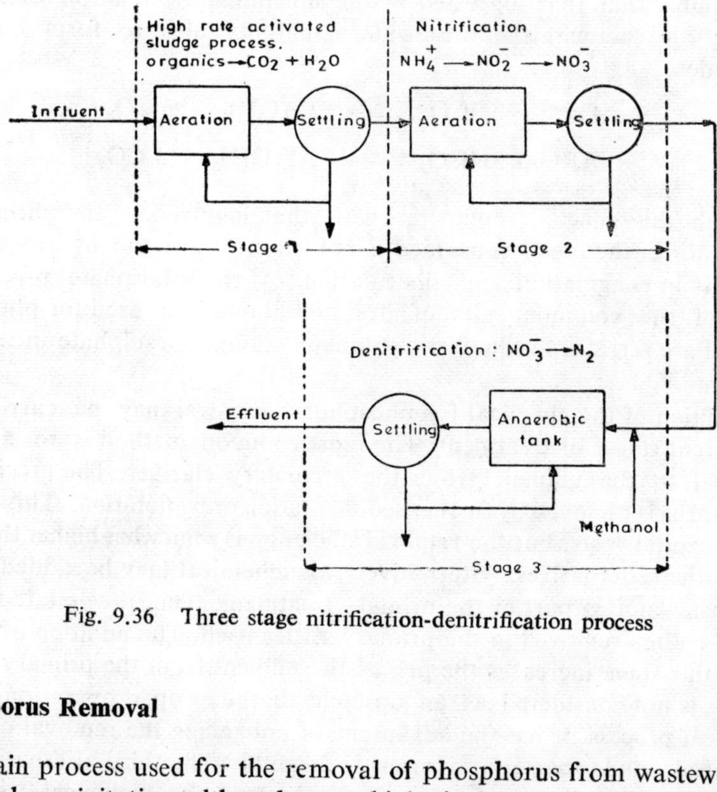

Fig. 9.36 Three stage nitrification-denitrification process

Phospborus Removal

The main process used for the removal of phosphorus from wastewater is chemical precipitation, although some biological methods have also been suggested. Many chemicals are effective in the precipitation of phosphorus from solution. Most commonly used chemicals are lime, iron salts, and aluminium salts. Lime precipitation can be carried out after secondary treatment where high levels of *p*H, in the range of 10.5 to 11.5, are required. In the process, phosphate reacts with calcium to form the insoluble compound hydroxyapatite as follows:

$$3HPO_4^{3-} + 5\,Ca^{2+} + 4OH^- \rightarrow Ca_5OH(PO_4)_3 \downarrow + 3\,H_2O \qquad (9.63)$$

In addition to this precipitation, lime also reacts with the bicarbonate alkalinity of wastewater to form calcium carbonate:

$$CaH(CO_3)_2 + Ca(OH)_2 \rightarrow 2\,CaCO_3 \downarrow + 2\,H_2O \qquad (9.64)$$

Trivalent iron and aluminium salts may be used to precipitate soluble phosphorus as ferric or aluminium phosphate according to the reactions:

$$Fe^{3+} + PO_3^{3-} \rightarrow FePO_4 \downarrow \qquad (9.65)$$

$$Al^{3+} + PO_4^{3-} \rightarrow AlPO_4 \downarrow \qquad (9.66)$$

These reactions indicate that 1 mole of the cation is required for each mole of phosphorus in the wastewater. However, the dose of the metallic salt is higher than that suggested by the stoichiometric relation because the metallic salt react with the wastewater alkalinity also to form insoluble hydroxide:

$$Fe^{3+} + 3HCO_3^- \rightarrow Fe(OH)_3 + 3\,CO_2 \qquad (9.67)$$

$$Al^{3+} + 3HCO_3^- \rightarrow Al\,(OH)_3 + 3\,CO_2 \qquad (9.68)$$

Although this reaction competes with that involved in the phosphorus precipitation, the metal consumed is not wasted since the hydroxide formed aids in coagulation and flocculation of the phosphate precipitate. Some of the common salts of iron and aluminium used for phosphate removal are ferric chloride, ferric sulphate, aluminium sulphate and sodium aluminate.

Addition of the chemical for phosphorus removal may be carried out at different stages of treatment. The most common method is to add the chemical to the effluent from the secondary clarifier. The precipitated phosphorus is removed by further sedimentation or by flotation. This increases the capital costs, but the removal efficiency is somewhat higher than that in the other alternatives. Alternatively, the chemical may be added to the raw wastewater as part of the primary treatment. The precipitated phosphorus is then removed in the primary settler itself. The addition of chemical at this stage increases the *p*H of the effluent from the primary settler, but this is not considered as an obstacle to the proper operation of the biological process. It has the advantage of enhancing the removal of BOD, suspended solids, bacteria, viruses and metals. The third alternative is to add the chemical directly to the aeration tank of the biological treatment system. In this method, the precipitate settles in the secondary settler together with the activated sludge. It has the advantage of acting as a conditioner and aids in the dewatering of the activated sludge. However, addition of chemicals to the aeration tank can cause damage to the bacterial cells and impair their efficiency for the removal of organic constituents. More than 95% phosphate can be removed through a proper choice of the chemicals and location of their application[15].

The precipitation reaction induced by lime or alum or both has been used for the removal of phosphate from the wastewaters of the mining, fertilizer and chemical industries.

Biological Phosphorus Removal

Some bacteria have the ability to take up phosphorus in excess of their immediate nutritional requirements and store it within the bacterial cell in the form of polyphosphates. The polyphosphates can then be removed from the system. The 'Bardenpho' process[16] and the 'Phostrip' process[17] are two of the biological processes developed for the removal of phosphorus from wastewater streams.

Advanced Biological Systems

New biological methods are being investigated for wastewater treatment. The use of shallow oxidation ponds or lagoons has proved very effective for the treatment of domestic wastewater. These ponds are classified into four main types:

Aerobic Ponds

Wastewaters containing organic impurities are purified by the action of aerobic bacteria and algae. Oxygen is supplied by natural diffusion across the pond surface, and by algal photosynthesis. These are shallow ponds (depth 0.3 m or less) designed to maximize the growth of algae. Aerobic conditions are maintained throughout the depth of the pond for all the time.

Facultative Ponds

This is the most frequently encountered type. These ponds have an aerobic upper zone and an anaerobic lower zone. Operation of a typical facultative pond is shown in Fig. 9.37. The organic waste enters at one end of the pond where the suspended solids settle to the bottom. At the bottom an anaerobic layer develops and the settled sludge is degraded by anaerobic organisms to produce CO_2, NH_3 and CH_4. In the upper zone aerobic bacterial degradation of the waste takes place. A facultative zone exists between these two zones, which is generally variable. It can be either aerobic or anaerobic at various times so that growth of facultative organisms, which are able to adapt to either condition, is favoured.

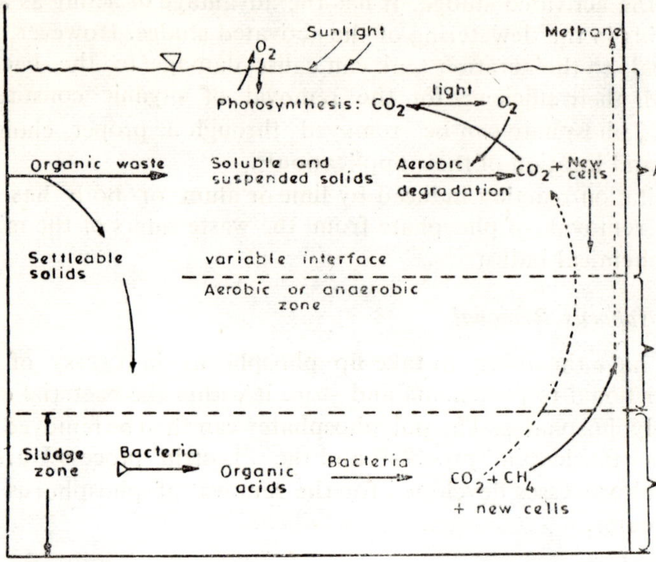

Fig. 9.37 Basic reactions in a facultative pond system

Aerated Ponds

These are similar to the activated sludge process units where oxygen is supplied by mechanical aerators installed on floats or on a permanent base. The basic difference between the aerated ponds or lagoons and the activated sludge system is that recycling and wasting of the sludge is provided in the latter as a means of controlling the solids in the aerator. In aerated ponds no recycle of the sludge is provided and the digested material leaves the system. The concentration of solids in this case is much lower than that in the activated sludge system.

Anaerobic Ponds

These ponds are maintained in anaerobic condition by applying a BOD load that exceeds oxygen production from photosynthesis. Anaerobic ponds are usually employed as pretreatment ponds for the treatment of high-temperature, high-strength wastewaters where the reduction in waste strength is more important than the effluent quality. Effluents from anaerobic ponds require additional treatment before they are discharged to water bodies.

Biological pond systems are traditionally designed on the basis of average liquid detention time and organic loading per unit surface area, or both. Typical design parameters are summarized in Table 9.7. The advantages of the biological pond systems are their simplicity of operation and

Table 9.7 Empirical pond design parameters (ref. 18)*

Pond type	Depth (m)	Loading (kg BOD/m^2. day)	Detention time (days)	% BOD removal
Aerobic	0.5-1.0	0.15	2-6	80-95
Facultative	1.0-2.0	0.05	5-30	70-85
Aerated	2.0-4.0	0.2	2-10	50-90
Anaerobic	2.5-5.0	0.5	30-60	50-70

*Reprinted with permission of Pergamon Press PLC, Copyright 1977

low costs. Their disadvantages include potential odour problems, extensive land area requirements and the adverse effects of seasonal weather changes on the pond's performance.

Chemical Oxidation

In the field of wastewater treatment, chemical oxidants such as chlorine, ozone and hydrogen peroxide are widely used for disinfection, removing organic materials that are resistant to biological or other treatment processes, and conversion of cyanides to innocuous products. Use of chlorine as a disinfectant destroys or inactivates bacteria present in the wastewater before it is discharged into receiving streams. Chlorine rapidly penetrates bacterial cells and kills the bacteria. However, the effectiveness of chlorine is influenced greatly by the physical and chemical characteristics of wastewater.

Chlorine

Initially when it is added to water, chlorine forms hypochlorous acid (HOCl):

$$Cl_2 + H_2O \rightleftharpoons HOCl + H^+ + Cl^- \qquad (9.69)$$

Hypochlorous acid is the disinfecting agent and is referred to as free residual or free available chlorine. However, if any reducing agents such as ferrous ions or hydrogen sulphide are present in wastewater, chlorine reacts with them, and the concentration of chlorine available to destroy pathogenic bacteria is reduced. The reduction reaction with hydrogen sulphide may be represented as:

$$H_2S + 4Cl_2 + 4H_2O \rightarrow H_2SO_4 + 8HCl \qquad (9.70)$$

Wastewaters usually contain ammonia. In the presence of ammonia, HOCl reacts to form, sequentially, monochloramine (NH_2Cl), dichloramine

($NHCl_2$) and trichloramine (NCl_3) according to the following reactions:

$$NH_3 + HOCl \rightarrow H_2O + NH_2Cl \tag{9.71}$$

$$HOCl + NH_2Cl \rightarrow H_2O + NHCl_2 \tag{9.72}$$

$$HOCl + NHCl_2 \rightarrow H_2O + NCl_3 \tag{9.73}$$

Monochloramine and dichloramine are referred to as combined residuals and are more stable than free residual, but are less effective as disinfectants. Once all ammonia has reacted, further addition of chlorine converts the combined residuals into a free residual, the conversion being proportional to the dose at the "break-point". This is the limit beyond which all the residual chlorine is available as free chlorine. Typical chlorine dosages required for disinfection are shown in Table 9.8.

Table 9.8 Typical chlorine dosages for disinfection (ref 5*)

Effluent from	Cl_2 dosage (mg/l)
Untreated wastewater	6-25
Primary sedimentation	5–20
Chemical precipitation	2–6
Trickling filter	3–15
Activated sludge plant	2–8
Multimedia filter following activated sludge plant	1–5

*Reprinted with permission of McGraw-Hill, N.Y., 1972

Chlorination is used to oxidize cyanide in industrial wastewaters to harmless carbon and nitrogen compounds. This is done in alkaline media at *p*H greater than 8.5 to prevent the generation of poisonous hydrogen cyanide gas. The overall reaction may be represented as

$$2CN^- + 5Cl_2 + 8OH^- \rightarrow 10Cl^- + 2CO_2 + N_2 + 4H_2O \tag{9.74}$$

The residual cyanide concentration, after a reasonable reaction time, is very small.

Ozone

Ozone is becoming more attractive as an oxidant, particularly as a substitute for chlorine. It is a powerful oxidizing agent and found to be an efficient disinfectant. It is also useful for the removal of colour, taste and odour. Like oxygen, it is only slightly soluble in water and because of its unstable form it leaves no undesirable byproducts such as organic chlorides. It is effective in the oxidation of many complex organic materials including pesticides, surfactants, cyanides and phenols. Ozonation in the presence of ultraviolet light has shown promise of reducing reaction times and improving ozone utilization.

9.5 RECOVERY OF MATERIALS FROM PROCESS EFFLUENTS

The possibility of reduction in costs of wastewater treatment processes by means of product or byproduct recovery should not be neglected in any industrial effluent treatment project. The methods employed for recovery are based on the same physical, chemical and biological techniques that were discussed previously. The methods of recovery of valuable materials from process effluents using the various techniques are summarized below.

Biological Reactions

Biological reactions are employed for chemical recovery as well as for waste disposal. Many wastewaters from paper and pulp mills, sugar plants and breweries are suitable as fermentation seeds. Drugs, vitamins, alcohols and acetone can be obtained by fermenting the above wastes.

Precipitation

Precipitation by alkalies is an excellent means of recovering heavy metals such as iron, copper, zinc and cadmium, while soluble barium salts, such as barium sulphide, have been used for the precipitation of chromium. The proteins present in wastewaters such as those from slaughterhouses can be precipitated by the addition of lignosulphonate and sulphuric acid.

Flotation

This technique is widely used in the meat processing industry for the recovery of renderable fats and greases. It is also used for the recovery of fibre in the paper industry.

Adsorption

Adsorption has been employed for the recovery of elemental sulphur from

SO_2 in paper and pulp industry, CS_2 and CCl_4 in petrochemical industry, and fluorides in fertilizer industry.

Ion Exchange

Ion exchange has been used in chemical industry for the recovery of chromic acid from cooling tower effluents, and recovery of phenol, and small amounts of dissolved electrolytes from solution. It has also been used in the recovery of metals like silver, gold and platinum from plating baths, and for the recovery of amino and other organic acids and alkaloids in petrochemical industry. Other uses of ion exchange process include the recovery of isotopes from high level radioactive wastes and the recovery and removal of ionized organics from aqueous solution.

Dialysis

Dialysis involves the separation of the solutes by means of unequal diffusion through membranes. The rate of diffusion is based on the concentration gradient and separations are made on the basis of molecular sizes, or diffusion rates of the solvents. It has been used to recover rayon from textile wastes and acids (H_2SO_4, H_3PO_4) from spent liquor in the metal treating wastes.

Electrodialysis

This process is extensively used in the recovery of acid and iron from pickle liquors in steel industry, and of lignosulphonic acids, wood sugars, pulping chemicals and acids in paper and pulp industry.

Reverse Osmosis

This process is applied to recover acid from pickle liquors, polyvinyl alcohols and mineral oils in textile industry, demineralized water in metal plating, hydrogen from petroleum effluents, and pigment-producing organometallic compounds in mining operations.

Solvent Extraction

Solvent exraction has been successfully used to recover phenol in steel industry, wool grease and pulping liquors in textile and paper industry, and iobutylene, paraffins and naphthalenes in petrochemical industry.

In India, the main focus of attention in the area of water pollution control is the need to control water-borne diseases such as typhoid and cholera. The relatively simple techniques of filtration and chlorination of drinking water can save thousands of lives. Water-borne diseases are not the only concern at present, as attention is now shifting towards the need to control oxygen-consuming wastes. Most of the municipalities still use the inadequate primary treatment, and there is an urgent need to design sewage treatment plants equipped with secondary treatment facilities so that the BOD load on receiving waters is reduced. In addition to the BOD load, significant amounts of plant nutrients are discharged into water bodies.

Industry must also learn to do a better job of controlling the release of its many pollutants—mercury, cadmium, acid mine wastes, pesticides, heat, etc.,—into receiving streams. The cost of the necessary programmes may be high initially, but is not at all unreasonable in view of the inherent dangers of taking these problems lightly.

REFERENCES

1. O. Barnes, P.J. Bliss, B.W. Gould and H.R. Vallentine, *Water and wastewater engineering systems* (Pitman Books Ltd., London), 1981, p. 163.
2. W.P. Talmadge and E.B. Fitch, "Determining thickener unit areas", *Ind. Engg. Chem.*, Vol. 47, 1955.
3. L.D. Banefield and C.W. Randall, *Biological process design for wastewater treatment* (Prentice-Hall, Inc., N.J.), 1980.
4. A.W. Lawrence and P.L. McCarty, "Unified basis for biological treatment design and operation", *J. Sanitary Engg. Div., ASCE*, Vol. 96 (SA3) (1970), p. 757.
5. Metcalf and Eddy, Inc., *Wastewater engineering-collection, treatment, disposal* (McGraw-Hill, New York), 1972.
6. A.W. Lawrence, "Modeling and simulation of slurry biological reactors", in *Mathematical Modeling for Water Pollution Control,* edited by T. Keinath and M. Wanielista (Ann Arbor Science Publishers, Mich.), 1975.
7. P. Mac Berthouex and D.F. Rudd, *Strategy of pollution control* (John Wiley & Sons, New York), 1977.
8. J.L. Pavoni (Ed.), *Handbook of water quality management planning* (Van Nostrand Reinhold Co., New York), 1977.
9. T.J. Mulligan and R.D. Fox, "Treatment of industrial wastewaters", *Chem. Engg.*, Vol. 83 (1976), Oct. 18.
10. P.B. Lederman, "Physical and chemical methods of wastewater treat-

ment", *Chem. Engg.*, Vol. 84 (1977), Aug. 15.

11. R.E. Treybal, *Mass-transfer operations*, 2nd Edn. (McGraw-Hill, New York), 1968.
12. "Cyanide tamer looks promising", *Chem. Week* (1976), April 14, p. 44.
13. R.W. Holcomb, "Wastewater treatment: the tide is turning" *Science*, Vol. 169 (1970), p. 457.
14. H.E. Wild, "Factors affecting nitrification kinetics", *J. of Water Pollution Control*, Vol. 43 (1971), p. 1845.
15. I.J. Kugelman, "Status of advanced waste treatment" in *Handbook of water resources and pollution control*, (Van Nostrand-Reinhold Co., New York), 1976, p. 593.
16. J.L. Barnard, "The bardenpho process", Chap. 6 in *Advances in water and wastewater treatment, biological nutrient removal,* edited by M.P. Wanielista and W.W. Eckenfelder (Ann Arbor Science Pub. Mich.), 1978, pp. 79-114.
17. L.C. Matsch and R.F. Drnevich, "Phostrip—A biological-chemical system for removing phosphorus", Chap. 7 in *Advances in water and wastewater treatment, biological nutrient removal,* edited by M.P. Wanielista and W.W. Eckenfelder (Ann Arbor Science Pub. Mich.), 1978, pp. 115-142.
18. T.H.Y. Tebbutt, *Principles of water quality control* (Pergamon Press, Oxford), 1977.

SUGGESTIONS FOR FURTHER READING

1. E.D. Schroeder, *Water and wastewater treatment* (McGraw-Hill, New York), 1977.
2. G.M. Fair, J.C. Geyer and D.A. Okun, *Water and wastewater engineering*, Vols. 1 and 2 (John Wiley & Sons, New York), 1971.
3. N.L. Nemerow, *Industrial water pollution* (Addison-Wesley, Reading, Mass.), 1978.
4. R.S. Ramalho, *Introduction to wastewater treatment processes* (Academic Press, New York), 1977.
5. M.A. Winkler, *Biological treatment of wastewater* (Ellis Horwood Ltd., Chichester, England), 1981.
6. R.L. Culp and G.L. Culp, *Advanced wastewater treatment* (Van Nostrand-Reinhold, New York), 1971.
7. P.A. Vesilind, *Treatment and disposal of wastewater sludges* (Ann Arbor Science Pub. Mich.), 1974.
8. J.N. English *et al.*, "Removal of organics from wastewater by activated carbon", *Chem. Engg. Progr. Symp. Series*, Vol. 67 (1970), p. 147.

9. E.C. Kaup, "Design factors in reverse osmosis", *Chem. Engg.*, Vol. 80 (1973), April 2.
10. T.P. O'Farrell *et al.*, "Nitrogen removal by ammonia stripping", *J. of Water Pollution Control*, Vol. 44 (1972), p. 1527.
11. E.F. Barth *et al.*, "Chemical-biological control of nitrogen and phosphorus in wastewater effluent", *J. of Water Pollution Control*, Vol. 40 (1968), p. 2040.
12. T.V. Arden, *Water purification by ion exchange* (Butterworth, London), 1968.

PROBLEMS

1. A settling-column analysis is performed on a dilute suspension of discrete particles. Data collected from samples taken at the depth of 1.2 m are as follows:

Settling time, min	15	30	45	60	90	180
Fraction of initial solids in sample	0.96	0.83	0.70	0.50	0.27	0.06

Determine the theoretical removal in solids from this suspension in a horizontal flow tank with surface overflow rate of (43.2 m^3/day)/m^2. (65.6%)

2. A secondary clarifier is to be designed to produce an underflow concentration of 30,000 mg/l from an influent with mixed liquor solids content of 4,500 mg/l. The wastewater flow rate is 0.06 m^3/s. Calculate the required clarifier area. The following data are obtained from a settling test in a 100 cm cylinder. (82.8 m^2).

Time, min	0	2	4	6	8	10	14	18	22	26	30
Interface height, cm	100	87.5	75	63	53	46	36	29	25	23	20

3. A wastewater treatment plant serves a population of 5,00,000. It receives domestic sewage at the per capita flow rate of 200 litres/day and per capita BOD loading of 50 g/day. Along with the domestic sewage, the plant also treats 0.2 m^3/s of industrial wastewater at a BOD concentration of 500 mg/l. The plant removes 90 per cent of the BOD, and the effluent is discharged into a stream which has a flow of

8 m^3/s and a BOD of 2 mg/l. Calculate the BOD concentration of the downstream immediately after discharge. (5.87 mg/l)

4. Ordinary municipal wastewater is to be treated by the conventional activated sludge process. The design loading is 0.1 m^3/s and 250 mg/l BOD. Fig. 9.16 defines process variables. The design criteria and operating values for the conventional process are as follows:

BOD removal per cent	= 90
Organic loading	= 0.5 kg BOD/day per kg MLVSS
The target for MLSS in aeration tank	= 2000 mg/l
Excess sludge production	= 0.5 kg sludge solids per kg BOD destroyed
Air flow rate	= 100 m^3/day per kg BOD/day
Final settling tank overflow rate	= (30 m^3/day) per m^2 surface
Solids concentration in the recycled sludge	= 10,000 mg/l
Solids concentration in the effluent leaving the system	= 25 mg/l

Calculate the following: (1) sludge recycle ratio, (2) aeration period (days), (3) amount of solids leaving the system (kg/day), (4) amount of BOD destroyed (kg BOD/day), (5) mean cell residence time (days), (6) waste sludge flow rate (m^3/day), (7) air flow requirement (m^3/day), and (8) surface area of the settling tank (m^2), (25%; 0.25; 972; 1944; 4.44; 75.8; 1, 94, 400; 288)

5. A 10^6 litres/day conventional activated sludge plant has an influent BOD of 200 mg/l. The primary settler removes 30% of that BOD. The plant is equipped with an aeration tank 18 m long × 5 m wide × 3 m deep. The mixed liquor volatile suspended solids (MLVSS) concentration is maintained at 1600 mg/l. Calculate the aeration period and the F/M ratio. (0.27 days; 0.32)

6. The success of an activated sludge system depends on the settling of the solids in the final settling tank. Suppose that the sludge in a system has started to "bulk" and experiments show that the volume of the settled sludge (30 minutes) is 300 ml both before and during bulking and the SVI is 100 and 250 respectively. What is the MLSS before and during the bulking problem? During bulking, what happens to BOD removal efficiency? Why?

7. 1000 kg of raw wet sludge with a solids content of 3% is fed into the

active digestion zone of an anaerobic digester as shown in Fig. 9.25. The volatile solids content of the sludge is 80% and the rest are inert solids. 580 kg of water is removed as a supernatant with negligible solids content. If the digester efficiency is 75%, what is the total solids content in the digested sludge? What is the maximum amount of volatile solids that could be converted as gas? (3%; 18 kg)

8. The chlorination of drinking and wastewater is the dominant method of disinfection in India and other countries. In *Chemical Engineering* [August 18, 1975], a review is presented of the problems of water disinfection and of new developments. Prepare a summary of that paper.

9. In *Environmental Science and Technology* [Vol. 9 (No. 1) (1975) 53–59] a reverse osmosis process has been proposed for the removal of pesticides from wastewater. Prepare a short review of that paper.

10. Mercury in waste effluents can be removed by an ion exchange process proposed in *Chemical Engineering* [August 23, 1971]. Discuss the process briefly.

11. 0.5 m^3/s of wastewater having a total phosphorus concentration of 20 mg/l as P is to be treated by precipitating with coagulant chemicals, alum [$Al_2(SO_4)_3.14H_2O$] and ferric chloride [$FeCl_3.6H_2O$]. Laboratory tests on the wastewater indicate that an alum dose of 2.0 mol Al(III)/mol P or a ferric chloride dose of 1.8 mol Fe(III)/mol P will be adequate for phosphorus removal. Calculate the coagulant doses of alum and ferric chloride needed in kg/day (12,658; 13,530).

10. Solid Waste Management

10.1 SOURCES AND CLASSIFICATION

Solid waste, often called the third pollution after air and water pollution, is that material which arises from various human activities and which is normally discarded as useless or unwanted. It consists of the highly heterogeneous mass of discarded materials from the urban community as well as the more homogeneous accumulation of agricultural, industrial and mining wastes.

Solid wastes may be classified based partly on content and partly on moisture and heating value. A typical classification is as follows:

(a) *Garbage*: Refers to the putrescible solid waste constituents produced during the preparation or storage of meat, fruit, vegetables, etc. These wastes have a moisture content of about 70% and a heating value of around 6×10^6 J/kg.

(b) *Rubbish*: Refers to non-putrescible solid waste constituents, either combustible or non-combustible. Combustible wastes would include paper, wood, scrap, rubber, leather etc. Non-combustible wastes are metals, glass ceramics etc. These wastes contain a moisture content of about 25% and the heating value of the waste is around 15×10^6 J/kg.

(c) *Pathological Wastes*: Dead animals, human waste, etc. The moisture content is 85% and there are 5% non-combustible solids. The heating value is around 2.5×10^6 J/kg.

(d) *Industrial Wastes*: Chemicals, paints, sand, metal ore processing, flyash, sewage treatment sludge, etc.

(e) *Agricultural Wastes*: Farm animal manure, crop residues, etc.

The principal sources of solid wastes are domestic, commercial, industrial and agricultural activities. Many times domestic and commercial wastes are considered together as the so-called urban wastes. The main constituents of urban wastes are similar throughout the world, but the weight generated, the density and the proportion of constituents vary widely from country to country, and from town to town within a country according to the level of economic development, geographic location, weather and social conditions. In general, it has been found that as the personal income rises, kitchen wastes decline but the paper, metals and glass wastes increase; the total weight generated rises but the density of the wastes declines.

In India authentic information regarding the composition of the urban wastes is not generally available as regular analysis of the refuse is not carried out by the municipalities. In fact, refuse is very heterogeneous in composition, and the geographical, temporal and seasonal variations in its composition make it difficult to define a "typical refuse". The solid refuse generated in urban areas contains articles of various sizes and types and consists of dust, vegetable leaves, waste paper, large paper-board cartons, glass bottles, worn out tyres, carcasses of animals and night soil. Table 10.1 gives the composition of refuse in various cities of India and a comparison is made with the urban refuse from a typical European city. As is seen from Table 10.1, the average paper content in the refuse of Indian

Table 10.1 Composition of city refuse (refs. 1, 2) (percentage by weight)

	Kanpur	Delhi	Calcutta	Bangalore	Bombay	Typical European city
Paper	1.35	5.88	0.14	1.5	3.20	27
Vegetable—putrescible matter	53.34	57.71	47.25	75.2	59.37	30
Dust, ash, etc.	25.93	22.95	33.58	12.0	15.90	16
Metals	0.18	0.59	0.66	0.1	0.13	7
Glass	0.38	0.31	0.24	0.2	0.52	11
Textiles	1.57	3.56	0.28	3.1	3.26	3
Plastics, leather, rubber, etc.	0.66	1.46	1.54	0.9	—	3
Other (stones, wooden matter, etc.)	18.59	6.4	16.98	18.9	16.4	3
Density, kg/m^3	500	—	540	578	—	132

cities is about 2 to 3 per cent as compared with about 27% for a typical European city. Similarly, the density of refuse in India is much higher than that of the refuse generated in the cities of Western countries because of the inclusion in it of the street sweepings. The amount of refuse collected from urban areas in India is of the order of 0.3 kg to 0.5 kg per person per day excluding night soil.

Manufacturing industries produce wastes which are solid or semi-solid. These wastes can be pyrophoric (self-igniting), explosive, toxic or radioactive. Chemical process industries (CPI) generate a variety of wastes, both organic and inorganic, which are mixtures with a wide range of component concentrations. Table 10.2 gives some of the physical and chemical characteristics of the wastes generated by the CPI.

Table 10.2 Solid wastes generated in the CPI (ref. 3)

Origin	Physical and chemical characteristics	Comments
Coke manufacture	Coke and coal fires	Generally contained in scrubber wastes
Dyes and pigments	Reaction or raw-material sludge, highly variable	—
Pharmaceuticals and fine-chemicals	Raw material solids, biological wastes	—
Inorganic chemicals	Insoluble salts, tailings, slimes	—
Metal processing (secondary)	Ash, scrubber wastes, metal hydroxide sludges	—
Petrochemicals	Oily, greasy, asphaltic	Usually float
Plastics and rubber	Latex or plastic crumbs, often alum-coagulated	—
Pulp and paper	Fibrous, often mixed with lime or alum	

Most of the industrial wastes generated in cities come from small scale operations and these are usually disposed off along with the city refuse. Larger industries are often located outside the cities and the disposal of their wastes is primarily the responsibility of the industries themselves. Some of the industrial wastes are often recycled (scrap metal and paper) while others can be utilized as an energy source for specific processing plants in some regions. Energy can be recovered from solid wastes by numerous thermal routes as well as by biochemical conversion. The toxic and radioactive wastes, often classified as hazardous wastes, need special consideration before their disposal.

Agricultural wastes comprise both crop residues and animal wastes such as manure and urine. Whereas urban wastes amount to between 0.3 and 0.5 kg per person per day in India, agricultural wastes amount to around 2 kg per person per day. Animal and vegetable wastes contain valuable minerals and nurients. Humus from agricultural wastes contains nitrogen, phosphorus, potash and trace elements which are vital to the fertility of the soil and optimum plant growth. Burning of wastes as fuel in the conventional manner makes poor use of the leaf content of the fuel burnt and, further, leads to loss of valuable nutrients.

10.2 PUBLIC HEALTH ASPECTS

The relationship between solid wastes and human disease is difficult to prove. Nevertheless, improper handling of solid wastes is a health hazard and causes damage to the environment. The main risks to human health arise from the breeding of disease vectors, primarily flies and rats. A common transmission route of bacillary dysentery, amoebic dysentery and diarrhoeal disease in India is from human faeces by flies to food or water and thence to humans[4]. It has been estimated that in warm climates, exposed garbage produces as many as 70,000 flies per 0.03 m^3 in a week.

The refuse dumps also serve as a source of food for rats and small rodents which quickly proliferate and spread to neighbouring areas. Rats destroy property, infect by direct bite and spread various diseases like plague, endemic typhus, salmonellosis, trichinosis, etc. Apart from diseases for which insects and rats are carriers, the handling and transfer of biological wastes poses a threat to the worker and those he contacts. Disease transmission may occur through direct contact with the waste, through infection of open sores or through vectors.

The hazardous wastes are injurious to human health; some have acute effects while others pose a health hazard after prolonged period of exposure. Improper disposal of such wastes has resulted in the death of humans and animals through contamination of crops or water supplies.

The environmental damage caused by solid wastes is mostly aesthetic in nature. Uncontrolled dumping of urban wastes destroys the beauty of the countryside; also, there is the danger of water pollution when the leachate from a refuse dump enters surface water or ground water resources. In addition, uncontrolled burning of open dumps can cause air pollution.

Solid waste disposal is at present limited to land and the ocean. Some of the wastes can be recovered and reprocessed, a procedure commonly known as recycling. However, before the wastes can be disposed of effectively, they must be collected efficiently. These activities, i.e., collection, disposal and/or recovery form a part of the solid waste management system.

10.3 METHODS OF COLLECTION

Efficient collection and transportation are essential parts of the overall solid waste management programme since these two activities constitute about 75% of the total cost. The basic mode of refuse collection in India is from communal storage points. The refuse is delivered to fixed storage bins usually built from concrete blocks, having capacities between 100 and

500 litres and placed at intervals of 50 to 200 metres. The refuse is stored in the bins till it is collected for disposal by a vehicle. Daily collection is essential because the organic matter in the refuse tends to decompose rapidly in the hot climate. The refuse is usually removed using a shovel and a basket from the bin and dumped into the waiting vehicle by the workers. The clearing of the bin is a slow process and the vehicle waiting time is excessive.

Other methods of refuse collection like block collection and kerbside collection are practised in developed countries. In block collection the waste is brought in containers by individuals to a waiting vehicle which travels a regular route twice or thrice a week. The containers are emptied by the vehicle crew and returned to the individual. In kerbside collection, the waste is brought in containers and placed on the footway in advance of the collection time to be retrieved later.

Transportation of the collected refuse constitutes a key stage in the overall waste management system. In India, no single mode of transport can prove effective, economical and efficient due to congested and narrow lanes and streets in towns and cities. Hence, various types of vehicles from hand carts to modern mechanized vehicles are used. Hand cart collection is the best mode of transport from congested and narrow places; the refuse is usually transferred from the carts to a waiting trailer for final transport and disposal. Large vehicles with some form of mechanical compression to increase the density and, hence, the pay load are suitable for some well-planned city centres in India, but generally smaller and simpler vehicles are better. Usually, the trailer or a truck is filled with refuse and driven to the disposal site. If the disposal site is situated far off from the loading points the collection crew waste a lot of time travelling to and from the disposal site. Their time could be optimally utilized by employing a relay system.

In a relay system the collection can be made in trailers employing a lesser number of tractors, or two vehicles can be employed for one unit of crew (2 : 1 relay). In the second alternative, the crew stays at the collection site and the unloading at the disposal site is done mechanically. This is particularly useful when the loading time is approximately equal to the truck travelling time to and from the disposal site. The 2 : 1 relay enables the crew to be busy throughout the day and avoids unnecessary waste of time. Similarly, when the transport time is approximately half of the time required to collect a full load, a 3 : 2 relay can be used.

Transfer Station

A transfer station may be described simply as a place for receiving refuse from a number of small collection vehicles such as hand carts and transferring it to larger line-haul type vehicles capable of undertaking a longer journey more economically. The best construction for a transfer station is a ramp leading to a concrete platform where the smaller vehicles discharge. The refuse is unloaded either by the force of gravity or manually into the transfer vehicles which wait on a roadway about 5 metres below. A typical two-level transfer station is shown in Fig. 10.1. If the refuse can be com-

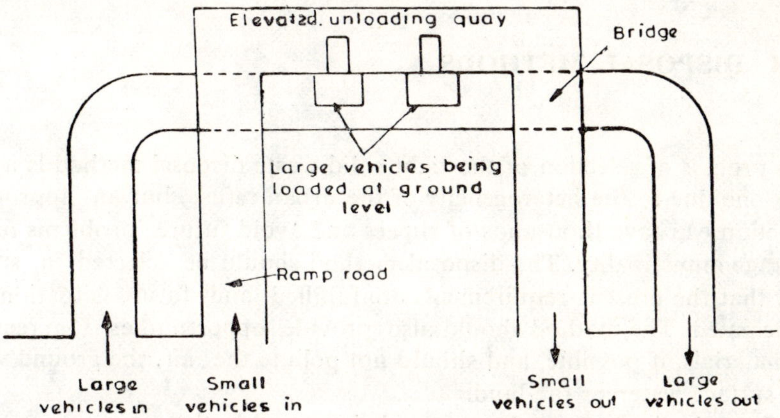

Fig. 10.1 Two-level transfer station (ref. 2, reproduced with permission of World Health Organization, New Delhi. 1976)

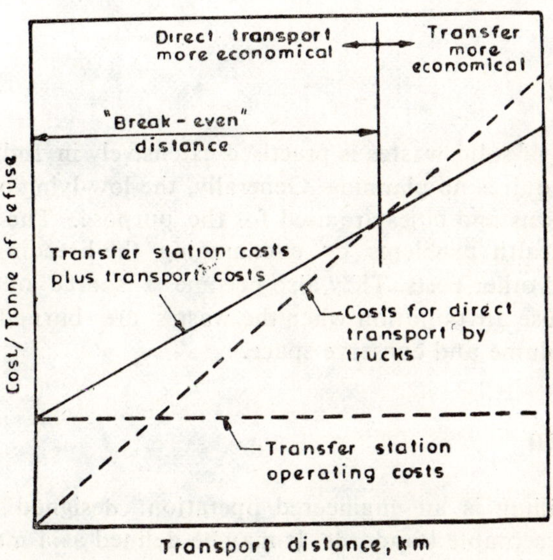

Fig. 10.2 Economic viability of a transfer station (ref. 5, reproduced with permission of Van Nostrand Reinhold, N.Y., copyright (c) 1973)

pressed at the transfer station, a heavy load can be carried economically. By using transfer loading it is possible to have disposal sites at long distances from city centres.

In order that the transfer station may be economically viable, the total cost of collection, transfer and disposal must be less than the total cost of collection, direct transport by collection trucks and disposal. This is graphically illustrated in Fig. 10.2.

10.4 DISPOSAL METHODS

The process of selection of the right solid waste disposal method is a complex one due to the heterogeneity of the urban refuse, but an appropriate selection can save thousands of rupees and avoid future problems for the average municipality. The disposal method should be selected in such a way that the present requirements are fulfilled and future situations are anticipated. The method should also provide opportunities for recycling of materials, if possible, and should not pollute the air, the ground water, the surface water or the land.

Several disposal methods are being used in the various parts of the world and the most prominent of these are: open dumping, sanitary land filling (controlled tipping), incineration and composting. Sanitary land filling is the main method used in the developed countries and open dumping is very common in India.

Open Dumping

Open dumping of solid wastes is practised extensively in India because it is cheap and requires no planning. Generally, the low-lying areas and outskirts of the towns and cities are used for the purpose. The open dumps cause public health problems by encouraging the breeding of flies, rats, mosquitoes and other pests. They also become a source of objectionable odours and cause air pollution when the wastes are burned in order to reduce their volume and conserve space.

Sanitary Landfill

Sanitary landfilling is an engineered operation, designed and operated according to acceptable standards. It may be defined as a method of disposing refuse on land without creating nuisances or hazards to public health or safety. The operation is carried out without environmental damage and in areas already spoiled or in need of restoration.

In sanitary landfill operation, refuse is spread and compacted in thin layers within a small area. This layered structure is usually referred to as a cell. To allow for proper compaction, the cell depth should not exceed about 2 metres. The cell is then covered with a layer of soil which is spread uniformly and then compacted. To provide an adequate seal the 'cover' should normally be at least 20 cm thick. If the refuse includes large irregular objects it may be necessary to increase the thickness of the cover. On the other hand, a cover thickness of less than 15 cm may be satisfactory if the refuse has been pulverized. When a number of cells reach the final desired elevation, a final cover of about one metre of earth is placed and it is again compacted. This final cover is necessary to prevent rodents from burrowing into the refuse. Fig. 10.3 shows the cross-sectional area of a typical sanitary landfill.

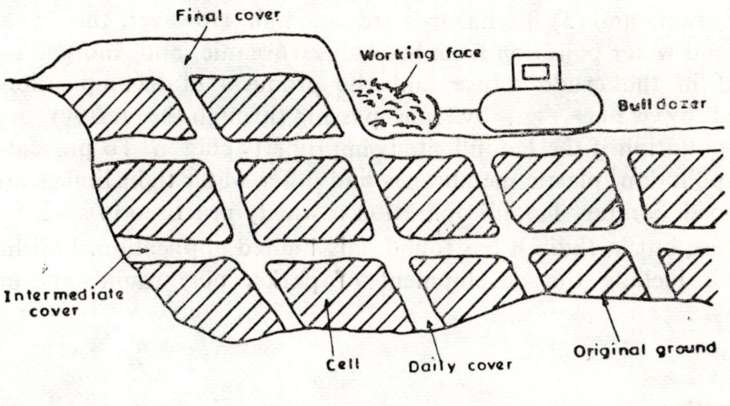

Fig. 10.3 Sanitary landfill operation

The landfill operation is essentially a biological method of waste treatment. The refuse stabilization may be divided into five distinct phases within the overall process[6]. During the first phase of operation, aerobic bacteria, which are dominant, deplete the available oxygen and as a result of aerobic respiration the temperature increases. In the second phase anaerobic conditions become established and hydrogen and carbon dioxide are evolved through acidogenic activity Phase three is characterized by the establishment of methanogenic population of bacteria and the beginning of methanogenic activity. In phase four the methanogenic activity becomes stabilized. In phase five the methanogenic activity decreases representing depletion of the organic matter, and ultimately the system returns to aerobic conditions within the landfill. The duration of each phase of landfill activity and degree of degradation varies with prevailing environmental conditions and the nature of the refuse.

The end products of decomposition during phases three and four are mostly CO_2 and CH_4, accompanied by small amounts of H_2S, NH_3 and water. Fig. 10.4 shows the typical gas composition profile during the overall process. During the first year the amount of N_2 decreases as air cannot penetrate the soil to the sealed cell. Methane production can usually be observed around 200 days after refuse placement, i.e., phase three of refuse digestion, but the period of production of CH_4 during phase four varies and is often difficult to predict. Escape vents for the gases are provided so that they do not build up to dangerous levels. Landfill gas is a rich and renewable soure of energy and attempts have been made to harness and exploit this potential.

The advantages of a sanitary landfill, as opposed to an open dump are: (1) the public health problems are minimized because flies, rats and other pests are unable to breed in the covered refuse, (2) there is no air pollution from burning and (3) fire hazards are minimal. However, there is a danger of ground water pollution from leachates (organic and inorganic matter present in the crude refuse and the products of decomposition can be leached by water, e.g., by rain passing through the refuse), or surface water pollution if the landfill site is improperly chosen. To prevent ground water pollution, plastic membranes and other water tight linings are being developed. Sanitary landfill operation is widely practised in some Western countries, but in India it has found only limited application. Finished landfill uses include the development of parks, playgrounds and even golf courses.

Incineration

Incineration involves the burning of solid wastes at high temperatures; leftover ashes, glass, metals and unburned combustibles amount to perhaps 25% of the original waste. This residue must still be disposed of in some manner. Incineration leads to air pollution unless the plant is designed, equipped and operated to comply with air pollution standards. Typical air pollutants from incineration are flyash, SO_2, hydrogen chloride, and organic acids. New techniques of handling the waste have been developed. Thus, materials which are not combustible are removed from the waste by gravity or magnetic separation. Many of the separated materials like glass or metals can be recycled. Air pollution can be controlled by installation of proper control equipment.

Industrial solid wastes that are incinerated are mostly cellulose type, and more often industries have to handle certain kinds of chemical wastes in the form of either solids or sludges. In the chemical process industries,

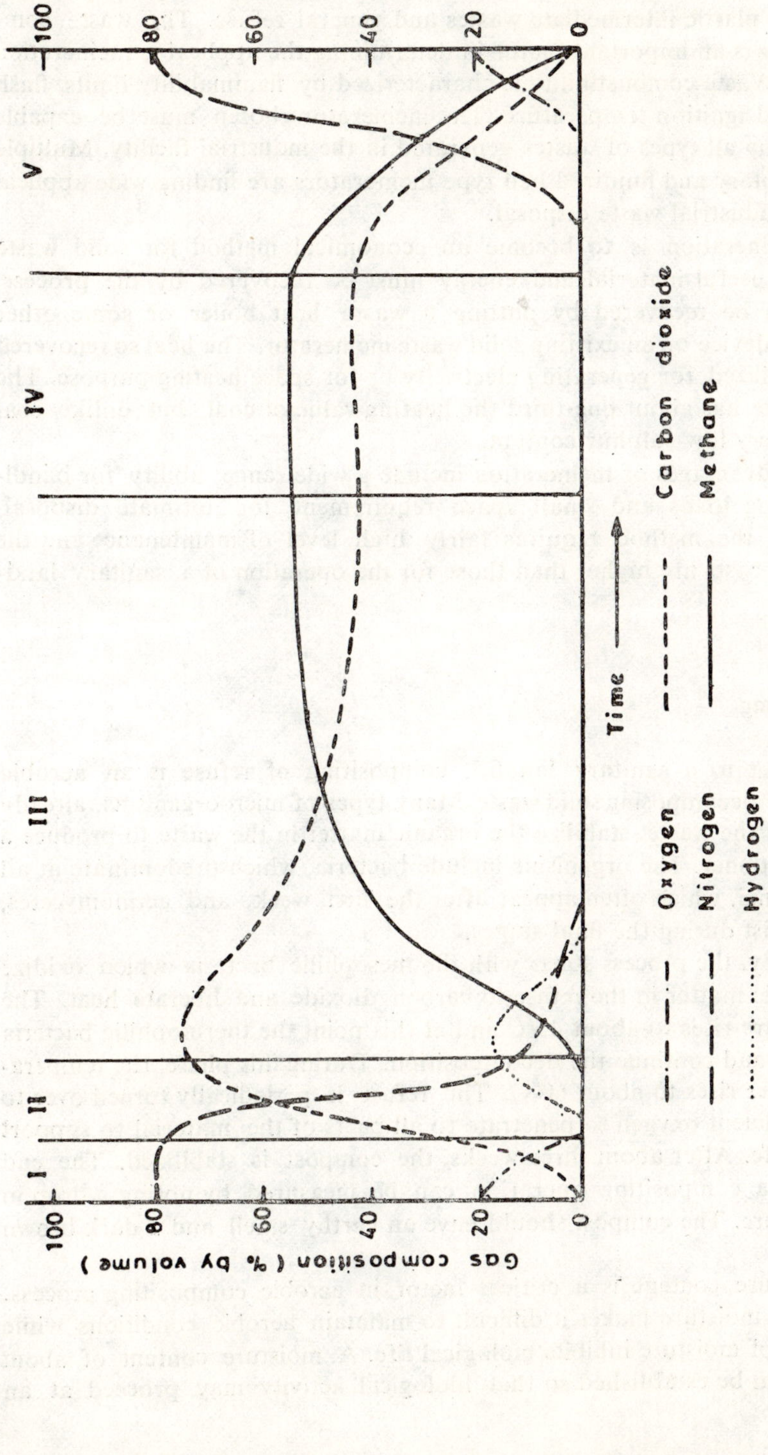

Fig. 10.4 Composition profile of main gases during refuge digestion in a landfill (ref. 7)

incineration is most frequently used to dispose of tarry and gummy petroleum and plastic intermediate wastes and general refuse. The waste combustibility is an important factor in determining the applicable incineration process. Waste combustibility is characterized by flammability limits, flash point, and ignition temperature. The incinerator chosen must be capable of handling all types of wastes generated in the industrial facility. Multiple hearth, rotary and fluidized bed type incinerators are finding wide applications in industrial waste disposal.

If incineration is to become an economical method for solid waste disposal, useful material and energy must be recovered by the process. Heat can be recovered by putting a waste heat boiler or some other recovery device on an existing solid waste incinerator. The heat so recovered can be utilized for generating electricity or for space heating purpose. The solid waste has about one-third the heating value of coal, but unlike coal it has a very low sulphur content.

The advantages of incineration include a wide range ability for handling varying loads and small space requirement for ultimate disposal. However, the method requires fairly high level of maintenance and the operating costs are higher than those for the operation of a sanitary landfill.

Compositing

In contrast to a sanitary landfill, compositing of refuse is an aerobic method of decomposing solid waste. Many types of microorganisms, already present in the waste, stabilize the organic matter in the waste to produce a soil conditioner. The organisms include bacteria, which predominate at all stages, fungi, which often appear after the first week, and actinomycetes, which assist during the final stages.

Initially, the process starts with the mesophilic bacteria which oxidize the organic matter in the refuse to carbon dioxide and liberate heat. The temperature rises to about 45°C and at this point the thermophilic bacteria take over and continue the decomposition. During this phase, the temperature further rises to about 60°C. The refuse is periodically turned over to allow sufficient oxygen to penetrate to all parts of the material to support aerobic life. After about three weeks, the compost is stabilized. The end point of a compositing operation can be measured by noting a drop in temperature. The compost should have an earthy smell and a dark brown colour.

Moisture content is a critical factor in aerobic compositing process. Excessive moisture makes it difficult to maintain aerobic conditions while a dearth of moisture inhibits biological life. A moisture content of about 55% should be established so that biological activity may proceed at an optimum rate. It may be necessary to add water to maintain a satisfactory moisture content.

Modern methods of composting fall into two broad categories: manual and mechanical. For towns with a rural bias, the manual window system is more attractive. In India two methods of composting refuse/night-soil mixtures have been developed. In the Indore process, layers of vegetable waste and night-soil are alternated, each about three inches thick to a depth of upto 1.5 metres in a trench or form a mound above ground called a windrow. The mixture is kept aerobic by turning regularly for two or three months. The compost is then left for another month or so without turning, the whole process thus takes about four months.

A modified process, known as the Bangalore process, is now widely adopted by municipal authorities throughout the country. The material is placed in layers, as in the Indore process, in a trench about a metre deep. The material is not turned but is digested under essentially anaerobic conditions whereby decomposition is complete in four to five months. Though the process avoids turnings altogether, it lays great emphasis on the initial C/N ratio of the compost heap and initial moisture content. The compost is free from pathogenic organisms and contains 1.5% N, 1.1% P (P_2O_5) and 1.5% K (K_2O) on dry basis and proves a valuable nutrient for the soil.

Fully mechanized plants involve shredding, grinding and machanical separation of high-density solids. There are basically four processes of mechanical compositing available in India. These are: (a) the Buhler process, (b) the Dano process, (c) the Tollemache process, and (d) the Nusoil process.

In the Buhler process the material is ground in two stages in hammer mills; the non-compostable inorganic materials are separated by strong sifting action on circular swinging sieves. The material is then decomposed aerobically in open windrows; stabilization may take about two to three months.

The Dano process uses a long rotating drum, called a bio-stabilizer unit, for decomposing the refuse. The rotating drum is inclined so that the waste flows from one end to the other. The refuse is partially decomposed in the drum and the outcoming refuse is generally free from odour and pathogenic organisms. It is then completely decomposed in windrows; the entire operation may take about four weeks.

In the Tollemache process the refuse is pulverized in a vertical pulveriser and then passed through a screening plant to screen out paper, plastics, etc. The pulverized-screened refuse is allowed to decompose in the windrows for three weeks, with three to four turnings. The compost is then cured for four to five weeks; the complete stabilization thus takes about two months.

In the Nusoil process the noncompostable material is separated from the refuse which is then pulverized in a hammer mill. The pulverized matter then goes to a vertical digester where the decomposition takes place. The digester is a circular unit having seven sections; the refuse moves downward through each section of the digester. It is kept for about a day in each section, and air flow rate and water addition are regulated so that decomposition takes place under optimum conditions. The digestion process is completed in seven days and the resultant compost is satisfactory for direct field application without the addition of supplementary nutrients.

Compost is not widely used in the U.S.A., because inorganic fertilizers are cheap and easy to apply, and most farms are located in areas where soil conditions are good. Compost, on the other hand, is in great demand in many European countries. Farmers in Spain, Austria and France apply compost to their farms, and in Netherlands composting is used to dispose of about 1/6th of all municipal waste. In India, during monsoon season, high-intensity rainfall and rapid run-off result in the erosion of valuable humus from a large area of crop land. This may be remedied by applying compost, which has good moisture retaining capacity and is not so easily washed away as the chemical fertilizer.

10.5 POTENTIAL METHODS OF DISPOSAL

In recent years new and better methods for solid waste management have been suggested and/or developed. Some of the ideas are discussed in this section.

Utilization

Many solid wastes generated by industry can be utilized directly. Fly and bottom ash from power plants can be used commercially, largely as cement substitute. New uses are being developed for fly ash, e.g., to make bricks, to dewater industrial wastewater sludge, as a land cover, etc. India produces about 6×10^6 tonnes of bagasse from sugar cane annually. This bagasse can be utilized for the manufacture of paper pulp which can displace hardwood pulp of similar quality and cost. There some novel uses of sugar cane bagasse, one sugar cane company in South Africa has integrated its operation with the production of eggs and dairy products.

Recovery and Recycling

Solid wastes contain significant amounts of valuable materials like steel, aluminium, copper and other metals which, if they are recovered and reused, would reduce the volume of the wastes to be collected and at the same time would yield significant salvage and resale income. In addition, better reclamation techniques will help to save valuable natural resources and turn wastes, which could be dangerous, into useful products. Some important solid wastes that have been successfully reclaimed are paper, glass, metals and plastics. These are discussed briefly in the following sections.

Paper

The paper industry treats the incoming waste paper according to its source. Some paper goes directly to the mills, while other paper needs considerable processing and sorting. Waste paper contains grit, sand, ink, tar, paper clips, plastic coatings, rubber bands, etc. These have to be separated before paper can be treated further. Only a small fraction of recycled paper ends up as virgin paper and, more commonly, used paper is made into paper-board, insulation, paper bags, etc. Recycled paper can be used for printing only a few times because with each recycle the fibres become somewhat shorter and more frayed, with the consequence that the recycled product is weaker than similar product made from virgin fibres.

The manufacture of paper board is the main use of waste paper. Waste paper after sorting into different grades is fed on to a conveyor belt in the proportions required by the mix. The conveyor belt feeds the paper to a hydropulper unit where it is repulped in water with a mixture of 3 parts of paper to 97 parts of water at about 38°C. The stirring force in the hydro-pulper is so great that the sheets of paper are rapidly broken down to a slurry. The fibres in the paper are separated and heavy unacceptable materials are drained from the bottom. The slushed waste paper is passed over a riffler system and light material is floated off. The slush is then screened to remove any undissolved matter and is then fed to a bitumen dispersal unit for the removal of tar. The slushed paper slurry is thickened and the stock is then ready for the paperboard machines.

Perhaps the most promising use or reuse of waste paper is the conversion of the material to enesgy. It has a high calorific value, approximately half that of coal and it does not cause air pollution. The plastic and wax coatings enhance the value of the paper as a fuel and contribute to its heating value.

Glass

Glass is the perfect product for recycling. It is clean, easy to reprocess, and can be used in many ways. At present, market exists in India for almost all of the waste glass which can be collected. Scrap glass, which is known as cullet, can be used in each new batch of glass produced. However, the main problem in any reuse of glass is its separation from other materials in the solid waste. The typical glass separation methods include froth flotation, dense media separation, and colour sorting. These methods are summarized in Table 10.3.

Table 10.3 Glass separation methods (ref. 8)*

Method	Separation procedure	Comments
Froth flotation	Separation by surface characteristics	Potential for 95% recovery, glass not colour-sorted
Dense media separation	Material separated in stages	Particles are large enough for colour sorting
Colour sorting	After removal of stones, ceramics, and residual material by a a transparency sorting device, a colour sorter matches particles with the background for flint, amber and green. Air jets deflect particles into bins	Marketability of recovered glass increased, unit requires constant calibration, organic residuals may cloud readings

*Reprinted with permission of Van Nostrand Reinhold, N.Y., Copyright (c) 1980.

Waste glass can be used as an aggregate in asphalt instead of crushed stone. The finished glasphalt product exposes no jagged edges and is skid resistant. Several industries in the U.S.A. have experimented with a water-soluble glass that can be treated with a thin protective coating to keep the glass from dissolving until the container is broken. Experimental use of ground glass as an ingredient of light-weight building materials has been technically successful.

Metals

Metals can be easily recycled from industrial scrap. The metals arising as scrap include both ferrous as well as nonferrous types. The ferrous metals are separated first from nonferrous metals; this can be achieved magnetically. While it is a simple matter to separate cast iron and steel, much skill is needed in the recognition of special steels so that their scrap may be usefully applied.

The principal nonferrous metals which are subject to recycling are aluminium, copper and lead. The methods used for the recovery of aluminium from solid waste include gravity separation, electric or magnetic separation, and chemical or thermal separation. A typical gravity separator is the zigzag separator (Fig. 10.5) where high pressure air is used to separate aluminium from heavier materials. In the magnetic separation an aluminium

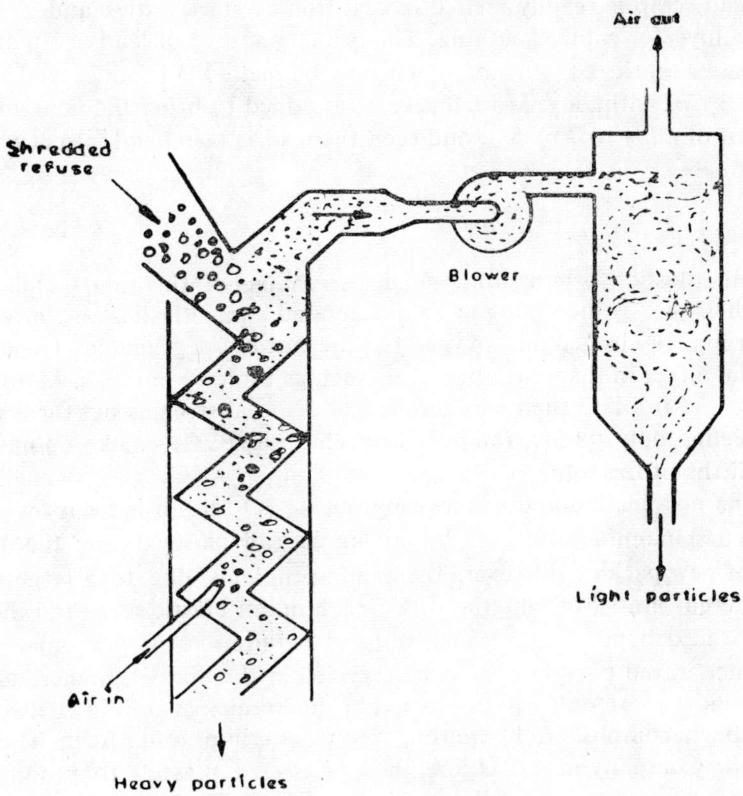

Fig. 10.5 Zigzag separator

magnet produces eddy currents and the interaction of these currents with the magnetic field is responsible for exerting a repellent force on the metal and causing its separation from other materials like copper and zinc. Chemical methods are also used for the separation of aluminium from other materials. For example, "Sweating" which exploits the differences in the melting points of metals has been used to separate aluminium from lead and zinc.

High purity recycled aluminium can be reused as a substitute for the pure virgin metal. Aluminium of less purity such as that recycled from municipal waste contains some other metals also like iron, zinc, copper and silicon which can add strength to the end product. The recycled aluminium is usually melted in oil fired reverberatory or rotary furnaces and converted into alloys for use in the die casting industry.

Copper is usually found in industrial wastes in the form of wire and is often covered with insulation. The insulation can be removed either by mechanical or thermal methods. Pure copper and its alloys may be melted for reprocessing and cast into billets or ingots for various applications.

Lead scrap is readily melted in cast iron or steel kettles and converted into alloys for cable sheathing. The primary source of lead scrap are storage batteries, the contents of which may be melted to produce lead carrying about 3% of antimony. More antimony is added to bring the final antimony content of alloy to 7 or 8%, and then the lead is re-refined into new battery plates.

Plastics

In India plastics form a rather small percentage of the total solid wastes, but their use is increasing at an accelerated pace. Plastics basically consist of carbon, hydrogen and oxygen but are not bio-degradable. Hence, they remain intact in disposal operations such as sanitary landfills. Many varieties of plastics find their way into waste disposal streams but three types—polyolefins, polystyrene, and poly vinyl chloride (PVC)—make up more than four-fifths of the total.

One possible method of recycling plastic is to use it in the form in which it was first manufactured, i.e., by mixing the raw plastic during the production of new packs. However, there are problems due to a large number of different grades of material within each major plastic type, and the grade employed depends on the characteristics required in the particular pack.

Since mixed plastics cannot be recycled in the above manner, and this is the most common form of plastics in the municipal solid waste, recycling may be accomplished by reusing the plastic in another form. The Union Carbide Company in the U.S.A. has pyrolyzed plastics to produce hard and soft waxes, greases, adhesives and tars as well as a range of gases.

Another idea which being carefully investigated is the total decomposition of plastic materials. When exposed to ultra-violet light, certain chemical groups absorb the light and release energy. This energy breaks the polymer chain so that the plastic loses its strength, becomes brittle and is broken up by the wind and rain. The small pieces then mix with the soil and can be decomposed by normal soil bacteria into basic chemicals.

Once the economic and technological problems are overcome, recycling of waste materials will become an important way of solid waste management. Due to resources depletion, what is now being discarded may become valuable in the future. In addition, proper handling and recycling could provide energy, and would lower the waste burden on the natural environment.

REFERENCES

1. *Report of the Committee of Urban Wastes, Govt. of India* (Ministry of Works and Housing. New Delhi), Dec. 1975.
2. F. Flintoff, *Management of solid wastes in developing countries* (World Health Organisation, New Delhi), 1976.
3. R.W. Okey, D. Digregorio and E.G. Kominek, "Waste-sludge treatment in the CPI" *Chem. Engg.*, Vol. 86 (1979), Jan. 29.
4. S. Rajagopalan, *Guide to simple sanitary measures for the control of enteric diseases* (WHO, Geneva), 1974.
5. D.J. Hagerty, J.L. Pavoni and J.E. Heer (Jr.) *Solid waste management* (Van Nostrand Reinhold Co., New York), 1973.
6. G.T. Farquhar and F.A. Rovers, "Gas production during refuse decomposition", *Water, Air & Soil Pollution*, Vol. 2 (1973), pp. 483-495.
7. *A basic study of landfill microbiology and biochemistry* (Dept. of Energy, ETSU B 1159, U.K.), 1988.
8. R.A. Conway and R.D. Ross, *Handbook of 'industrial waste disposal* (Van Nostrand Reinhold Co., New York), 1980.

SUGGESTIONS FOR FURTHER READING

1. S. Rao *et al.*, "Character and quality of refuse in rural homes", *Institution of Engineers (India), J. of Public Health*, Vol. 51 (1970), pp. 23-26.
2. P.A. Witt, "Disposal of solid wastes", *Chem. Engg.*, Vol. 78 (1971). p. 65.
3. Committee on Sanitary Engineering Research, "Survey of sanitary landfill practices : thirtieth progress report". *J. San. Engg. Div. ASCE* Vol. 87 (SA4) (1961), pp. 65-84.
4. G.M. Mallan, "A total recycling process for municipal solid wastes", paper 46C, *National AIChE meeting*, Atlantic city, 1971.
5. D.V. Jackson *et al.*, "Recovery from industrial and domestic wastes in centralized treatment plants", in *Chem. Engg. in a changing world*, edited by W.T. Koetsier (Elsevier Scientific Pub. Co., Amsterdam), 1976.

PROBLEMS

1. Discuss how you would run a sanitary landfilling operation in your city. Be sure to include the technical, economic and psychological considerations.
2. List in a tabular form the advantages and disadvantages of the following methods of solid wastes disposal : incineration, sanitary landfill, composting and pyrolysis.
3. Helmut W. Schulz [*Environ Science & Technology,* Vol. 8 (1975) 423 -27] reported on a study on the recovery of energy, glass and metals from municipal solid wastes, which can be separated economically. Discuss the several separation schemes that seem to be the most effective in processing the solid waste.
4. A biological method of producing sugar and protein from trash has been reported [*Eviron Science & Technology* Vol. 8 (9) (1975), pp. 784-85]. Report on the latest developments.
5. Suppose the solid waste from a city of 3 million people is used to fuel a 40% efficient steam-electric power plant. Assuming that half of the waste is combustible with a heating value of 15×10^6 J/kg, and assuming that each person's contribution to the solid waste is 0.4 kg/day, calculate the size of the power plant.

Appendix A—Conversion Factors and Constants

Table A1 Conversion factors

Length (m)		
	1 inch	= 0.0254 m
	1 foot	= 0.3048 m
	1 mile	= 1.609 km
	1 micron (μ)	= 10^{-6} m
	1 Angstrom (Å)	= 10^{-10} m
Mass (kg)		
	1 ounce	= 28.35 g
	1 ounce	= 0.02835 kg
	1 pound	= 0.4536 kg
	1 ton	= 1000 kg
Area (m^2)		
	1 in^2	= 6.4516 cm^2
	1 ft^2	= 929.03 cm^2
	1 ft^2	= 0.0929 m^2
	1 Acre	= 4046.85 m^2
	1 Acre	= 0.4046 hectare
	1 $mile^2$	= 2.59 km^2
Volume (m^3)		
	1 $inch^3$	= 16.387 cm^3
	1 $inch^3$	= 0.01639 *l*
	1 ft^3	= 0·0283 m^3
	1 ft^3	= 28.316 *l*

1 gallon (U.S)	= 3.785 *l*
1 gallon (U.S)	= 0.003785 m^3
1 litre	= 10^{-3} m^3

Energy (*J* = N.m)

1 joule	= 1 kg-m^2/s^2
1 erg	= 1 g-cm^2/s^2
1 joule	= 10^7 ergs
1 cal	= 4.184 J
1 ft-lb_f	= 1.3558 J
1 Btu	= 1.055×10^3 J
1 hp-hr	= 2.6845×10^6 J
1 kW-hr	= 3.600×10^6 J

Power (*W* = *J*/*s*)

1 hp	= 745.7 W
1 Btu/s	= 1055 W
1 ft-lb_f/s	= 1.3558 W
1 kW	= 1.341 hp

Pressure (*Pa* = *N*/m^2)

1 kg/m.s^2	= 1 N/m^2
1 lb_f/ft^2	= 47.88 Pa
1 atm	= 1.0133×10^5 Pa
1 mm Hg	= 1.3332×10^2 Pa
1 in Hg	= 3.3864×10^3 Pa
1 bar	= 10^5 Pa
1 mm H_2O	= 9.806 Pa
1 in H_2O	= 249.09 Pa

Density (*kg*/m^3)

1 g/cc	= 10^3 kg/m^3
1 mg/1	= 10^{-3} kg/m^3
1 lbm/ft^3	= 16 kg/m^3

Viscosity ($N\ s/m^2$)

1 kg/m.s	= 1 N.s/m^2
1 cP	= 10^{-3} N.s/m^2
1 lbm/ft.s	= 1.488 N.s/m^2
1 lbm/ft.hr	= 4.1338 N.s/m^2

Thermal conductivity ($W/m\ K$)

1 cal/s.cm. K	= 418.4 W/m K
1 kcal/hr.m.K	= 1.1623 W/m K
1 Btu/hr.ft, °F	= 1.7307 W/m K

Table A2 Frequently used constants

Universal gas constant	$R_u = 8314$ J/kg-mol K
Standard acceleration of gravity	$g_c = 9.8067$ m/s^2
Standard atmospheric pressure	$p = 101{,}326$ N/m^2
Joule's constant	$J_c = 4.184$ J/cal
Boltzmann's constant	$k = 1.38 \times 10^{-23}$ J/K
Faraday's constant	$F = 9.6487 \times 10^7$ C/kg–mol
Planck's constant	$h = 6.626 \times 10^{-34}$ Js
Mathematical constants	$e = 2.71828...$
	$\ln 10 = 2.30259...$
	$\pi = 3.14159...$

Appendix B—Physical Properties of Fluids

Table B1—Properties of dry air at atmospheric pressure

T (°C)	ρ (kg/m³)	$\mu \times 10^5$ (N.s/m²)	k (W/m.K)	C_p (J/kg.K)
0	1.293	1.72	0.0244	1005
10	1.247	1.77	0.0251	1005
20	1.205	1.81	0.0259	1005
30	1.165	1.86	0.0267	1005
40	1.128	1.91	0.0276	1005
50	1.093	1.96	0.0283	1005
60	1.060	2.01	0.0290	1005
70	1.029	2.06	0.0297	1009
80	1.000	2.11	0.0305	1009
90	0.972	2.15	0.0313	1009
100	0.946	2.19	0.0321	1009

Table B2—Properties of gases at low pressures and 300 K

Gas	Molec. wt	Gas constant, R (J/kg. K)	C_p (J/kg. K)	Ratio of specific heats γ	k (W/m.K)
Carbon monoxide	28.01	298	1042	1.40	0.0251
Helium	4.00	2080	5230	1.66	0.1507
Hydrogen	2.02	4120	14350	1.40	0.1867
Nitrogen	28.02	297	1038	1.40	0.0261
Oxygen	32.00	262	916	1.40	0.0266
Ammonia	17.04	515	2177	1.31	0.0545
Hydrogen sulphide	34.08	245	1009	1.32	—
Methane	16.04	535	2261	1.31	0.0342
Nitric oxide	30.01	175	963	1.40	0.0259
Sulphur dioxide	64.07	141	628	1.29	—

NOTATION INDEX

A = Area of the spot, m^2

A = Interfacial area, m^2

A = Surface area of settling zone, m^2

A_c = Collector surface area, m^2

A_c = Clarification area, m^2

A_i = Inlet cross-sectional area, m^2

A_{sc} = Tower cross-sectional area, m^2

A_t = Surface area required for thickening, m^2

a = Interfacial area per unit of tower volume, m^2/m^3

b = Birth rate

b = Inlet width of cyclone, m

C = Particle concentration, kg/m^3

C = Cunningham correction factor

C_A = Component-A concentration in liquid phase, kg-mol/m^3

C_D = Drag coefficient

C_{Df} = Fibre drag coefficient

C_i = Solution concentration, kg-mol/m^3

C_L = Actual concentration of oxygen in water, kg/m^3 or mg/l

C_p = Specific heat at constant pressure per unit mass of air parcel, J/kg. K (kg air parcel)

C_S = Saturation concentration of oxygen in water, kg/m^2 or mg/l

C_u = Underflow concentration, kg/m^3 or mg/l

C_V = Specific heat at constant volume, J/kg. K.

C_v = Specific heat at constant volume per unit mass of air parcel, J/kg. K (kg air parcel)

C_y, C_z = Diffusion parameters

c = Concentration of solute in the bulk of the liquid, kg. mol/m^3

c_i = Concentration at the interface, kg-mol/m^3

c^* = Equilibrium concentration of solute, kg-mol/m^3

D = Dissolved oxygen deficit, kg/m^3 or mg/1

D^* = Dilution factor

$\mathcal{D}$ = Particle diffusivity, m^2/s

D_{AM} = Mass diffusivity of A, m^2/s

D_G = Gas diffusivity of solute absorbed, m^2/s

D_h = Hydraulic diameter, m

D_L = Liquid diffusivity of solute absorbed, m^2/s

d = average droplet size, m or μm

d_f = Fibre diameter, m or μm

d_p = Particle diameter, m or μm

d_{pc} = Particle cut size, m or μm

E = Activation energy, J/kg-mol or Cal/g-mol

E = Flow rate of extract, kg/s

F = Buoyancy flux parameter, m^4/s^3

F = Drag force on particle, N

F = Packing factor

F = Feed rate, kg/s

F_l = Solute flux, kg/s. m^2

F_m = Solvent flux, kg/s. m^2

f = Coriolis parameter

f = Fractional ability of adsorbent

G = Flow rate of gas phase, kg-mol/s. m^2

G_S = Flow rate of non-diffusing material in gas phase, kg-mol/s. m^2

g = Acceleration of gravity, m/s^2

H' = Henry's law constant, N/m^2

H_G = Height of gas transfer unit, m

H_L = Height of liquid transfer unit, m

H_{OG} = Over-all height of transfer unit, m

I = Intensity of radiation

$\vec{j}_A$ = Mass flux vector of A, kg/s. m^2

$K(T)$ = Equilibrium constant

K = Distribution coefficient

K_d = Microbial decay coefficient, s^{-1}

K_G = Overall mass transfer coefficient based on partial pressure driving force, kg-mol/(s) (m^2) (Δp units of concentration)

K_l = Overall mass transfer coefficient as defined in Eq. (9.58), m/s

K_L = Overall mass transfer coefficient based on liquid phase concentration driving force, m/s

K_s = Saturation constant as defined in Eq. (9.15), mg/l

K_{xx}, K_{yy}, K_{zz} = Eddy diffusivities, m^2/s

$K_y a$ = Overall capacity coefficient, kg-mol/(s) (m^3) (mol fraction)

k = Von Karman's constant

k = Boltzmann's constant, J/K

k = Maximum specific substrate utilization. s^{-1}

k_G = Gas phase mass transfer coefficient, kg-mol/(s) (m^2) (Δp units of concentration)

k_L = Liquid phase mass transfer coefficient, m/s

k_1 = Deoxygenation rate constant, d^{-1} or s^{-1}

k_2 = Reaeration constant

L = Flow rate of liquid phase, kg-mol/s.m^2

L = BOD remaining in time 't', kg/m^3 or mg/l

L_s = Flow rate of non-diffusing material in liquid phase, kg-mol/s.m^2

L_u = Ultimate BOD, kg/m^3 or mg/l

M = Molecular weight of gas mixture

$\hat{M}_p$ = Molecular weight of pollutant

m = Net migration rate

m_a = Mass of air, kg or μg

m_p = Mass of pollutant, kg or μg

m_s = Mass flow rate of spray liquid, kg/s

N = Oxygen transfer rate, kg/s.m^2

N_A = Mass transfer rate of component A, kg-mol/s.m^2

N_e = Effective number of turns of the gas inside cyclone

N_{OG} = Number of overall transfer units

N_t = Number of transfer units
n' = Turbulence index as defined in Eq. (3.62)
P = Total pressure, N/m^2
P = Power consumption in a scrubber, W
Pe = Peclet number
P_0 = Reference pressure, N/m^2
P_0 = Osmotic pressure, N/m^2
p = Air pressure, N/m^2
p = Atmospheric pressure, N/m^2
p = Perimeter, m
p_G = Partial pressure of solute in bulk of gas stream, N/m^2
p_i = Partial pressure at the interface, N/m^2
p_0 = Pressure at ground level, N/m^2
p^* = Partial pressure of solute in equilibrium with bulk concentration in liquid phase, N/m^2
Q = Source strength, kg/s or g/s
Q = Volumetric flow rate, m^3/s
Q_H = Heat emission rate, J/s
Q_w = Waste sludge flow rate, m^3/s
$\vec{q}$ = Mass average velocity, m/s
$\vec{q}_A$ = Absolute velocity of A relative to stationary coordinate axes, m/s
q_G = Geostrophic wind speed, m/s
R = Gas constant, J/kg.K
R = Overall fraction removed from suspension
R = Raffinate flow rate, kg/hr or kg/s
Re_f = Reynolds number based on fibre diameter
Re_p = Particle Reynolds number
R_u = Universal gas constant, J/kg-mol, K
r = Radius of rotation, m
r = Reaeration rate constant, s^{-1}
r_A = Rate of production of A by chemical reaction, $kg/s.m^3$
S = Stability parameter, $1/s^2$

S = Substrate concentration, kg/m³ or mg/l

S = Solvent flow rate, kg/hr or kg/s

Sc = Schmidt number

T = Temperature, K

T = Percentage of light transmission through spot

T_a = Ambient air temperature at stack height, K

T_d = Doubling time, years

T_s = Stack gas temperature, K

T_0 = Temperature at ground level, K

t_B = Time to break-through point, hr or s

t_E = Time to bed exhaustion, hr or s

t_F = Adsorption zone formation time, hr or s

t_0 = Retention time, s

U = Internal energy, J

u = Wind velocity, m/s

u_* = Friction velocity, m/s

V = Precipitator volume, m³

V = Aeration tank volume, m³

V_s = Stack gas exit velocity, m/s

v = Specific volume, kg/m³

v = Velocity inside settling chamber, m/s

v_h = Hindered settling velocity, m/s

v_i = Gas inlet velocity, m/s

v_p = Volume of pollutant, m³

v_p = Particle velocity, m/s

v_r = Radial velocity of gas, m/s

v_r = Relative velocity of gas to liquid at ventury throat, m/s

v_s = Terminal velocity of a particle, m/s

v_t = Terminal settling velocity, m/s

v_{td} = Droplet terminal free falling velocity, m/s

v_0 = Gas free stream velocity, m/s

v_θ = Tangential velocity of particle (and gas), m/s

W = Work, J

W = Membrane coefficient

X = Mol solute/mol non-solute in liquid phase

X = Concentration of biological solids, kg/m^3 or mg/l

X_A = Mole fraction of component A in liquid

X_e = Effluent solids concentration, kg/m^3 or mg/l

X_r = Return sludge suspended solids concentration, kg/m^3 or mg/l

X_T = Bed saturation concentration, kg/m^3

x = Mole fraction of solute in liquid phase

x = Weight fraction solute in raffinate

x_t = Fraction of particles with settling velocity, v_s

Y = Mol solute/mol non-solute in gas phase

Y = Amount of BOD utilised in time 't' kg/m^3 or mg/l

Y = Growth coefficient

y = Mole fraction of solute in gas phase

y = Weight fraction solute in extract

y_{ppm} = Volume concentration, ppm

y_s = Weight fraction solute in solvent

Z = Tower height, m

Z = Height of adsorbent bed, m

z_0 = Aerodynamic roughness

α = Exponent as defined in Eq. (3.37)

Γ = Dry adiabatic lapse rate as defined by Eq. (3.16)

γ = Ratio of specific heats

η = Efficiency

η = Degree of saturation at break-through point

η_d = Drop collection efficiency

η_T = Gravimetric efficiency

θ = Temperature coefficient

θ = Hydraulic retention time, s

θ_c = Mean cell residence time, s

λ = Latent heat, J/kg or cal/g

λ = Mean free path of gas molecules, m

λ = Concentration of coliforms per ml

μ = Viscosity of gas, N.s/m^2

μ = Specific growth rate, s^{-1}

μ_a = Viscosity of air, N.s/m^2

μ_p = Mass of particulate matter, kg

ν = Kinematic viscosity, m^2/s

ν = Stoichiometric coefficient

ρ = Density, kg/m^3

ρ = Total mass concentration of mixture, kg/m^3 or μg/m^3

ρ_A = Mass concentration of species *A*, kg/m^3

ρ_{air} = Air density, kg/m^3

ρ_b = Bulk density, kg/m^3

ρ_g = Gas density, kg/m^3

ρ_L = Spray liquid density, kg/m^3

ρ_p = Mass-volume concentration, kg/m^3 or μg/m^3

ρ_p = Particle density, kg/m^3

σ = Spreading coefficient, m

ϕ_a = Adherence factor

ϕ_c = Collision factor defined by Eq. (5.30)

ψ = Stokes or inertial impaction parameter

ω_A = Mass fraction of species *A*

ω_p = Mass concentration

Index

Absorption, 200
Absorption towers, 204
Activated sludge system:
  design criteria, 356
  design of, 350
  efficiency, 360
  loading criteria, 355
  material balance, 352
  process control, 354
  process modification, 356
Activation energy, 293
Adherence factor, 176
Adiabatic lapse rate, 79
Adsorbers, 371
Adsorbents:
  properties and uses, 222
Adsorption, 219–229, 389
  principles of, 221
Advanced wastewater treatment, 368–389
Aerated ponds, 386
Aerobic ponds, 385
Aerodynamic roughness, 87
Aerosol:
  definition, 35
  deposition model, 113
Air-fuel ratio, 271
Air pollutants:
  analysis, 142
  behaviour and fate of, 46
  classification, 34, 39
  collection devices, 124, 126
  concentration units, 30
  dispersion of, 94–114
  global emissions, 41
  in photochemical reactions, 50
  primary, 34
  properties, 34–38
  sampling, 122, 132
    freeze-out, 126
    grab, 124
    Hi-vol, 127
  secondary, 34
  sources, 38–46
  synergistic action, 59
  topographical effects on, 88
Air pollution:
  control of, 157, 160, 200
  definition, 30
  effects of, 55–70
    on human health, 55
    on materials, 69
    on vegetation, 68
  laws and standards, 71
  major disasters, 56
  mobile sources, 270
Air equality:
  in the United States, 72
  measurement, 142
Alkalinity, 323
Alkalized alumina process, 244
Anaerobic ponds, 386
Asbestos, 62
Asphyxiation, 64
Atmosphere:
  polytropic, 78
  stability of, 81
  turbulent, 85
Atmospheric stability regions, 82
Automobile emissions, control of, 270

Benz-α-pyrene, 39
Beryllium, 62
Biochemical oxygen demand (BOD), 289
  measurement of, 316
Biological pond systems, 385
  design parameters, 387
Biological treatment:
  activated sludge process, 349
  trickling filters, 361
Biosphere, definition, 1
BOD levels in industrial effluents, 295
Boltzmann constant, 179, 417
Break-through curve, 221
Break-through point, 221
Buoyancy flux parameter, 115

Cadmium in atmosphere, 61
Capacity coefficient, 208
Carbon analyzer, 318
Carbon dioxide, global increases in, 6
Carbon monoxide:
  control of, 268
  effects, 64, 65
  emission sources, 43, 44
  fate in atmosphere, 48
  measurement of, 148
  properties, 38
Carboxyhemoglobin, 64
Catalytic:
  decomposition of NO, 267
  reduction of NO, 267
Characteristic settling curve, 344
Chemical oxygen demand (COD), 317
Chemical pollution, 23
Chemico process, 249
Chlorosis, 68
Claus process, 240
Clostridium botulinum, 303
Coal:
  gasification, 239
  hydrodesulphurization, 239
  resources, 14, 16
Coliform bacteria, 299, 326
Colour and odour, 324
Combustion methods:
  catalytic oxidation, 231
  direct, 229
  thermal incineration, 230
Comminutor, 335
Composting of solid wastes, 406
  modern methods of, 407
Coning plume, 90, 91
Continuity equation, 95
Coriolis parameter, 100
Cunningham corrector factor, 176, 195
Cyclone separators:
  efficiency of collection, 173
  particles cut-size, 172
  pressure drop, 173
  separation factor, 170

DCDA method, 253
DDT levels in man, 301
Deoxygenation rate constant, 289
Desulphurization:
  Coal, 239
  flue gases, 243–253
  fuel oils, 241
Detergents, 301
Diffusion coefficient, 95
Dilution factor, 317
Dissolved oxygen (DO):
  analyzer, 316
  deficit, 288
  factors affecting, 286
  measurement of, 315
Direct interception, 176, 178
Distribution coefficient, 374
Doubling time, 11, 12
Drag coefficient, 166, 178
Drag force on particle, 165
Drop collection efficiency, 190
Dust, definition, 35
Dust fall, measurement of, 127

Ecosystems, 1, 2
Effective stack height, 112
Electrostatic precipitator:
  Advantages and disadvantages, 187
  collection efficiency, 183
  migration velocity, 185
  particulate sampling, 131
  principle, 182
Equilibrium constant, 257, 258
Equilibrium curve, 206, 207
Escherichia coli, 299
Eutrophication, 6, 302

Fabric filter:
  Advantages, 182
  choice of, 179
  collection efficiency, 179
  fibre diameter, 176
  interception and inertial impaction, 176, 178
Facultative ponds, 385
Fanning plume, 90, 91
Fibre materials, properties of, 181
Fick's law, 95
Flame combustion, 229
Flue gas scrubbing, 243, 263
Fluidized bed combustion, 243, 263
Fog, definition, 35
Fossil fuels:
  coal resources, 14, 16
  oil resources, 14, 16
  impact on environment, 17
Friction velocity, 100
Fuel oils:
  desulphurization, 241
  gasification, 242
Fume, definition, 35
Fumigation, 92

Gas cleaning devices:
efficiencies of various types, 200
Gas constant, 77, 80, 293
Gasification:
coal, 239
fuel oils, 242
Gaussian plume model, 101–109
Geostrophic wind speed, 100
Glass separation methods, 410
Greenburg-Smith impinger, 124
Grit chambers, 334

Height of a transfer unit, 209, 210, 225
Henry's law, 202
Hindered settling, 343
Hydraulic:
diameter, 164
retention time, 353
Hydrocarbons:
control of, 269
effects, 66
emission sources, 43
measurement of, 151, 152
properties, 38
Hydro-electric energy, 20

Impaction parameter, 176
Impingement plate scrubber, 193
Incineration, 404
Indicator organisms, 326
Inertial impaction, 176, 188
Interfacial area, 208
Internal energy, 79
Ion exchange process, 377, 390
Isokinetic sampling, 135–140
Inversion:
advective, 84
radiation, 83
subsidence, 83

Kjeldahl nitrogen, 319

Laminar flow, 165
Lapple's correlation, 172
Lapse rate, 78
adiabatic, 79
environmental, 78
Laws and standards:
air pollution, 71
water pollution, 308, 328
Lead, 59, 60, 304
Limestone process, 242
Lofting plume, 92
Looping plume, 90

Magnesium oxide process, 249
Manganese oxide process, 244
Mass average velocity, 95
Mass of pollutant, 30
Mass concentration, 30, 95
Mass flux, definition, 95
Mass fraction, 95
Mass-volume concentration, 32
Mass transfer coefficient, 205, 207
Mean cell residence time, 351
Mean free path of gas molecules, 176
Membrane coefficient, 379
Mercury, 61, 304
Methaemoglobin anemia, 303
Microstraining, 368
Migration velocity of particles, 185
Mist, definition, 35,
Mobile sources:
Hydrocarbon emissions control, 272
NO_x emissions control, 273
Mole fraction, 207, 208
Molecular weight, 32
MPN method, 326

Natural cycles:
carbon, 4
hydrologic, 3
nitrogen, 6
phosphorus, 7
sulphur, 8
Nessler's reagent, 320
Nickel, 61
Nitrogen oxides:
control of, 257–268
design conditions modification, 263
effects, 65
effluent gas treatment, 263
emission sources, 37, 43
fate in atmosphere, 49, 50
kinetics of formation, 257
measurement of, 146
operating conditions modification, 260
Nuclear energy:
India's resources, 21
environmental effects, 21, 305
radioactive pollution, 22
wastes generated, 22
Number of transfer units, 209, 225
Number of turns, 172

Oil:
pollution, 307
resources, 14, 15
Oil slick, 307

Operating line, 205, 206, 225
Optical haze, coefficient, 129
Origin of :
air pollutants, 38, 41, 44
water pollutants, 282, 286
solid wastes, 397, 398
Osmotic pressure, 378
Oxidants, 66, 150
Oxidation of wastewater sludges, 366
Oxygen sag curve, 296, 298
Ozonation, 389
Ozone, 67, 69, 70, 388
Ozone depletion, 25

Packed tower, 192, 204
diameter, 211
height or packing, 208
Packing factor, 212, 213
Paper board manufacture, 409
Particles :
capture mechanism, 177
diffusivity, 178
Particulate matter :
capture efficiency, 179
emission sources, 41, 45
fate in atmosphere, 46, 47
measurement of, 127
properties, 34
sampling, 126, 135
toxic effects, 58–62
Particulate collector selection, 198
Parts per million, 30
Peclet number, 178
Permeable membranes, 379
Peroxyacylnitrate (PAN), 52
Pesticides, 24, 299
limiting standards in drinking water, 300
Phosphorus, 7, 302, 320,
removal from wastewaters, 383, 385
Photochemical smog :
behaviour, 53
principal reactions, 51
Photosynthesis, 288
Phytoplankton, 1, 2, 305
Plant nutrients, 2, 302
Plume behaviour, 90
Plume rise estimate, 114–116
buoyant, 114
non-buoyant, 116
Polytropic atmosphere, 78
Population :
doubling time, 12
environmental impact, 13
growth in world, 10
urban growth in India, 12
Power consumption in scrubber, 197
Pressure drop :
cyclone separator, 173
venturi scrubber, 196

Radial velocity, 171
Radiation inversion, 83
Radius of rotation, 170
Reaeration, 296
coefficient, 297
Reinluft process, 246
Respiration, 289
Reverse osmosis, 378
Reynolds number, 164, 176
Ringelmann method, 153
Rossby number, 100

Scavenging mechanism, 46
Schmidt number, 178, 210
Scrubbers :
centrifugal, 191
efficiency correlation, 200
impingement plate, 193
packed bed, 192
spray tower, 189
venturi, 194
Sedimentation, 335–345
Settling types .
discrete, 337
flocculent, 341
zone, 343
Settling chambers :
collection efficiency 165
laminar flow, 165
turbulent flow, 167
Settling velocity, 166
spherical particles, 167
Sludge volume index (SVI), 359
Smoke, definition, 35
Solid wastes:
agricultural, 398
classification, 396
collection and disposal, 399, 402
health hazards, 399
industrial, 398
incineration of, 404
recovery and recycling, 408–412
refuse composition, 397
sanitary landfill, 402–404
ground water pollution from, 404
Specific heat, 80
Specific volume, 80

Spectrophotometer, 321
Spreading coefficients, 102, 105, 108
Stability parameter, 115
Stack sampling, 132
Stratosphere, 79
Subsidence inversion, 83
Sulphur, extraction from fuel oil, 241
Sulphur dioxide :
  effects, 63
  emission control methods, 238, 243
  emission sourcess, 43, 45
  fate in atmosphere, 48, 49
  measurement cf, 142–145
    ambient air, 142
    stack gases, 144
  properties, 36
Tangential velocity 170
Tape sampler, 128
Terminal particle velocity, 166, 337
Thermal discharges, 306
Thermal incineration, 230
Threshold odour number, 325
Total organic carbon (TOC), 318
Toxic metals, 59–62, 304
  effects, 60, 304
Trickling filters, 361
Troposphere, 78, 79
Turbulence :
  closure problem, 96
  mechanical, 85
  power law, 85
  thermal, 85
Turbulent diffusion, 94
Two-resistance theory, 201
Volatile suspended solids, 324
Volume concentration, 30
Von karman's constant, 100

Wastewater :
  characteristics, 282
  composition, 283, 284
  materials recovery, 389, 390
  methods of analysis, 315
  sampling, 313–315
Wastewater treatment processes :
  adsorption, 370
  ammonia stripping, 381
  biological, 348
  coagulation, 369
  electrodialysis, 380
  flotation, 347
  ion exchange, 377
  nitrification-denitrification, 382
  reverse osmosis, 378
  sedimentation, 335
  solvent extraction, 373
Wastewater sludges :
  dewatering, 365
  digestion, 364
  disposal methods, 367
  oxidation, 366
Water :
  domestic supply 279
  Quality standards, 308, 328
  resources, 277
  utilization of, 279
Water pollutants :
  classification, 286
  disease causing agents, 298
  inorganic chemicals and minerals, 303
  oil, 307
  oxygen demanding wastes, 286
  plant nutrients, 302
  radioactive substances, 305
  sediments, 305
  sources of, 282
  synthetic organic compounds, 299
  thermal discharges, 306
  trace elements, 304
Welman-Lord process, 251
Westvaco process, 248
Wet collectors, advantages and disadvantages, 189
Wet oxidation, 366
Windrows, 407
Winkler's method, 315